T0342077

Large-Scale Inverse Problems and Quantification of Uncertainty

Wiley Series in Computational Statistics

Consulting Editors:

Wiley Series in Computational Statistics is comprised of practical guides and cutting edge research books on new developments in computational statistics. It features quality authors with a strong applications focus. The texts in the series provide detailed coverage of statistical concepts, methods and case studies in areas at the interface of statistics, computing, and numerics.

With sound motivation and a wealth of practical examples, the books show in concrete terms how to select and to use appropriate ranges of statistical computing techniques in particular fields of study. Readers are assumed to have a basic understanding of introductory terminology.

The series concentrates on applications of computational methods in statistics to fields of bioinformatics, genomics, epidemiology, business, engineering, finance and applied statistics.

Titles in the Series

Billard and Diday – Symbolic Data Analysis: Conceptual Statistics and Data Mining
Bolstad – Understanding Computational Bayesian Statistics
Borgelt, Steinbrecher and Kruse – Graphical Models, 2e
Dunne – A Statistical Approach to Neutral Networks for Pattern Recognition
Liang, Liu and Carroll – Advanced Markov Chain Monte Carlo Methods
Ntzoufras – Bayesian Modeling Using WinBUGS

Large-Scale Inverse Problems and Quantification of Uncertainty

Edited by

Lorenz Biegler
Carnegie Mellon University, USA

George Biros
Georgia Institute of Technology, USA

Omar Ghattas
University of Texas at Austin, USA

Matthias Heinkenschloss
Rice University, USA

David Keyes
KAUST and Columbia University, USA

Bani Mallick,
Texas A&M University, USA

Youssef Marzouk
Massachusetts Institute of Technology, USA

Luis Tenorio
Colorado School of Mines, USA

Bart van Bloemen Waanders
Sandia National Laboratories, USA

Karen Willcox
Massachusetts Institute of Technology, USA

A John Wiley and Sons, Ltd., Publication

This edition first published 2011

Registered office
John Wiley & Sons Ltd, The Atrium, Southern Gate, Chichester, West Sussex, PO19 8SQ, United Kingdom

For details of our global editorial offices, for customer services and for information about how to apply for permission to reuse the copyright material in this book please see our website at www.wiley.com.

Library of Congress Cataloging-in-Publication Data

Large-scale inverse problems and quantification of uncertainty / edited by Lorenz Biegler ... [et al.].
p. cm.
Includes bibliographical references and index.
ISBN 978-0-470-69743-6 (cloth)
1. Bayesian statistical decision theory. 2. Inverse problems (Differential equations) 3. Mathematical optimization. I. Biegler, Lorenz T.
QA279.5.L37 2010
515′.357 – dc22

2010026297

A catalogue record for this book is available from the British Library.

Sandia National Laboratories is a multi-program laboratory operated by Sandia Corporation, a wholly owned subsidiary of Lockheed Martin Corporation, for the U.S. Department of Energy's National Nuclear Security Administration under contract DE-AC04-94AL85000.

Print ISBN: 978-0-470-69743-6
ePDF ISBN: 978-0-470-68586-0
oBook ISBN: 978-0-470-68585-3

Typeset in 10/12pt Computer Modern font by Laserwords Private Limited, Chennai, India

Contents

List of Contributors

D. Calvetti
Department of Mathematics
Case Western Reserve University
Cleveland, OH, USA

A. Datta-Gupta
Department of Petroleum Engineering
Texas A&M University
College Station, TX, USA

F. Delbos
Institut Francais du petrole
Rueil-Malmaison, France

C. Duffet
Institut Francais du petrole
Rueil-Malmaison, France

Y. Efendiev
Department of Mathematics
Texas A&M University
College Station, TX, USA

G. Evensen
Statoil Research Centre
Bergen, Norway

M. Frangos
Department of Aeronautics and Astronautics
Massachusetts Institute of Technology
Cambridge, MA, USA

E. Haber
Department of Mathematics
University of British Columbia
Vancouver, Canada

S. Habib
Nuclear & Particle Physics, Astrophysics, and Cosmology Group
Los Alamos National Laboratory
Los Alamos, NM, USA

K. Heitmann
International, Space and Response Division
Los Alamos National Laboratory
Los Alamos, NM, USA

D. Higdon
Statistical Sciences
Los Alamos National Laboratory
Los Alamos, NM, USA

L. Horesh
IBM T. J. Watson Research Center
Yorktown Heights, NY, USA

J. Hove
Statoil Research Centre
Bergen, Norway

D.B.P. Huynh
Department of Mechanical Engineering
Massachusetts Institute of Technology
Cambridge, MA, USA

K. Hwang
Department of Statistics
Texas A&M University
College Station, TX, USA

P.K. Kitanidis
Department of Civil and Environmental Engineering
Stanford University
Stanford, CA, USA

E. Lawrence
Statistical Sciences
Los Alamos National Laboratory
Los Alamos, NM, USA

X. Ma
Department of Petroleum Engineering
Texas A&M University
College Station, TX, USA

J. McFarland
Southwest Research Institute
San Antonio, TX, USA

B. Mallick
Department of Statistics
Texas A&M University
College Station, TX, USA

Y. Marzouk
Department of Aeronautics and Astronautics
Massachusetts Institute of Technology
Cambridge, MA, USA

I. Myrseth
Department of Mathematical Sciences
Norwegian University of Science and Technology
Trondheim, Norway

N.C. Nguyen
Department of Mechanical Engineering
Massachusetts Institute of Technology
Cambridge, MA, USA

H. Omre
Department of Mathematical Sciences
Norwegian University of Science and Technology
Trondheim, Norway

A.T. Patera
Department of Mechanical Engineering
Massachusetts Institute of Technology
Cambridge, MA, USA

G. Rozza
Department of Mechanical Engineering
Massachusetts Institute of Technology
Cambridge, MA, USA

A. Sandu
Department of Computer Science
Virginia Polytechnic Institute and State University
Blacksburg, VA, USA

A. Seiler
Statoil Research Centre
Bergen, Norway

D. Sinoquet
Institut Francais du petrole
Rueil-Malmaison, France

J.-A. Skjervheim
Statoil Research Centre
Bergen, Norway

E. Somersalo
Department of Mathematics
Case Western Reserve University
Cleveland, OH, USA

P.B. Stark
Department of Statistics
University of California
Berkeley, CA, USA

L. Swiler
Optimization and Uncertainty Quantification
Sandia National Laboratories
Albuquerque, NM, USA

L. Tenorio
Mathematical and Computer Sciences
Colorado School of Mines
Golden, CO, USA

J.G. Vabø
Statoil Research Centre
Bergen, Norway

B. van BloemenWaanders
Applied Mathematics and Applications
Sandia National Laboratories
Albuquerque, NM, USA

R.D. Wilkinson
School of Mathematical Sciences
University of Nottingham
Nottingham, United Kingdom

K. Willcox
Department of Aeronautics and Astronautics
Massachusetts Institute of Technology
Cambridge, MA, USA

N. Zabaras
Sibley School of Mechanical and Aerospace Engineering
Cornell University
Ithaca, NY, USA

Chapter 1

Introduction

1.1 Introduction

To solve an inverse problem means to estimate unknown objects (e.g. parameters and functions) from indirect noisy observations. A classic example is the mapping of the Earth's subsurface using seismic waves. Inverse problems are often ill-posed, in the sense that small perturbations in the data may lead to large errors in the inversion estimates. Furthermore, many practical and important inverse problems are large-scale; they involve large amounts of data and high-dimensional parameter spaces. For example, in the case of seismic inversion, typically millions of parameters are needed to describe the material properties of the Earth's subsurface. Large-scale, ill-posed inverse problems are ubiquitous in science and engineering, and are important precursors to the quantification of uncertainties underpinning prediction and decision-making.

In the absence of measurement noise, the connection between the parameters (input) and the data (output) defines the forward operator. In the inverse problem we determine an input corresponding to given noisy data. It is often the case that there is no explicit formula for the inversion estimate that maps output to input. We thus often rely on forward calculations to compare the output of different plausible input models to choose one model, or a distribution of models, that is consistent with the data. A further complication is that the forward operator itself may not be perfectly known, as it may depend on unknown tuning parameters. For example, the forward operator in seismic inversion may depend on a velocity model that is only partly known.

It is then clear that there is a need in inverse problems for a framework that includes computationally efficient methods capable of incorporating

Large-Scale Inverse Problems and Quantification of Uncertainty Edited by L. Biegler, G. Biros, O. Ghattas, M. Heinkenschloss, D. Keyes, B. Mallick, Y. Marzouk, L. Tenorio, B. van Bloemen Waanders and K. Willcox

prior information and accounting for uncertainties at different stages of the modeling procedure, as well as methods that can be used to provide measures of the reliability of the final estimates. However, efficient modeling of the uncertainties in the inputs for high-dimensional parameter spaces and expensive forward simulations remain a tremendous challenge for many problems today – there is a crucial unmet need for the development of scalable numerical algorithms for the solution of large-scale inverse problems. While complete quantification of uncertainty in inverse problems for very large-scale nonlinear systems has often been intractable, several recent developments are making it viable: (1) the maturing state of algorithms and software for forward simulation for many classes of problems; (2) the arrival of the petascale computing era; and (3) the explosion of available observational data in many scientific areas.

This book is focused on computational methods for large-scale inverse problems. It includes methods for uncertainty quantification in input and output parameters and for efficient forward calculations, as well as methodologies to incorporate different types of prior information to improve the inversion estimates. The aim is to cross-fertilize the perspectives of researchers in the areas of data assimilation, statistics, large-scale optimization, applied and computational mathematics, high performance computing, and cutting-edge applications.

Given the different types of uncertainties and prior information that have to be consolidated, it is not surprising that many of the methods in the following chapters are defined in a Bayesian framework with solution techniques ranging from deterministic optimization-based approaches to variants of Markov chain Monte Carlo (MCMC) solvers. The choice of algorithms to solve a large-scale problem depends on the problem formulation and a balance of computational resources and a complete statistical characterization of the inversion estimates. If time to solution is the priority, deterministic optimization methods offer a computationally efficient strategy but at the cost of statistical inflexibility. If a complete statistical characterization is required, large computational resources must accommodate Monte Carlo type solvers. It is often necessary to consider surrogate modeling to reduce the cost of forward solves and/or the dimension of the state and inversion spaces.

The book is organized in four general categories: (1) an introduction to statistical Bayesian and frequentist methodologies, (2) approximation methods, (3) Kalman filtering methods, and (4) deterministic optimization based approaches.

1.2 Statistical Methods

Inverse problems require statistical characterizations because uncertainties and/or prior information are modeled as random. Such stochastic structure helps us deal with the complex nature of the uncertainties that plague many aspects of inverse problems including the underlying simulation model, data measurements and the prior information. A variety of methods must be considered to achieve an acceptable statistical description at a practical computational cost. Several chapters discuss different approaches in an attempt to reduce the computational requirements. Most of these methods are centered around a Bayesian framework in which uncertainty quantification is achieved by exploring a posterior probability distribution. The first three chapters introduce key concepts of the frequentist and Bayesian framework through algorithmic explanations and simple numerical examples. Along the way they also explain how the two different frameworks are used in geostatistical applications and regularization of inverse problems. The subsequent two chapters propose new methods to further build on the Bayesian framework.

Brief summaries of these chapters follow:

- Stark and Tenorio present frequentist and Bayesian methods for inverse problems. They discuss the different ways prior information is used by each school and explain basic statistical procedures such as estimators, confidence intervals and credible regions. They also show how decision theory is used to compare statistical procedures. For example, a frequentist estimator can be compared to a Bayesian one by computing the frequentist mean squared error of each. Credible regions can be compared to frequentist regions via their frequentist coverage. Stark and Tenorio provide illustrative examples of these and other types of comparisons.

- Calvetti and Somersalo clarify where the subjectivity in the Bayesian approach lies and what it really amounts to. The focus is on the interpretation of the probability and on its role in setting up likelihoods and priors. They show how to use hierarchical Bayesian methods to incorporate prior information and uncertainty at different levels of the mathematical model. Algorithms to compute the maximum a-posteriori estimate and sampling methods to explore the posterior distribution are discussed. Dynamic updating and the classic Kalman filter algorithm are introduced as a prelude to the chapters on Kalman and Bayesian filtering.

- Kitanidis presents the Bayesian framework as the appropriate methodology to solve inverse problems. He explains how the Bayesian approach differs from the frequentist approach, both in terms of methodology and in terms of the meaning of the results one obtains, and discusses some disagreements between Bayesians and non-Bayesians in the selection of prior distributions. Bayesian methods for geostatiscal analysis are also discussed.

- Higdon *et al.* consider the problem of making predictions based on computer simulations. They present a Bayesian framework to combine available data to estimate unknown parameters for the computer model and assess prediction uncertainties. In addition, their methodology can account for uncertainties due to limitations on the number of simulations. This chapter also serves as an introduction to Gaussian processes and Markov chain Monte Carlo (MCMC) methods.

- Efendiev *et al.* present a strategy to efficiently sample from a surrogate model obtained from a Bayesian Partition Model (BPM) which uses Voronoi Tessellations to decompose the entire parameter space into homogeneous regions and use the same probability distribution within each region. The technique is demonstrated on an inversion of permeability fields and fractional flow simulation from the Darcy and continuity equations. The high dimensional permeability field is approximated by a Karhunen-Loeve expansion and then combined using regression techniques on different BPM regions. A two-stage MCMC method has been employed, where at the first stage the BPM approximation has been used thereby creating a more efficient MCMC method.

1.3 Approximation Methods

The solution of large-scale inverse problems critically depends on methods to reduce computational cost. Solving the inverse problem typically requires the evaluation of many thousands of plausible input models through the forward problem; thus, finding computationally efficient methods to solve the forward problem is one important component of achieving the desired cost reductions. Advances in linear solver and preconditioning techniques, in addition to parallelization, can provide significant efficiency gains; however, in many cases these gains are not sufficient to meet all the computational needs of large-scale inversion problems. We must therefore appeal to approximation techniques that seek to replace the forward model with an inexpensive surrogate. In addition to yielding a dramatic decrease in forward problem solution time, approximations can reduce the dimension of the input space and entail more efficient sampling strategies, targeting a reduction in the number of forward solves required to find solutions and assess uncertainties. Below we describe a number of chapters that employ combinations of these approximation approaches in a Bayesian inference framework.

- Frangos *et al.* summarize the state of the art in methods to reduce the computational cost of solving statistical inverse problems. A literature survey is provided for three classes of approaches – approximating the forward model, reducing the size of the input space, and reducing the number of samples required to compute statistics of the posterior. A simple example demonstrates the relative advantages of polynomial chaos-based surrogates and projection-based model reduction of the forward simulator.

- Nguyen *et al.* present a reduced basis approximation approach to solve a real time Bayesian parameter estimation problem. The approach uses Galerkin projection of the nonlinear partial differential equations onto a low-dimensional space that is identified through adaptive sampling. Decomposition into 'Offline' and 'Online' computational tasks achieves solution of the Bayesian estimation problem in real time. A posteriori error estimation for linear functionals yields rigorous bounds on the results computed using the reduced basis approximation.

- Swiler *et al.* present a Bayesian solution strategy to solve a model calibration in which the underlying simulation is expensive and observational data contains errors or uncertainty. They demonstrate the use of Gaussian surrogates to reduce the computational expense on a complex thermal simulation of decomposing foam dataset.

- Wilkinson discusses emulation techniques to manage multivariate output from expensive models in the context of calibration using observational data. A focus of this work is on calibration of models with long run time. Consequently an ensemble comprising of only a limited number of forward runs can be considered. A strategy is presented for selecting the design points that are used to define the ensemble. When the simulation model is computationally expensive, emulation is required and here a Bayesian approach is used.

1.4 Kalman Filtering

The next two chapters discuss filtering methods to solve large-scale statistical inverse problems. In particular, they focus on the ensemble Kalman filter, which searches for a solution in the space spanned by a collection of ensembles. Analogous to reducing the order of a high dimensional parameter space using a stochastic spectral approximation or a projection-based reduced-order model, the ensemble Kalman filter assumes that the variability of the parameters can be well approximated by a small number of modes. As such, large numbers of inversion parameters can be accommodated in combination with complex and large-scale dynamics. This comes however, at the cost of less statistical flexibility, since approximate solutions are needed for nonlinear non-Gaussian problems.

- Myrseth *et al.* provide an overview of the ensemble Kalman filter in addition to various other filters. Under very specific assumptions about linearity and Gaussianity, exact analytical solutions can be determined for the Bayesian inversion, but for any deviation from these assumptions, one has to rely on approximations. The filter relies on simulation based inference and utilizes a linearization in the data conditioning. These approximations make the ensemble Kalman filter computationally efficient and well suited for high-dimensional hidden Markov models.

- Seiler *et al.* discuss the use of the ensemble Kalman filter to solve a large inverse problem. This approach uses a Monte Carlo process for calculating the joint probability density function for the model and state parameters, and it computes the recursive update steps by approximating the first and second moments of the predicted PDF. The recursive Bayesian formulation can be solved using the ensemble Kalman filter under the assumption that predicted error statistics are nearly Gaussian. Instead of working with the high-dimensional parameter space, the inverse problem is reduced to the number of realizations included in the ensemble. The approach is demonstrated on a petroleum reservoir simulation dataset in which a history matching problem is solved.

1.5 Optimization

An alternative strategy to a Bayesian formulation is to pose the statistical inverse problem as an optimization problem. Computational frameworks that use this approach build upon state-of-the-art methods for simulation of the forward problem, as well as the machinery for large-scale deterministic optimization. Typically, Gaussian assumptions are made regarding various components of the data and models. This reduces the statistical flexibility and perhaps compromises the quality of the solution. However as some of these chapters will demonstrate, these methods are capable of addressing very large inversion spaces while still providing statistical descriptions of the solution.

- Horesh *et al.* present optimal experimental design strategies for large-scale nonlinear ill-posed inverse problems. In particular, strategies for a nonlinear impedance imaging problem are presented. Optimal selection of source and receiver locations is achieved by solving an optimization problem that controls the performance of the inversion estimate subject to sparsity constraints.

- Zabaras outlines a methodology to perform estimation under multiple sources of uncertainties. By relying on the use of the deterministic simulator, the solution of the stochastic problem is constructed using sparse grid collocation. Furthermore, the stochastic solution is converted to a deterministic optimization problem in a higher dimensional space. Stochastic sensitivities are calculated using deterministic calculations. The evolution of a PDF requires the solution of a billion DOFs at each stochastic optimization iteration. The technique is demonstrated on a heat flux problem.

- Frederic *et al.* present an uncertainty analysis approach for seismic inversion and discuss tomography and ray tracing, which are efficient methods to predict travel time. The deterministic inversion approach makes the connection between the Hessian and the covariance. The prior and posterior PDFs are assumed to be Gaussian. By estimating just the diagonal terms of the covariance matrix for the prior and therefore assuming the errors are uncorrelated, the computational expense is reduced considerably. By assuming Gaussianity and calculating the diagonal terms, the inversion formulation has a specific form that involves the Hessian. Two strategies are presented, one where the full Hessian is inverted to give the covariance and the other where a multi-parameter approach is used to reduce the computational expense of the Hessian inversion.

- Sandu discusses the use of adjoint methods, which are at the core of many optimization strategies for large-scale inverse problems. This chapter presents an analysis of the properties of Runge-Kutta and linear multistep methods in the context of solving ODEs that arise in adjoint equations. An example shows the use of discrete adjoint methods in the solution of large-scale data assimilation problems for air quality modeling.

Chapter 2

A Primer of Frequentist and Bayesian Inference in Inverse Problems

P.B. Stark[1] and L. Tenorio[2]
[1] *University of California at Berkeley, USA*
[2] *Colorado School of Mines, USA*

2.1 Introduction

Inverse problems seek to learn about the world from indirect, noisy data. They can be cast as statistical estimation problems and studied using statistical decision theory, a framework that subsumes Bayesian and frequentist methods. Both Bayesian and frequentist methods require a stochastic model for the data, and both can incorporate constraints on the possible states of the world. Bayesian methods require that constraints be re-formulated as a *prior probability distribution*, a stochastic model for the unknown state of the world. If the state of the world is a random variable with a known distribution, Bayesian methods are in some sense optimal. Frequentist methods are easier to justify when the state of the world is unknown but not necessarily random, or random but with an unknown distribution; some frequentist methods are then optimal – in other senses.

Parameters are numerical properties of the state of the world. *Estimators* are quantities that can be computed from the data without knowing the state of the world. Estimators can be compared using *risk functions*, which quantify

Large-Scale Inverse Problems and Quantification of Uncertainty Edited by L. Biegler, G. Biros, O. Ghattas, M. Heinkenschloss, D. Keyes, B. Mallick, Y. Marzouk, L. Tenorio, B. van Bloemen Waanders and K. Willcox

the expected 'cost' of using a particular estimator when the world is in a given state. (The definition of cost should be dictated by the scientific context, but some definitions, such as mean squared error, are used because they lead to tractable computations.)

Generally, no estimator has the smallest risk for all possible states of the world: there are tradeoffs, which Bayesian and frequentist methods address differently. Bayesian methods seek to minimize the expected risk when the state of the world is drawn at random according to the prior probability distribution. One frequentist approach – *minimax estimation* – seeks to minimize the maximum risk over all states of the world that satisfy the constraints.

The performance of frequentist and Bayesian estimators can be compared both with and without a prior probability distribution for the state of the world. Bayesian estimators can be evaluated from a frequentist perspective, and vice versa. Comparing the minimax risk with the Bayes risk measures how much information the prior probability distribution adds that is not present in the constraints or the data.

This chapter sketches frequentist and Bayesian approaches to estimation and inference, including some differences and connections. The treatment is expository, not rigorous. We illustrate the approaches with two examples: a concrete one-dimensional problem (estimating the mean of a Normal distribution when that mean is known to lie in the interval $[-\tau, \tau]$), and an abstract linear inverse problem.

For a more philosophical perspective on the frequentist and Bayesian interpretations of probability and models, see Freedman and Stark (2003); Freedman (1995). For more careful treatments of the technical aspects, see Berger (1985); Evans and Stark (2002); Le Cam (1986).

2.2 Prior Information and Parameters: What Do You Know, and What Do You Want to Know?

This section lays out some of the basic terms, most of which are shared by frequentist and Bayesian methods. Both schools seek to learn about the state of the world – to estimate parameters – from noisy data. Both consider constraints on the possible states of the world, and both require a stochastic measurement model for the data.

2.2.1 The State of the World, Measurement Model, Parameters and Likelihoods

The term 'model' is used in many ways in different communities. In the interest of clarity, we distinguish among three things sometimes called 'models,' namely, the state of the world, the measurement model, and parameters.

The state of the world, denoted by θ, is a mathematical representation of the physical system of interest, for example, a parametrized representation of seismic velocity as a function of position in the Earth, of the angular velocity of material in the Sun, or of the temperature of the cosmic microwave background radiation as a function of direction. Often in physical problems, some states of the world can be ruled out by physical theory or prior experiment. For instance, mass densities and energies are necessarily nonnegative. Transmission coefficients are between 0 and 1. Particle velocities are less than the speed of light. The rest mass of the energy stored in Earth's magnetic field is less than the mass of the Earth (Backus 1989). The set Θ will represent the possible states of the world (values of θ) that satisfy the constraints. That is, we know a priori that $\theta \in \Theta$.[1]

The observations Y are related to the particular state of the world θ through a *measurement model* that relates the probability distribution of the observations to θ. The set of possible observations is denoted $\mathcal{Y}$, called the *sample space*. Typically, Y is an n-dimensional vector of real numbers; then, $\mathcal{Y}$ is $\mathbb{R}^n$. Depending on θ, Y is more likely to take some values in $\mathcal{Y}$ than others. The probability distribution of the data Y when the state of the world is η (i.e., when $\theta = \eta$) is denoted $\mathbb{P}_\eta$; we write $Y \sim \mathbb{P}_\eta$. (The 'true' state of the world is some particular – but unknown – $\theta \in \Theta$; η is a generic element of Θ that might or might not be equal to θ.) We shall assume that the set of distributions $\mathcal{P} \equiv \{\mathbb{P}_\eta : \eta \in \Theta\}$ is dominated by a common measure μ.[2] (In the special case that μ is Lebesgue measure, that just means that all the probability distributions $\mathcal{P} \equiv \{\mathbb{P}_\eta : \eta \in \Theta\}$ have densities in the ordinary sense.) The *density of* $\mathbb{P}_\eta$ *(with respect to* μ*) at* y is

$$p_\eta(y) \equiv d\mathbb{P}_\eta / d\mu|_y. \tag{2.1}$$

The *likelihood of* η *given* $Y = y$ is $p_\eta(y)$ viewed as a function of η, with y fixed.

For example, suppose that for the purposes of our experiment, the state of the world can be described by a single number $\theta \in \mathbb{R}$ that is known to be in the interval $[-\tau, \tau]$, and that our experiment measures θ with additive Gaussian noise that has mean zero and variance 1. Then the measurement model is

$$Y = \theta + Z, \tag{2.2}$$

where Z is a standard Gaussian random variable (we write $Z \sim N(0, 1)$). The set $\Theta = [-\tau, \tau]$. Equivalently, we may write $Y \sim N(\theta, 1)$ with $\theta \in [-\tau, \tau]$.

---

[1]There is a difference between the physical state of the world and a numerical discretization or approximation of the state of the world for computational convenience. The numerical approximation to the underlying physical process contributes uncertainty that is generally ignored. For discussion, see, e.g., Stark (1992b).

[2]By defining μ suitably, this can allow us to work with 'densities' even if the family $\mathcal{P}$ contains measures that assign positive probability to individual points. Assuming that there is a dominating measure is a technical convenience that permits a general definition of likelihoods.

(The symbol $\sim$ means 'has the probability distribution' or 'has the same probability distribution as.') Thus $\mathbb{P}_\eta$ is the Gaussian distribution with mean η and variance 1. This is called the *bounded normal mean* (BNM) problem. The BNM problem is of theoretical interest, and is a building block for the study of more complicated problems in higher dimensions; see, for example, Donoho (1994). The dominating measure μ in this problem can be taken to be Lebesgue measure. Then the density of $\mathbb{P}_\theta$ at y is

$$\varphi_\theta(y) \equiv \frac{1}{\sqrt{2\pi}} e^{-(y-\theta)^2/2}, \tag{2.3}$$

and the likelihood of η given $Y = y$ is $\varphi_\eta(y)$ viewed as a function of η with y fixed.

As a more general example, consider the following canonical linear inverse problem. The set Θ is a ball in a norm or semi-norm in a separable, infinite-dimensional Banach space (for example, Θ might be a set of functions whose integrated squared second derivative is less than some constant $C < \infty$, or a set of nonnegative functions that are continuous and bounded).[3] The data Y are related to the state of the world θ through the action of a linear operator K from Θ to $\mathcal{Y} = \mathbb{R}^n$, with additive noise:

$$Y = K\theta + \varepsilon, \quad \theta \in \Theta, \tag{2.4}$$

where the probability distribution of the noise vector ε is known. We assume that $K\eta = (K_1\eta, K_2\eta, \ldots, K_n\eta)$ for $\eta \in \Theta$, where $\{K_j\}_{j=1}^n$ are linearly independent bounded linear functionals on Θ. Let $f(y)$ denote the density of ε with respect to a dominating measure μ. Then $p_\eta(y) = f(y - K\eta)$. The BNM problem is an example of a linear inverse problem with K the identity operator, $\mathcal{Y} = \mathbb{R}$, $\Theta \equiv [-\tau, \tau]$, $\varepsilon \sim N(0, 1)$, $f(y) = \varphi_0(y)$.

A *parameter* $\lambda = \lambda[\theta]$ is a property of the state of the world. The entire description of the state of the world, θ, could be considered to be a parameter; then λ is the identity operator. Alternatively, we might be interested in a simpler property of θ. For example, in gravimetry, the state of the world θ might be mass density as a function of position in Earth's interior, and the parameter of interest, $\lambda[\theta]$, might be the average mass density in some region below the surface. In that case, the rest of θ is a *nuisance parameter*: it can affect the (probability distribution of the) measurements, but it is not of primary interest.

Our lead examples in this paper are the 'bounded normal mean' problem and the canonical linear inverse problem just described.

2.2.2 Prior and Posterior Probability Distributions

In the present framework, there is prior information about the state of the world θ expressed as a constraint $\theta \in \Theta$. Frequentists use such constraints

[3] Separable Banach spaces are measurable with respect to the σ-algebra induced by the norm topology, a fact that ensures that prior probability distributions – required by Bayesian methods – can be defined.

as-is. Bayesians augment the constraints using *prior probability distributions*. Bayesians treat the value of θ as a realization of a random variable that takes values in Θ according to a prior probability distribution π.[4] The constraint $\theta \in \Theta$ is reflected in the fact that π assigns probability 1 (or at least high probability) to the set Θ.[5] In the BNM problem, a Bayesian would use a prior probability distribution π that assigns probability 1 to the interval $[-\tau, \tau]$. For instance, she might use as the prior π the 'uninformative'[6]

---

[4]To use prior probability distributions, Θ must be a measurable space; frequentist methods generally do not need the set of parameters to be measurable. We will not worry about technical details here. For rigor, see Le Cam (1986).

[5]Freedman (1995) writes, 'My own experience suggests that neither decision-makers nor their statisticians do in fact have prior probabilities. A large part of Bayesian statistics is about what you would do if you had a prior. For the rest, statisticians make up priors that are mathematically convenient or attractive. Once used, priors become familiar; therefore, they come to be accepted as "natural" and are liable to be used again; such priors may eventually generate their own technical literature.' And, 'Similarly, a large part of [frequentist] statistics is about what you would do if you had a model; and all of us spend enormous amounts of energy finding out what would happen if the data kept pouring in.' We agree. One argument made in favor of Bayesian methods is that if one does not use Bayes' rule, an opponent can make 'Dutch book' against him – his beliefs are not consistent in some sense. ('Dutch book' is a collection of bets that acts as a money pump: no matter what the outcome, the bettor loses money.) This is superficially appealing, but there are problems: First, beliefs have to be characterized as probabilities in the first place. Second, the argument only applies if the prior is 'proper,' that is, has total mass equal to one. So, the argument does not help if one wishes to use the 'uninformative (uniform) prior' on an unbounded set, such as the real line. See also Eaton (2008); Eaton and Freeman (2004).

[6]Selecting a prior is not a trivial task, although most applications of Bayesian methods take the prior as given. See footnote 5. Many studies simply take the prior to be a multivariate normal distribution on a discretized set of states of the world. Some researchers tacitly invoke Laplace's Principle of Insufficient Reason to posit 'uninformative' priors. Laplace's Principle of Insufficient Reason says that if there is no reason to believe that outcomes are not equally likely, assume that they are equally likely. For a discussion, see Freedman and Stark (2003). When the state of the world represents a physical quantity, some researchers use priors that are invariant under a group with respect to which the physics is invariant; similarly, when an estimation problem is invariant with respect to a group, some researchers use priors that are invariant with respect to that group. See, e.g., Lehmann and Casella (1998, p. 245ff). This is intuitively appealing and mathematically elegant, but there are difficulties. First, not all uncertainties are expressible as probabilities, so insisting on the use of any prior can be a leap. Second, why should the prior be invariant under the same group as the physics? The argument seems to be that if the physics does not distinguish between elements of an orbit of the group, the prior should not either: a change of coordinate systems should not affect the mathematical expression of our uncertainties. That is tantamount to applying Laplace's principle of insufficient reason to orbits of the group: if there is no reason to think any element of an orbit is more likely than any other, assume that they are equally likely. But this is a fallacy because 'no reason to believe that something is false' is not reason to believe that the thing is true. Absence of evidence is not evidence of absence; moreover, uncertainties have a calculus of their own, separate from the laws of physics. Third, invariance can be insufficient to determine a unique prior – and can be too restrictive to permit a prior. For example, there are infinitely many probability distributions on $\mathbb{R}^2$ that are invariant with respect to rotation about the origin, and there is no probability distribution on $\mathbb{R}^2$ that is invariant with respect to translation.

Moreover, without a prior probability distribution there can be no posterior distribution. There is interest in 'nonparametric' priors and hierarchical priors as well. Allowing a more general functional form for a prior does not alleviate the difficulties: hierarchical priors are uniform distribution on $[-\tau, \tau]$, which has probability density function

$$U_\tau(\eta) \equiv \begin{cases} \frac{1}{2\tau}, & -\tau \leq \eta \leq \tau \\ 0, & \text{otherwise} \end{cases}$$
$$\equiv \frac{1}{2\tau} 1_{[-\tau,\tau]}(\eta). \quad (2.5)$$

(In this example, $\pi(d\eta) = \frac{1}{2\tau} 1_{[-\tau,\tau]} d\eta$.) There are infinitely many probability distributions that assign probability 1 to $[-\tau, \tau]$; this is just one of them – one way to capture the prior information, and more. The constraint $\theta \in \Theta$ limits the *support* of π but says nothing about the probabilities π assigned to subsets of Θ. Every particular assignment – every particular choice of π – has more information than the constraint $\theta \in \Theta$, because it has information about the chance that θ is in each (measurable) subset of Θ. It expresses more than the fact that θ is an element of Θ. Below we will show how the assumption $\theta \sim \pi$ (with π the uniform distribution) reduces the apparent uncertainty (but not the true uncertainty) in estimates of θ from Y when what we really know is just $\theta \in \Theta$.

In the linear inverse problem, the prior π is a probability distribution on Θ, which is assumed to be a measurable space. In most real inverse problems, θ is a function of position – an element of an infinite-dimensional space. Constraints on θ might involve its norm or seminorm, non-negativity or other pointwise restrictions, for example Backus (1988, 1989); Evans and Stark (2002); Parker (1994); Stark (1992b,c). There are serious technical difficulties defining probability distributions on infinite-dimensional spaces in an 'uninformative' way that still captures the constraint $\theta \in \Theta$. For example, rotationally invariant priors on separable Hilbert spaces are degenerate: they either assign probability 1 to the origin, or they assign probability 1 to the event that the norm of θ is infinite – contrary to what a constraint on the norm is supposed to capture (Backus 1987).[7]

Recall that $p_\eta(y)$ is the likelihood of $\eta \in \Theta$ given y. We assume that $p_\eta(y)$ is jointly measurable with respect to η and y. The prior distribution π and the distribution $\mathbb{P}_\eta$ of the data Y given η determine the joint distribution of θ and Y. The *marginal distribution of* Y averages the distributions $\{\mathbb{P}_\eta : \eta \in \Theta\}$, weighted by π, the probability that θ is near η. The density of the marginal distribution of Y (with respect to μ) is

$$m(y) = \int_\Theta p_\eta(y)\, \pi(d\eta). \quad (2.6)$$

The marginal distribution of Y is also called the *predictive distribution* of Y. The information in the data Y can be used to update the prior π through

still priors, though priors built by putting priors on priors are rather harder to think about than priors that directly state the distribution of the parameters.

[7]Discretizing an infinite-dimensional inverse problem then positing a prior distribution on the discretization hides the difficulty, but does not resolve it.

Bayes' rule; the result is the *posterior distribution of θ given $Y = y$*:

$$\pi(d\eta|Y = y) = \frac{p_\eta(y)\,\pi(d\eta)}{m(y)}. \tag{2.7}$$

(The marginal density $m(y)$ can vanish; this happens with probability zero.)

In the BNM problem with uniform prior, the density of the predictive distribution of Y is

$$m(y) = \frac{1}{2\tau}\int_{-\tau}^{\tau} \varphi_\eta(y)d\eta = \frac{1}{2\tau}(\Phi(y+\tau) - \Phi(y-\tau)). \tag{2.8}$$

The posterior distribution of θ given $Y = y$ is

$$\pi(d\eta|Y = y) = \frac{\varphi_\eta(y)\frac{1}{2\tau}1_{\eta\in[-\tau,\tau]}(\eta)}{m(y)} = \frac{\varphi_\eta(y)1_{\eta\in[-\tau,\tau]}(\eta)}{\Phi(y+\tau) - \Phi(y-\tau)}. \tag{2.9}$$

Figure 2.1 shows the posterior density $f_\theta(\eta|y)$ of θ (the density of $\pi(d\eta|Y = y)$) for four values of y using a uniform prior on the interval $[-3, 3]$. The posterior is a re-scaled normal density with mean y, restricted to Θ. It is unimodal with mode y, but symmetric only when $y = 0$. The mode of the posterior density is the closest point in Θ to y.

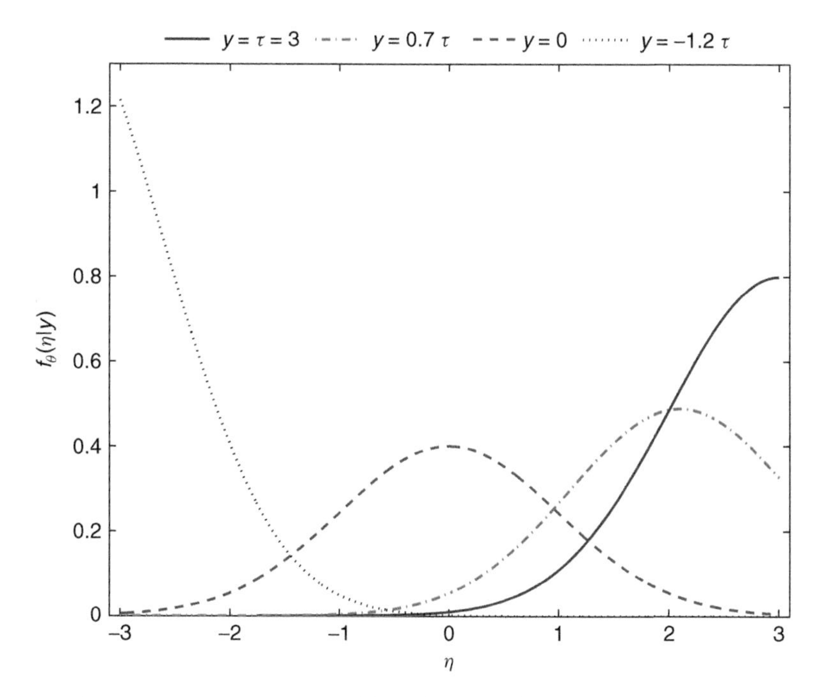

Figure 2.1 Posterior densities of θ for a bounded normal mean with a uniform prior on the interval $[-3, 3]$ for four values of y.

In the linear inverse problem, the density of the predictive distribution is

$$m(y) = \int_{\Theta} f(y - K\eta)\pi(d\eta), \tag{2.10}$$

and the posterior distribution of θ given $Y = y$ is

$$\pi(\eta|Y = y) = \frac{f(y - K\eta)\pi(d\eta)}{m(y)}. \tag{2.11}$$

Any parameter $\lambda[\theta]$ has a *marginal posterior distribution given* $Y = y$, $\pi_\lambda(d\ell|Y = y)$, induced by the posterior distribution of θ given $Y = y$. It is defined by

$$\int_{\Lambda} \pi_\lambda(d\ell|Y = y) \equiv \int_{\eta:\lambda[\eta]\in\Lambda} \pi(d\eta|Y = y) \tag{2.12}$$

for suitably measurable sets Λ.

2.3 Estimators: What Can You Do with What You Measure?

Estimators are measurable mappings from the set $\mathcal{Y}$ of possible observations to some other set. Let $\mathcal{L}$ denote the set of possible values of the parameter $\lambda[\eta]$ as η ranges over Θ. *Point estimators* of $\lambda[\theta]$ assign an element $\ell \in \mathcal{L}$ to each possible observation $y \in \mathcal{Y}$. For example, in the BNM problem, $\mathcal{L} = [-\tau, \tau]$ is the set of possible values of the parameter θ, and we might consider the *truncation estimator*

$$\begin{aligned} \widehat{\theta}_\tau(y) &\equiv \begin{cases} -\tau, & y \leq -\tau, \\ y, & -\tau < y < \tau \\ \tau, & y \geq \tau \end{cases} \\ &\equiv -\tau 1_{(-\infty,-\tau]}(y) + y 1_{(-\tau,\tau)}(y) + \tau 1_{[\tau,\infty)}(y). \end{aligned} \tag{2.13}$$

We will consider other point estimators for the BNM problem below.

A *set estimator* of a parameter assigns a subset of $\mathcal{L}$ to each possible observation $y \in \mathcal{Y}$. Confidence intervals are set estimators. For example, in the BNM problem we might consider the truncated naive interval

$$\mathcal{I}_\tau(y) \equiv [-\tau, \tau] \cap [y - 1.96, y + 1.96]. \tag{2.14}$$

(This interval is empty if $y < -\tau - 1.96$ or $y > \tau + 1.96$.) This interval has the property that

$$\mathbb{P}_\eta\{\mathcal{I}_\tau(Y) \ni \eta\} = 0.95, \tag{2.15}$$

whenever $\eta \in [-\tau, \tau]$. Below we will study the *truncated Pratt interval*, a confidence interval that uses the constraint in a similar way, but tends to be shorter – especially when τ is small.

More generally, a set estimator S for the parameter $\lambda[\theta]$ that has the property

$$\mathbb{P}_\eta\{S(Y) \ni \lambda[\eta]\} \geq 1 - \alpha \tag{2.16}$$

for all $\eta \in \Theta$ is called a $1 - \alpha$ *confidence set for* $\lambda[\theta]$.

A $1 - \alpha$ *Bayesian credible region* for a parameter $\lambda[\theta]$ is a set that has posterior probability at least $1 - \alpha$ of containing $\lambda[\theta]$. That is, if $S_\pi(y)$ satisfies

$$\int_{S_\pi(y)} \pi_\lambda(d\ell | Y = y) \geq 1 - \alpha, \tag{2.17}$$

then $S_\pi(y)$ is a $1 - \alpha$ credible region. Among sets with the property (2.17), we may choose as the credible region the one with smallest volume: that is, the smallest region that contains θ with posterior probability at least $1 - \alpha$. If the posterior distribution of $\lambda[\theta]$ has a density, that credible region is the *highest posterior density credible set*, a set of the form $S_\pi(y) = \{\eta : \pi_\lambda(d\ell | Y = y) \geq c\}$ with c chosen so that $S_\pi(y)$ satisfies (2.17). See, e.g., Berger (1985).

In the bounded normal mean problem with uniform prior, the highest posterior density credible region for θ is of the form

$$\{\eta \in [-\tau, \tau] : e^{-(y-\eta)^2/2} 1_{\eta \in [-\tau,\tau]}(\eta) \geq c\}, \tag{2.18}$$

where c depends on y and is chosen so that (2.17) holds. Figure 2.2 shows examples of the truncated naive confidence interval, the truncated Pratt confidence interval and credible regions for several values of y and τ. The truncated naive and truncated Pratt intervals are at 95 % confidence level; the credible regions are at 95 % credible level. When y is near 0, the truncated Pratt interval is the shortest. When $|y|$ is sufficiently large, the truncated naive interval is empty. Section 2.5 shows the confidence level of the credible region in the BNM problem.

2.4 Performance of Estimators: How Well Can You Do?

This section presents several properties of estimators that can be used to define what it means for an estimator to be 'good,' or for one estimator to be 'better' than another.

2.4.1 Bias, Variance

Statisticians commonly consider the bias and variance of point estimators. The bias is the expected difference between the parameter and the estimate

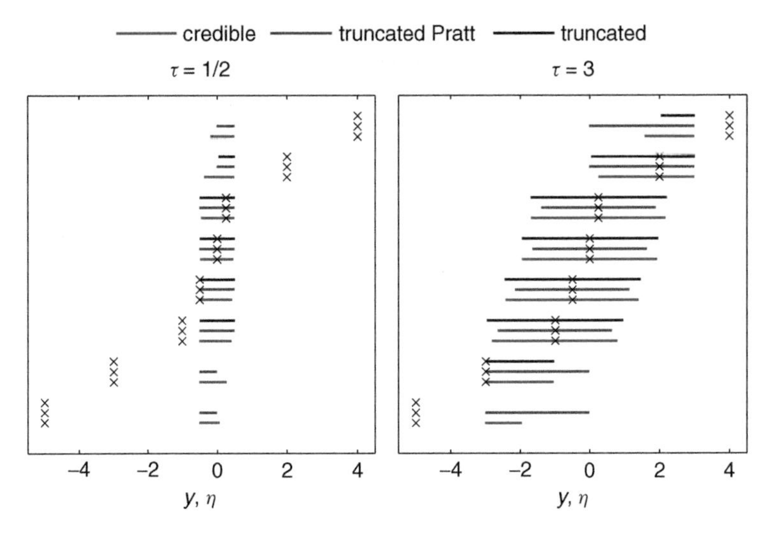

Figure 2.2 Examples of confidence intervals and credible sets for the BNM with $\tau = 1/2$ and $\tau = 3$, for eight different values of y. The horizontal axis represents the datum $Y = y$, indicated by an x, and the values η in the confidence interval for the observed value of y, which are plotted as horizontal bars. The red bars (bottom bar in each triple) are 95 % maximum posterior density credible regions using a uniform prior on $[-\tau, \tau]$. The blue bars (middle bar in each triple) are 95 % truncated Pratt intervals and the black bars (top bar in each triple) are 95 % truncated naive intervals. The truncated naive intervals are empty when $|y| > \tau + 1.96$.

of the parameter. For example, let $\widehat{\lambda}(Y)$ be an estimator of $\lambda[\theta]$. The *bias at* η *of* $\widehat{\lambda}$ *is*

$$\text{Bias}_\eta(\widehat{\lambda}) \equiv \mathbb{E}_\eta(\widehat{\lambda}(Y) - \lambda[\eta]) = \mathbb{E}_\eta\widehat{\lambda}(Y) - \lambda[\eta]. \tag{2.19}$$

when the expectation exists. The estimator $\widehat{\lambda}$ is *unbiased for* $\lambda[\eta]$ *at* η if $\text{Bias}_\eta(\widehat{\lambda}) = 0$. It is *unbiased* if $\text{Bias}_\eta(\widehat{\lambda}) = 0$ for all $\eta \in \Theta$. If there exists an unbiased estimator of $\lambda[\theta]$ then we say $\lambda[\theta]$ *is unbiasedly estimable* (or *U-estimable*). If $\lambda[\theta]$ is U-estimable, there is an estimator that, on average across possible samples, is equal to $\lambda[\theta]$. (Not every parameter is U-estimable.)

For example, the bias at η of the truncation estimator $\widehat{\theta}_\tau$ in the BNM problem is

$$\text{Bias}_\eta(\widehat{\theta}_\tau) = \mathbb{E}_\eta\widehat{\theta}_\tau(Y) - \eta, \tag{2.20}$$

where

$$\begin{aligned}\mathbb{E}_\eta\widehat{\theta}_\tau(Y) &= \int_{-\infty}^{\infty} \widehat{\theta}_\tau(y)\varphi_\eta(y)dy \\ &= (-\tau - \eta)\Phi(-\tau - \eta) + (\tau - \eta)\Phi(-\tau + \eta) + \\ &\quad + \varphi(-\tau - \eta) - \varphi(\tau - \eta),\end{aligned} \tag{2.21}$$

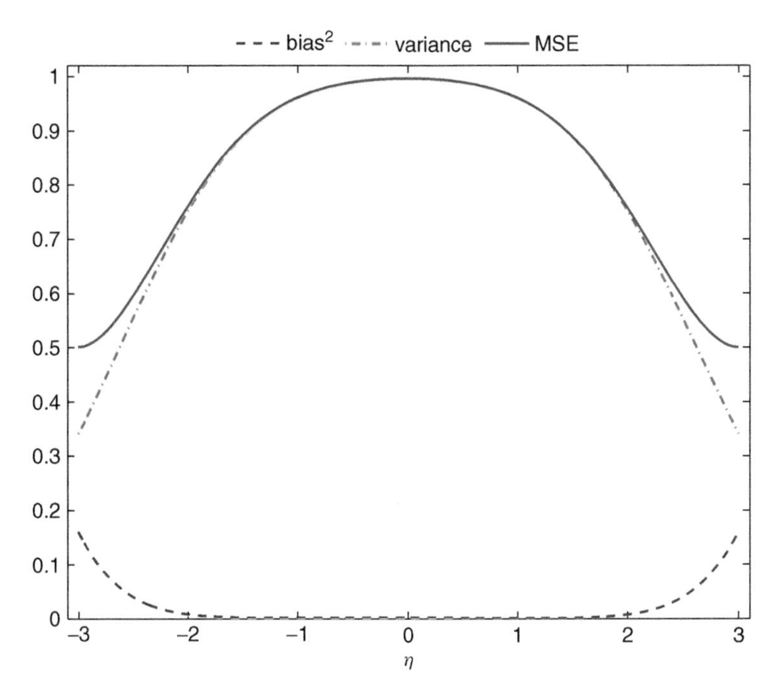

Figure 2.3 Squared bias, variance and MSE of the truncation estimator $\widehat{\theta}_\tau$ as a function of $\eta \in [-3, 3]$.

and $\varphi(y) = \varphi_0(y)$ is the standard Gaussian density (Equation (2.3) with $\eta = 0$) and $\Phi(y) = \int_{-\infty}^{y} \varphi(x)dx$ is the standard Gaussian cumulative distribution function (cdf). Note that if $\eta = 0$, the bias at η of $\widehat{\theta}_\tau$ is zero, as one would expect from symmetry. Figure 2.3 shows $\text{Bias}^2_\eta(\widehat{\theta}_\tau)$ as a function of η. Note that the squared bias increases as $|\eta|$ approaches τ. There, the truncation acts asymmetrically, so that the estimator is much less likely to be on one side of η than the other. For example, when $\eta = -\tau$, the estimator is never below η but often above, so the expected value of $\widehat{\theta}_\tau$ is rather larger than η – the estimator is biased.

The *variance at* η of an estimator $\widehat{\lambda}$ of a real-valued parameter $\lambda[\theta]$ is

$$\text{Var}_\eta(\widehat{\lambda}) \equiv \mathbb{E}_\eta \left(\widehat{\lambda}(Y) - \mathbb{E}_\eta(\widehat{\lambda}(Y)) \right)^2 \tag{2.22}$$

when the two expectations on the right exist. For example, the variance at η of the truncation estimator $\widehat{\theta}_\tau$ in the BNM problem is

$$\begin{aligned}\text{Var}_\eta \widehat{\theta}_\tau &= \int_{-\infty}^{\infty} \left(\widehat{\theta}_\tau(y) - \left(\int_{-\infty}^{\infty} \widehat{\theta}_\tau(x)\varphi(x-\eta)dx \right) \right)^2 \varphi(y-\eta)dy \\ &= (\tau^2 - \eta^2 - 1)\Phi(-\tau-\eta) + (\eta^2 - \tau^2 + 1)\Phi(\tau-\eta) + (\eta-\tau)\varphi(\tau+\eta) \\ &\quad -(\tau+\eta)\varphi(\tau-\eta) + \tau^2 - (\,\mathbb{E}_\eta\widehat{\theta}_\tau(Y)\,)^2. \end{aligned} \tag{2.23}$$

Figure 2.3 shows $\text{Bias}^2_\eta(\widehat{\theta}_\tau)$ and $\text{Var}_\eta(\widehat{\theta}_\tau)$ as functions of η for $\tau = 3$. The variance is biggest near the origin, where truncation helps least – if τ is large, the variance at $\eta = 0$ approaches the variance of the additive Normal noise, $\sigma^2 = 1$. Near $\pm\tau$, the truncation reduces the variance because it keeps the estimator from being very far from η on one side. But as we have seen, that introduces bias.

Neither bias nor variance alone suffices to quantify the error of an estimator: even if an estimator is unbiased, its variance can be so large that it is rarely close to the parameter it estimates.[8] Conversely, an estimator can have small or zero variance, and yet its bias can be so large that it is never close to the parameter. Small variance means small scatter; small bias means that on average the estimator is right.[9] Insisting that estimators be unbiased can be counter-productive, for a number of reasons. For instance, there are problems in which there is no estimator that is unbiased for all $\eta \in \Theta$. And allowing some bias can reduce the overall MSE, or another measure of the accuracy of an estimator. We would agree that it is important to have a bound on the magnitude of the bias – but that bound need not be zero.

The next subsection presents a general approach to defining the quality of estimators.

2.4.2 Loss and Risk

A more general way to evaluate and compare estimators is through a *loss function* that measures the loss we incur if we take a particular *action* ℓ when the true value of the parameter is $\lambda[\eta]$. The *risk at* η of an estimator is the expected value of the loss when that estimator is used and the world is in state η:

$$\rho_\eta(\widehat{\lambda}, \lambda[\eta]) \equiv \mathbb{E}_\eta \operatorname{loss}(\widehat{\lambda}(Y), \lambda[\eta]). \tag{2.24}$$

For example, consider estimating a real-valued parameter $\lambda[\theta]$ from data Y. For point estimates, actions are possible parameter values – real numbers in this case. We could define the loss for taking the action $\ell \in \mathbb{R}$ when the true parameter value is $\lambda[\eta]$ to be $\operatorname{loss}(\ell, \lambda[\eta]) = |\ell - \lambda[\eta]|^p$, for example. Scientific context should dictate the loss function,[10] but most theoretical work

[8]Three statisticians go deer hunting. They spot a deer. The first shoots; the shot goes a yard to the left of the deer. The second shoots; the shot goes a yard to the right of the deer. The third statistician shouts, 'we got him!'

[9]Estimation is like shooting a rifle. If the rifle is well made, the shots will hit nearly the same mark; but if its sights are not adjusted well, that mark could be far from the bullseye. Conversely, shots from a poorly made rifle will be scattered, no matter how well adjusted its sights are. The quality of the rifle is analogous to the variance: small variance corresponds to high quality – low scatter. The adjustment of the sights is analogous to the bias: low bias corresponds to well adjusted sights – on average, the estimate and the shots land where they should.

[10]For example, consider the problem of estimating how deep to set the pilings for a bridge. If the pilings are set too deep, the bridge will cost more than necessary. But if the pilings

on point estimation uses *squared-error loss*: $\text{loss}(\ell, \lambda[\eta]) = |\ell - \lambda[\eta]|^2$. For squared-error loss, the risk at η of the estimator $\hat{\lambda}$ is the *mean squared error*:

$$\text{MSE}_\eta(\widehat{\lambda}) \equiv \mathbb{E}_\eta|\widehat{\lambda}(Y) - \lambda[\eta]|^2, \tag{2.25}$$

when the expectation exists. The MSE of a point estimate can be written as the square of the bias plus the variance:

$$\text{MSE}_\eta(\widehat{\lambda}) = \text{Bias}^2_\eta(\widehat{\lambda}) + \text{Var}_\eta(\widehat{\lambda}). \tag{2.26}$$

In the BNM problem, the MSE at η of the truncation estimator $\widehat{\theta}_\tau$ of θ is

$$\text{MSE}_\eta(\widehat{\theta}) = \mathbb{E}_\eta|\widehat{\theta} - \eta|^2 = \text{Bias}^2_\eta\widehat{\theta}_\tau + \text{Var}_\eta(\widehat{\theta}_\tau). \tag{2.27}$$

Figure 2.3 plots $\text{MSE}_\eta(\widehat{\theta})$ as a function of η for $\tau = 3$. Near the origin the MSE is dominated by the variance; near the boundaries the squared bias becomes important, too.

For set estimators, one measure of loss is the size (Lebesgue measure) of the set; the risk at η is then the expected measure of the set when the true state of the world is η. For example, let $\mathcal{I}$ be a (suitably measurable) set of confidence set estimators – estimators that, with probability at least $1 - \alpha$ produce a (Lebesgue-measurable) set that contains $\lambda[\eta]$ when η is the true state of the world. That is, if $I \in \mathcal{I}$ then

$$\mathbb{P}_\eta\{I(Y) \ni \lambda[\eta]\} \geq 1 - \alpha \tag{2.28}$$

for all $\eta \in \Theta$. If $I(Y)$ is always Lebesgue-measurable,

$$\rho_\eta(I, \lambda[\eta]) \equiv \mathbb{E}_\eta \int_\ell I(Y) d\ell \tag{2.29}$$

is the expected measure of the set, a reasonable risk function for set estimates; see Evans *et al.* (2005). For example, Figure 2.4 shows the expected length of the truncated naive interval (and some other intervals) for the bounded normal mean, as a function of η. We will discuss this plot below.

2.4.3 Decision Theory

Decision theory provides a framework for selecting estimators. Generally, the risk of an estimator $\widehat{\lambda}$ of a parameter $\lambda[\theta]$ depends on the state θ that the world is in: some estimators do better when θ has one value; some do better when θ takes a different value. Some estimators do best when θ is a random

are set too shallow, the bridge will collapse, and lives could be lost. Such a loss function is asymmetric and highly nonlinear. Some theoretical results in decision theory depend on details of the loss function (for instance, differentiability, convexity, or symmetry), and squared error is used frequently because it is analytically tractable.

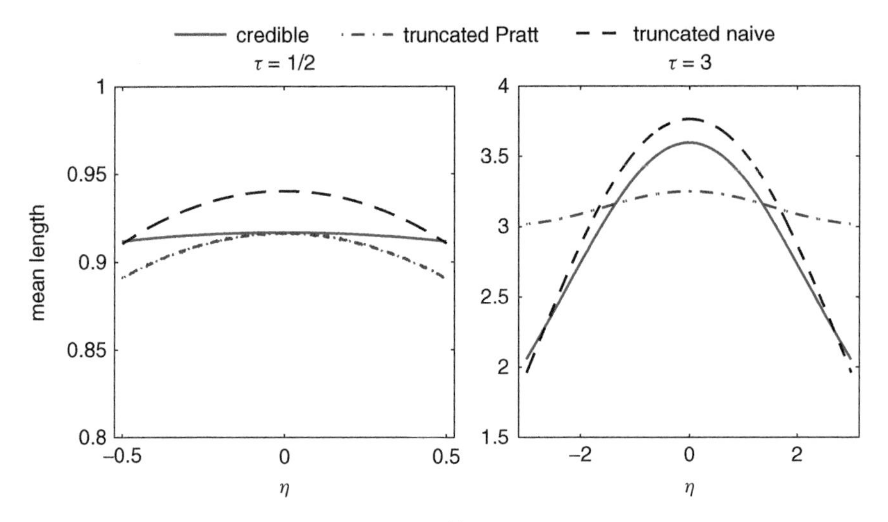

Figure 2.4 Expected length of the 95 % credible, truncated naive and truncated Pratt interval for the BNM as a function of η for $\tau = 1/2$ and $\tau = 3$. For reference, the length of the naive confidence interval, which does not use the information $\theta \in \Theta$, is $2 \times 1.96 = 3.92$ for all y, so its expected length is 3.92 for all η.

variable with a known distribution π. If we knew θ (or π), we could pick the estimator that had the smallest risk. But if we knew θ, we would not need to *estimate* $\lambda[\theta]$: we could just *calculate* it.

Decision theory casts estimation as a two-player game: Nature versus statistician. Frequentists and Bayesians play similar – but not identical – games. In both games, Nature and the statistician know Θ; they know how data will be generated from θ; they know how to calculate $\lambda[\theta]$ from any $\theta \in \Theta$; and they know the loss function ρ that will be used to determine the payoff of the game. Nature selects θ from Θ without knowing what estimator $\widehat{\lambda}$ the statistician plans to use. The statistician, ignorant of θ, selects an estimator $\widehat{\lambda}$. The referee repeatedly generates Y from the θ that Nature selected and calculates $\widehat{\lambda}(Y)$. The statistician has to pay the average value of $\text{loss}(\widehat{\lambda}(Y), \lambda[\theta])$ over all those values of Y; that is, $\rho_\theta(\widehat{\lambda}, \lambda[\theta])$.

The difference between the Bayesian and frequentist views is in how Nature selects θ from Θ. Bayesians suppose that Nature selects θ at random according to the prior distribution π – and that the statistician knows *which* distribution π Nature is using.[11] Frequentists do not suppose that Nature draws θ from π, or else they do not claim to know π. In particular, frequentists do

[11] Some Bayesians do not claim to know π, but claim to know that π itself was drawn from a probability distribution on probability distributions. Such *hierarchical priors* do not really add any generality: they are just a more complicated way of specifying a probability distribution on states of the world. A weighted mixture of priors is a prior.

not rule out the possibility that Nature might select θ from Θ to maximize the amount the statistician has to pay on average. The difference between Bayesians and frequentists is that Bayesians claim to know more than frequentists: both incorporate constraints into their estimates and inferences, but Bayesians model uncertainty about the state of the world using a particular probability distribution in situations where frequentists would not. Bayesians speak of the probability that the state of the world is in a given subset of Θ. For frequentists, such statements generally do not make sense: the state of the world is not random with a known probability distribution – it is simply unknown. The difference in the amount of information in 'θ is a random element of Θ drawn from a known probability distribution' and 'θ is an unknown element of Θ' can be small or large, depending on the problem and the probability distribution.

So, which estimator should the statistician use? The next two subsections show those differences in the rules of the game lead to different strategies Bayesians and frequentists use to select estimators.

Minimax Estimation

One common strategy for frequentists is to minimize the maximum they might be required to pay. That is, they choose the estimator (among those in a suitable class) that has the smallest worst-case risk over all possible states of the world $\eta \in \Theta$.

For example, consider the BNM problem with MSE risk. Restrict attention to the class of estimators of θ that are affine functions of the datum Y, that is, the set of estimators $\widehat{\theta}$ of the form

$$\widehat{\theta}_{ab}(y) = a + by, \tag{2.30}$$

where a and b are real numbers. We can choose a and b to minimize the maximum MSE for $\eta \in \Theta$. The resulting estimator, $\widehat{\theta}_A$, is the *minimax affine estimator* (for MSE loss). To find $\widehat{\theta}_A$ we first calculate the MSE of $\widehat{\theta}_{ab}$. Let $Z \sim N(0, 1)$ be a standard Gaussian random variable, so that $Y \sim \eta + Z$ if $\theta = \eta$.

$$\begin{aligned} MSE_\eta(\widehat{\theta}_{ab}) &= \mathbb{E}_\eta(\widehat{\theta}_{ab}(Y) - \eta)^2 \\ &= \mathbb{E}_\eta(a + b(\eta + Z) - \eta)^2 \\ &= \mathbb{E}_\eta(a + (b-1)\eta + bZ)^2 \\ &= (a + (b-1)\eta)^2 + b^2 \\ &= a^2 + 2a(b-1)\eta + (b-1)^2\eta^2 + b^2. \end{aligned} \tag{2.31}$$

This is quadratic in η with positive leading coefficient, so the maximum will occur either at $\eta = -\tau$ or $\eta = \tau$, whichever has the same sign as $a(b-1)$. Since the leading term is nonnegative and the second term will be positive if $a \neq 0$, it follows that $a = 0$ for the minimax affine estimator. We just have to find the optimal value of b. The MSE of the minimax affine estimator is thus $\tau^2(b-1)^2 + b^2$, a quadratic in b with positive leading coefficient. So the minimum MSE will occur at a stationary point with respect to b, and we find

$$2\tau^2(b-1) + 2b = 0, \tag{2.32}$$

i.e. the minimax affine estimator is

$$\widehat{\theta}_A(y) = \frac{\tau^2}{\tau^2+1} \cdot y. \tag{2.33}$$

This is an example of a *shrinkage estimator*: it takes the observation and shrinks it towards the origin, since $\tau^2/(\tau^2+1) \in (0,1)$. Multiplying the datum by a number less than 1 makes the variance of the estimator less than the variance of Y – the variance of the minimax affine estimator is $\tau^2/(\tau^2+1)^2$, while the variance of Y is one. Shrinkage also introduces bias: the bias is $\eta(\tau^2/(\tau^2+1)-1)$. But because we know that $\eta \in [-\tau, \tau]$, the square of bias is at most $\tau^2(\tau^2/(\tau^2+1)-1)^2$, and so the MSE of the minimax affine estimator is at most

$$\frac{\tau^2}{(\tau^2+1)^2} + \tau^2\left(1 - \frac{\tau^2}{\tau^2+1}\right)^2 = \frac{\tau^2}{\tau^2+1} < 1. \tag{2.34}$$

By allowing some bias, the variance can be reduced enough that the MSE is smaller than the MSE of the raw estimator Y of θ (the MSE of Y as an estimator of θ is 1 for all $\eta \in \Theta$)

Affine estimators use the data in a very simple way. The truncation estimator $\widehat{\theta}_\tau$ is not affine. What about even more complicated estimators? If we allowed an estimator to depend on Y in an arbitrary (but measurable) way, how small could its maximum MSE be than the maximum MSE of the minimax affine estimator?

In the BNM problem, the answer is 'not much smaller': the maximum MSE of the minimax (nonlinear) estimator of θ is no less than 4/5 the maximum MSE of the minimax affine estimator of θ (Donoho *et al.* 1990). Let $a \vee b \equiv \max(a, b)$ and $a \wedge b \equiv \min(a, b)$. Figure 2.6 also shows the risk of the *truncated minimax affine estimator*

$$\widehat{\theta}_{A\tau}(y) = \tau \wedge (-\tau \vee \tau(\tau^2+1)^{-1}y) \tag{2.35}$$

as a function of η. This estimator improves the minimax affine estimator by ensuring that the estimate is in $[-\tau, \tau]$ even when $|Y|$ is very large. For $\tau = 1/2$, the maximum risk of the truncated minimax affine estimator is about 25 % larger than the lower bound on the maximum risk of any nonlinear estimator. For $\tau = 3$, its maximum risk is about 12 % larger than the lower bound.

We can also find confidence set estimators in various classes that are minimax for a risk function, such as expected length. Stark (1992a) studies the minimax length of fixed-length confidence intervals for a BNM when the intervals are centered at an affine function of Y. Zeytinoglu and Mintz (1984) study the minimax length of fixed-length confidence intervals for a BNM when the intervals are centered at nonlinear functions of Y. Evans *et al.* (2005) study minimax expected size confidence sets where the size can vary with Y. They show that when $\tau \le 2\Phi^{-1}(1-\alpha)$, the minimax expected size confidence interval for the BNM problem is the *truncated Pratt interval*:

$$\mathcal{I}_{\mathrm{P}\tau}(Y) \equiv \mathcal{I}_P(Y) \cap [-\tau, \tau], \tag{2.36}$$

where $\mathcal{I}_P(Y)$ is the Pratt interval (Pratt 1961):

$$\mathcal{I}_{\mathrm{P}}(Y) \equiv \begin{cases} [(Y-\Phi^{-1}(1-\alpha)), 0 \vee (Y+\Phi^{-1}(1-\alpha))], & Y \le 0 \\ [0 \wedge (Y-\Phi^{-1}(1-\alpha)), Y+\Phi^{-1}(1-\alpha)], & Y > 0. \end{cases} \tag{2.37}$$

The truncated Pratt interval $\mathcal{I}_{\mathrm{P}\tau}$ has minimax expected length among all $1-\alpha$ confidence intervals for the BNM when $\tau \le 2\Phi^{-1}(1-\alpha)$. (For $\alpha = 0.05$, that is $\tau \le 3.29$.) For larger values of τ, the minimax expected length confidence procedure can be approximated numerically; see Evans *et al.* (2005); Schafer and Stark (2009). Figure 2.4 compares the expected length of the minimax expected length confidence interval for a BNM with the expected length of the truncated naive interval as a function of η. It also shows the expected length of the Bayes credible region using a uniform prior (see the next section). Note that the minimax expected length interval has the smallest expected length when η is near zero.

Bayes Risk

The *average ρ-risk of an estimator for prior π* is the mean risk of the estimator when the parameter θ is chosen from the prior π:

$$\rho(\widehat{\lambda}, \lambda; \pi) \equiv \int_\eta \rho_\eta(\widehat{\lambda}, \lambda[\eta])\pi(d\eta). \tag{2.38}$$

The Bayes ρ-risk for prior π is the smallest average ρ-risk of any estimator for prior π:

$$\rho(\lambda; \pi) \equiv \inf_{\widehat{\lambda}} \rho(\widehat{\lambda}, \lambda; \pi), \tag{2.39}$$

where the infimum is over some suitable class of estimators. An estimator whose average ρ-risk for prior π is equal to the Bayes ρ-risk for prior π is called a *Bayes estimator*.

Consider estimating a real-valued parameter $\lambda[\theta]$ under MSE risk. It is well known that the Bayes estimator is the mean of the marginal posterior distribution of $\lambda[\theta]$ given Y, $\pi_\lambda(d\ell|Y = y)$:

$$\hat{\lambda}_\pi(Y) \equiv \int \ell \pi_\lambda(d\ell|Y = y). \tag{2.40}$$

See e.g. Berger (1985); Lehmann and Casella (1998).[12]

For example, in the BNM problem the average MSE risk of the estimator $\widehat{\lambda}(Y) = Y$ of θ for the uniform prior on $[-\tau, \tau]$ is

$$\rho(Y, \theta; \pi) \equiv \frac{1}{2\pi} \int_{-\tau}^{\tau} \mathbb{E}_\eta (Y - \eta)^2 d\eta = 1. \tag{2.41}$$

The Bayes risk is

$$\rho(\theta, \pi) = \inf_{\widehat{\lambda}} \rho(\widehat{\lambda}, \theta; \pi). \tag{2.42}$$

And the Bayes estimator is

$$\hat{\theta}_\pi(Y) \equiv Y - \frac{\varphi(\tau - Y) - \varphi(\tau + Y)}{\Phi(\tau - Y) - \Phi(-\tau - Y)}. \tag{2.43}$$

Because the posterior distribution of θ has a unimodal density, a level set of the posterior density of the form $S(Y) = \{\eta : \pi(d\eta|Y = y) \geq c\}$ is an interval. If c is chosen so that the posterior probability of $S(Y)$ is $1 - \alpha$, S is a $1 - \alpha$ credible interval. Figure 2.4 shows the expected length of the Bayes credible region for the BNM using a uniform prior.

Bayes/Minimax Duality

The Bayes risk depends on the parameter to be estimated and the loss function – and also on the prior. Consider a set of priors that includes point masses at each element of Θ (or at least, priors that can assign probability close to 1 to small neighborhoods of each $\eta \in \Theta$). Vary the prior over that set to find a prior for which the Bayes risk is largest. Such priors are called 'least favorable.'

The frequentist minimax problem finds an $\eta \in \Theta$ for which estimating $\lambda[\eta]$ is hardest. The least favorable prior is a distribution on Θ for which estimating $\lambda[\eta]$ is hardest *on average* when θ is drawn at random from that prior. Since the set of priors in the optimization problem includes distributions concentrated near the $\eta \in \Theta$ for which the minimax risk is attained, the Bayes risk for the least favorable prior is no smaller than the minimax risk. Perhaps surprisingly, the Bayes risk for the least favorable prior is in fact equal to

[12]This follows from the general result that if X is a real-valued random variable with probability measure μ, then among all constants c, the smallest value of $\int (x - c)^2 \mu(dx)$ is attained when $c = \int x\mu(dx) = \mathbb{E}(X)$. Let $\mu(dx) = \pi_\lambda(dx|Y = y)$.

the minimax risk under mild conditions. See, for example, Berger (1985); Lehmann and Casella (1998).[13]

Because the Bayes risk for the least favorable prior is a lower bound on the minimax risk, comparing the Bayes risk for any particular prior with the minimax risk measures how much information the prior adds: whenever the Bayes risk is smaller than the minimax risk, it is because the prior is adding more information than simply the constraint $\theta \in \Theta$ or the data. When that occurs, it is prudent to ask whether the statistician *knows* that θ is drawn from the distribution π, or has adopted π to capture the constraint that $\theta \in \Theta$. If the latter, the Bayes risk understates the true uncertainty.

2.5 Frequentist Performance of Bayes Estimators for a BNM

In this section, we examine Bayes estimates in the BNM problem with uniform prior from a frequentist perspective.[14] In particular, we look at the maximum MSE of the Bayes estimator of θ for MSE risk, and at the frequentist coverage probability and expected length of the 95 % Bayes credible interval for θ.[15]

2.5.1 MSE of the Bayes Estimator for BNM

Figure 2.5 shows the squared bias, variance and MSE of the Bayes estimator of a bounded normal mean for MSE risk using a uniform prior, for $\tau = 3$. The figure can be compared with Figure 2.3, which plots the squared bias, variance and MSE of the truncation estimator for the same problem. Note that the squared bias of the Bayes estimate is comparatively larger, and the variance is smaller. The Bayes estimate has its largest MSE when $|\eta|$ is close to τ, which is where the truncation estimate has its smallest MSE. Figure 2.6 compares the MSE of the Bayes estimator of θ with the MSE of the truncation estimator $\widehat{\theta}_\tau$ and the minimax affine estimator $\widehat{\theta}_A$, as a function of θ, for $\tau = 1/2$ and $\tau = 3$. When $\tau = 1/2$, the truncation estimator is *dominated* by both of the others: the other two have risk functions that are smaller for every $\eta \in \Theta$. The risk of the Bayes estimator is smaller than that of the nonlinear minimax estimator except when $|\eta|$ is close to τ, although the two are quite close everywhere.

[13]Casella and Strawderman (1981) derive the (nonlinear) minimax MSE estimator for the BNM problem for small τ by showing that a particular 2-point prior is least favorable when τ is sufficiently small and that a 3-point prior is least favorable when τ is a little larger. They show that the number of points of support of the least favorable prior grows as τ grows. Similarly, Evans *et al.* (2005) construct minimax expected measure confidence intervals by characterizing the least favorable prior distributions; see also Schafer and Stark (2009).

[14]Generally, Bayes point estimators are biased. For more about (frequentist) risk functions of Bayes estimators, see, e.g., Lehmann and Casella (1998, p. 241ff).

[15]For a discussion of the frequentist consistency of Bayes estimates, see Diaconis and Freedman (1986, 1998).

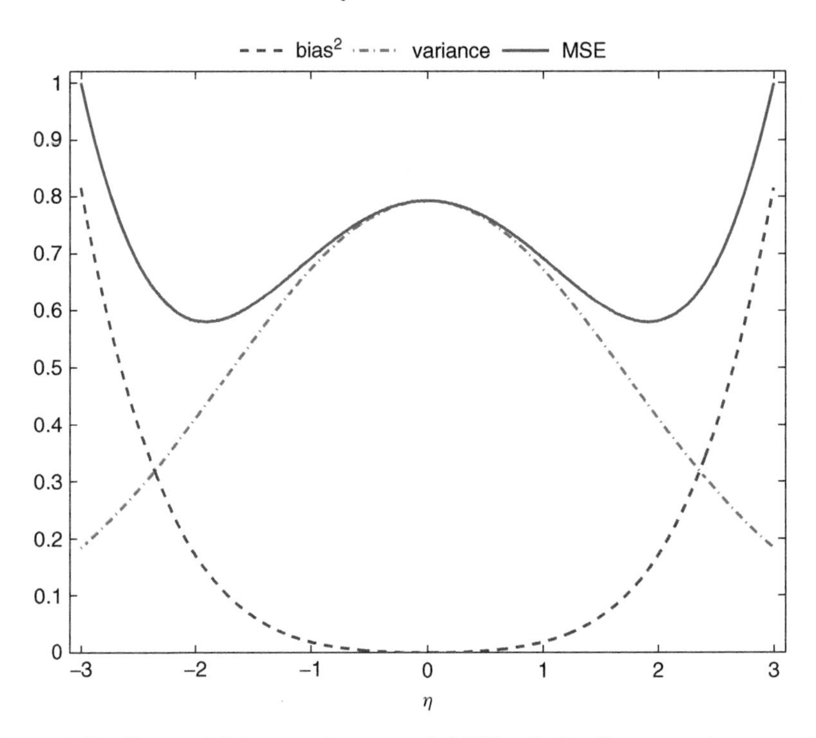

Figure 2.5 Squared bias, variance and MSE of the Bayes estimator of a bounded normal mean using a uniform prior on $[-3, 3]$.

When $\tau = 3$, none of the estimators dominates either of the others: there are regions of $\eta \in \Theta$ where each of the three has the smallest risk. The truncation estimator does best when $|\eta|$ is close to τ and worst when $|\eta|$ is near zero. The Bayes estimator generally has the smallest risk of the three, except when $|\eta|$ is close to τ, where its risk is much larger than that of the truncation estimator and noticeably larger than that of the minimax estimator.

2.5.2 Frequentist Coverage of the Bayesian Credible Regions for BNM

Figure 2.7 plots the coverage probability of the 95 % Bayes maximum posterior density credible region as a function of $\eta \in \Theta$, for $\tau = 1/2$ and $\tau = 3$. For $\tau = 1/2$, the coverage probability is nearly 100 % when η is in the middle of Θ, but drops precipitously as η approaches $\pm\tau$, where it is 68 %. For $\tau = 3$, the coverage probability is smallest when η is near zero, where it is 90.9 %. For most values of η, the coverage probability of the credible region is at least 95 %. On average over $\eta \in \Theta$, the coverage probability is about right, but for some values is far too large and for some it is far too small.

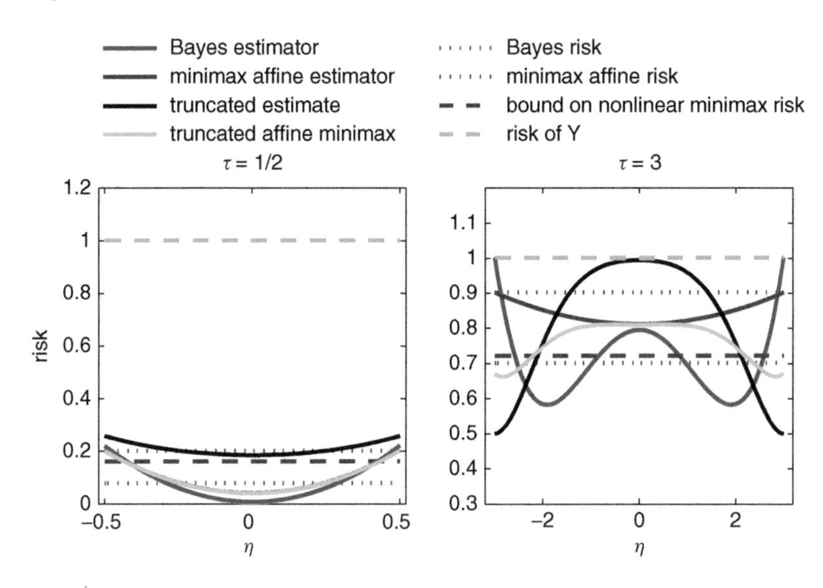

Figure 2.6 MSE risk of the naive estimator Y, the truncation estimator, the Bayes estimator (for uniform prior), the minimax affine estimator and the truncated minimax affine estimator for the BNM problem. For each estimator, the risk at η is plotted as a function of η for $\tau = 1/2$ and $\tau = 3$. The risk of Y is constant and equal to 1. The other three horizontal lines are the Bayes risk of the Bayes estimator, the minimax risk of the minimax affine estimator and a lower bound on the minimax risk of any nonlinear estimator. Some of the risks are computed analytically; others using 6×10^6 simulations. The fluctuations in the empirical approximation are less than the linewidth in the plots.

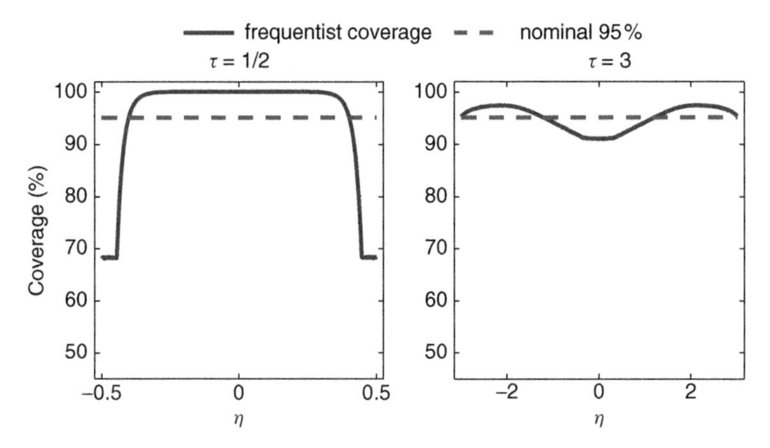

Figure 2.7 Frequentist coverage as a function of η of the 95 % credible interval for the BNM with a uniform prior on $[-\tau, \tau]$ for $\tau = 1/2$ and $\tau = 3$.

2.5.3 Expected Length of the Bayesian Credible Region for BNM

The frequentist coverage probability of the Bayesian 95 % credible region is generally close to 95%, but smaller for some values of θ. Does the loss in coverage probability buy a substantial decrease in the expected length of the interval? Figure 2.4 shows that is not the case: when τ is small, the expected length truncated Pratt interval is nowhere longer than that of the maximum posterior density credible region; when τ is moderate, the truncated naive interval has expected length only slightly larger than that of the credible region, and the truncated Pratt interval has smaller expected length for much of Θ.

2.6 Summary

Inverse problems involve estimating or drawing inferences about a *parameter* – a property of the unknown state of the world – from indirect, noisy data. Prior information about the state of the world can reduce the uncertainty in estimates and inferences. In physical science, prior information generally consists of constraints. For example, mass density is nonnegative; velocities are bounded; energies are finite. Frequentist methods can use such constraints directly. Bayesian methods require that constraints be re-formulated as prior probability distribution. That re-formulation inevitably adds information not present in the original constraint.

Quantities that can be computed from the data (without knowing the state of the world or the value of the parameter) are called *estimators*. Point estimators map the data into possible values of the parameter. The sample mean is an example of a point estimator. Set estimators map the data into sets of possible values of the parameter. Confidence intervals are examples of set estimators. There are also other kinds of estimators. And within types of estimators, there are classes with different kinds of functional dependence on the data. For example, one might consider point estimators that are affine functions of the data, or that are arbitrary measurable functions of the data.

Estimators can be compared using *risk functions*. Mean squared error is a common risk function for point estimators. Expected measure of a confidence set is a reasonable risk function for set estimators. Scientific considerations should enter the choice of risk functions, although risk functions are often selected for their mathematical tractability rather than their relevance.

The risk depends on the true state of the world. Typically, no estimator has smallest risk for all possible states of the world. Decision theory makes the risk tradeoff explicit. For example, one might consider the worst-case risk over all possible states of the world. The estimator with the smallest worst-case risk is the *minimax estimator*, a frequentist notion. Or one might consider the average risk if the state of the world were drawn at random from a given prior probability distribution. The estimator with the smallest average risk is the *Bayes estimator* for that prior distribution. If the state of the world is not random, or random but with unknown probability distribution, frequentist methods may be preferable to Bayesian methods.

Bayesian methods can be evaluated from a frequentist perspective, and vice versa. For example, one can calculate the Bayes risk of a frequentist estimator, or the maximum risk of a Bayes estimator. The frequentist notion of a confidence set is similar to the Bayesian notion of credible region, but the two are not identical. The frequentist coverage probability of a Bayesian $1 - \alpha$ credible set is typically not $1 - \alpha$ for all parameter values: it can be much higher for some and much lower for others. There is a duality between Bayesian and frequentist methods: under mild technical conditions, the risk of the Bayes estimator for the *least favorable prior* is equal to the risk of the minimax estimator. When the Bayes risk for a given prior is less than the minimax risk, the apparent uncertainty in the Bayes estimate has been reduced by the choice of prior: the prior adds information not present in the constraints.

References

Backus, G.E. 1987 Isotropic probability measures in infinite-dimensional spaces. *Proc. Natl. Acad. Sci.* **84**, 8755–8757.

Backus, G.E. 1988 Comparing hard and soft prior bounds in geophysical inverse problems. *Geophys. J.* **94**, 249–261.

Backus, G.E. 1989 Confidence set inference with a prior quadratic bound. *Geophys. J.* **97**, 119–150.

Berger, J.O. 1985 *Statistical Decision Theory and Bayesian Analysis. 2nd Edn.* Springer-Verlag.

Casella, G. and Strawderman, W.E. 1981 Estimating a bounded normal mean. *Ann. Stat.* **9**, 870–878.

Diaconis, P. and Freedman, D.A. 1986 On the consistency of Bayes estimates. *Ann. Stat.* **14**, 1–26.

Diaconis, P. and Freedman, D.A. 1998 Consistency of Bayes estimates for nonparametric regression: normal theory. *Bernoulli* **4**, 411–444.

Donoho, D.L. 1994 Statistical estimation and optimal recovery. *Ann. Stat.* **22**, 238–270.

Donoho, D.L., Liu, R.C. and MacGibbon, B. 1990 Minimax risk over hyper rectangles, and implications. *Ann. Stat.* **18**, 1416–1437.

Eaton, M.L. 2008 Dutch book in simple multivariate normal prediction: another look. In D. Nolan, and T. Speed,, editors. *Probability and Statistics: Essays in Honor of David A. Freedman.* Institute of Mathematical Statistics.

Eaton, M.L. and Freedman, D.A. 2004 Dutch book against some objective priors. *Bernoulli* **10**, 861–872.

Evans, S.N., Hansen, B. and Stark, P.B. 2005 Minimax expected measure confidence sets for restricted location parameters. *Bernoulli* **11**, 571–590.

Evans, S.N. and Stark, P.B. 2002 Inverse problems as statistics. *Inverse Problems* **18**, R1–R43.

Freedman, D.A. and Stark, P.B. 2003 What is the chance of an earthquake? in *NATO Science Series IV: Earth and Environmental Sciences* **32**, pp. 201–213.

Freedman, D.A. 1995 Some issues in the foundations of statistics. *Foundations of Science* **1**, 19–39.

Le Cam, L. 1986 *Asymptotic Methods in Statistical Decision Theory*. Springer-Verlag, New York.

Lehmann, E.L. and Casella, G. 1998 *Theory of Point Estimation 2nd Edn*. Springer-Verlag.

Parker, R.L. 1994 *Geophysical Inverse Theory*. Princeton University Press.

Pratt, J.W. 1961 Length of confidence intervals. *J. Am. Stat. Assoc.* **56**, 549–567.

Schafer, C.M. and Stark, P.B. 2009 Constructing confidence regions of optimal expected size. *J. Am. Stat. Assoc.* **104**, 1080–1089.

Stark, P.B. 1992a Affine minimax confidence intervals for a bounded normal mean. *Stat. Probab. Lett.* **13**, 39–44.

Stark, P.B. 1992b Inference in infinite-dimensional inverse problems: Discretization and duality. *J. Geophys. Res.* **97**, 14,055-14,082.

Stark, P.B. 1992c Minimax confidence intervals in geomagnetism. *Geophys. J. Intl.* **108**, 329–338.

Zeytinoglu, M. and Mintz, M. 1984 Optimal fixed size confidence procedures for a restricted parameter space. *Ann. Stat.* **12**, 945–957.

Chapter 3

Subjective Knowledge or Objective Belief? An Oblique Look to Bayesian Methods

D. Calvetti and E. Somersalo

Case Western Reserve University, Ohio, USA

3.1 Introduction

The idea of applying the Bayesian statistical formalism to solve inverse problems (Calvetti and Somersalo 2007; Kaipio and Somersalo 2004; Tarantola 2005) is almost as old as the Bayesian statistics itself (Laplace 1774). Despite its long tradition and demonstrated effectiveness, the approach continues to raise opposition and controversy. For its adversaries, the notion of subjective probability and the use of prior information are moot points, and they insist on the necessity of using techniques of either classical analysis, which avoids altogether the use of probability, or from frequentist statistics, with the obvious implicit assumption that it represents a less subjective framework to solve inverse problems.

Large-Scale Inverse Problems and Quantification of Uncertainty Edited by L. Biegler, G. Biros, O. Ghattas, M. Heinkenschloss, D. Keyes, B. Mallick, Y. Marzouk, L. Tenorio, B. van Bloemen Waanders and K. Willcox

The resistance to subjectivity in science which we still observe today is to a certain extent understandable, although based on a rather fundamental misinterpretation. The major philosophical achievement of Newtonian physics is its formulation of the laws of nature in a manner which is independent of the observer or his probability assignments. For the non-Bayesian scientists, the concept of subjectivity seems to be in conflict with the idea of the *detached observer*. Quantum physics added a new dimension to the discussion concerning the separation of the observer and the observed, and the intrinsic probabilistic nature of quantum physics has led many to believe that the quantum mechanical probability and subjective probability are in conflict in the sense that not all probabilities have a subjective interpretation. Despite this belief, the Bayesian interpretation of quantum mechanics is coherent and justified (Caves *et al.* 2002, 2007).

The goal of this article is to clarify where the subjectivity in the Bayesian approach lies and what it really amounts to. Characteristic to Bayesian statistics and its applications to inverse problems is the employment of certain computational techniques, such as Markov Chain Monte Carlo methods or variational Bayes methods, to explore the posterior densities, to the point that occasionally a work not relying on these methods is not even considered Bayesian. In this article, the focus is on the interpretation of the probability, and on its role in setting up likelihoods and priors, while the algorithms in the examples are mostly non-Bayesian. In fact, the exploration of posterior densities by sampling methods is restricted to one example. This rather eclectic choice is meant to emphasize the possible integration of statistical and traditional numerical methods to best take vantage of both of them; see Calvetti and Somersalo (2004) for further topics in this direction. An extensive repository of sampling techniques can be found in the reference Liu (2004).

3.2 Belief, Information and Probability

Consider a simple example that clarifies the meaning of probability as an expression of subjective belief.

■ **Example 3.1** Consider a sequence of integers,

26535 89793 23846 26433 83279 50288 41971 69399
37510 58209 74944 59230 78164 06286 20899 ...

The problem to predict the continuation of the sequence is presented to two agents A and B. Agent A, with the lack of further information, assumes that each number from 0 to 9 has an equal probability to appear, and generates the prediction with a random number generator.

Agent B, instead, observes that

$$3,14159\ 26535\ 89793\ 23846\ 26433\ 83279\ 50288\ 41971\ 69399$$
$$37510\ 58209\ 74944\ 59230\ 78164\ 06286\ 20899\ \ldots$$

is a decimal approximation of π and therefore predicts the continuation by looking up the continuation of the decimal expansion.

This rather trivial example requires a couple of quick observations. First, the need for using a probabilistic model (agent A) stems from subject's lack of information. Second, the concept of probability is a property of the subject, not of the object. Bruno de Finetti, one of the prominent advocates of the concept of subjective probability, went as far as claiming that *probability does not exist* (de Finetti 1974), referring to this nature of probability. The third observation is related to the certainty of an agent. Indeed, agent B may be convinced of having found a generative model that predicts correctly the continuation of the sequence, but this certainty does not imply that the prediction is correct. Thus, the proper way for agent B to express the certainty is to say, 'the sequence continues like a decimal approximation of π with probability one'. It is also important to understand that subjective certainty does not imply that the outcome is necessarily as stated.

An important concept related to subjective probability is calibration (O'Hagan and Forster 2004). Returning to the previous example, assume that the agents are shown a short segment of how the sequence continues, and it turns out that the continuation deviates from the decimal expansion of π. In this case, agent A may still keep the probabilistic model, while agent B needs to revise completely the model – in fact, one incorrectly predicted number is enough – and it would indeed be foolish for agent B to insist on his generative model (this is related to the concept of falsifiability of a scientific statement; see D'Agostini (2005); Tarantola (2006)). The calibration of the probability model is tantamount to updating as new information arrives. The concept of subjective probability is flexible enough to allow calibration. Hence, while probability assignments are subjective, they should not be arbitrary, and the agent should be receptive to all information available. Or, using the terminology of betting, odds should be placed in such a way that the subject feels comfortable whichever way the result turns out; this is the essence of the 'Dutch book argument' (Jeffrey 2004). Observe that in the previous example, agent B might not be following this principle: probability 1 is very committal, and one could argue that agent B should not even be interested in obtaining data. Frequencies of occurrence can be used to update probabilities, but there is no reason to argue that they should objectively determine the probability assignments. In the framework of subjective probability, the question whether a probability assignment is correct or not cannot be answered. In fact, it has no meaning. For further discussion, see D'Agostini (2003).

3.3 Bayes' Formula and Updating Probabilities

In the Bayesian framework, probability is used to express the subject's ignorance, or lack of information, of some of the quantities in the model describing the object. Therefore, deterministic models need to be extended as stochastic models concerning random variables and probability distributions.

It is commonly stated that the Bayesian statistics differs from the frequentist statistics in its use of prior probabilities, and that the subjective part is in the setup of the prior. While there is some truth to this, the statement is an oversimplification. Although the data is the same for all observers – and, in this sense, objective – the interpretation of the data is up to the observer, therefore necessarily subjective.

The Bayesian modeling of inverse problems starts by identifying the quantity of primary interest and the observed quantity. It is assumed here that both of these quantities take on values in finite dimensional real spaces, e.g. $\mathbb{R}^n$ and $\mathbb{R}^m$, respectively. We denote by X and B the multivariate random variables corresponding to the quantity of interest and the observable, respectively.

To establish the notations, assume that, a priori, the probability density of the variable X can be expressed in terms of a probability density $\pi_{\text{prior}} : \mathbb{R}^n \to \mathbb{R}_+$,

$$\mathrm{P}\{X \in \mathcal{B}\} = \int_{\mathcal{B}} \pi_{\text{prior}}(x)dx, \quad \mathcal{B} \subset \mathbb{R}^n \text{ measurable.}$$

Hence, the prior density π_{prior} expresses the degree of subject's information and ignorance of the variable X prior to the arrival of the information that the measurement of B conveys. Furthermore, we assume that the likelihood distribution, or the probability distribution of B conditional on $X = x \in \mathbb{R}^n$, is also expressed in terms of a probability density,

$$\mathrm{P}\{B \in \mathcal{C} \mid X = x\} = \int_{\mathcal{C}} \pi_{\text{likelihood}}(b \mid x)db, \quad \mathcal{C} \subset \mathbb{R}^m \text{ measurable.}$$

The essence of Bayesian inference is in the process of updating subject's prior probabilities by passing to the posterior probability using Bayes' formula. In terms of the prior and likelihood density, Bayes' formula gives

$$\pi_{\text{post}}(x) = \pi(x \mid b) = \frac{\pi_{\text{prior}}(x)\pi_{\text{likelihood}}(b \mid x)}{\pi(b)}, \quad b = b_{\text{observed}}, \tag{3.1}$$

where the denominator is the marginal density,

$$\pi(b) = \int_{\mathbb{R}^n} \pi_{\text{prior}}(x)\pi_{\text{likelihood}}(b \mid x)dx.$$

The posterior density (3.1) is the Bayesian solution of the inverse problem, and it expresses the subject's degree of information of X after the arrival of the data, i.e. the realization of the measurement $B = b_{\text{observed}}$.

In the sequel, the following notations are used: the symbol '$\propto$' is used to indicate identity up to an insignificant multiplicative constant, typically a normalizing factor. The notation $X \sim \mathcal{N}(x_0, \Gamma)$ means that X is a normally distributed random variable with mean x_0 and covariance matrix Γ. By the notation $X \sim \pi(x)$ it is meant that X is a random variable with probability density $\pi(x)$.

3.3.1 Subjective Nature of the Likelihood

The subjectivity of the probability is not only restricted to the prior modeling but applies to the likelihood as well. The subjective component in the likelihood model is related directly to the probability assignment of the noise, and indirectly to the selection of the model. The distribution of the observation noise can be calibrated, e.g. by independent repeated measurements, or can be justified by a generative theoretical model such as a counting model leading to Poisson noise, but ultimately the question of a correct noise model, and therefore a correct likelihood distribution, remains open.

The subjective nature of forward model selection is often overlooked, either because the models are in some sense obvious as is often the case in traditional statistics, or because the models are based on physics where the affinity between mathematical modeling and experimental methods is high due to the long tradition. A completely different situation occurs, e.g. in life sciences where extremely complex systems need to be modeled, and often no single model is beyond dispute. It is often a matter of subjective judgement which features should be included in the model, and the model validation is difficult because the interpretation of the data may be controversial.

A further challenge to the objectivity of the likelihood is related to the question how well a computational model describes the data. In fact the likelihood should account for the approximation error due to deficiencies of the model, including its incompleteness and discretization errors. The effect of modeling errors have been discussed, e.g. in Arridge *et al.* (2006); Kaipio and Somersalo (2007).

To clarify the point about noise, and to set up the computational model discussed later in this article, consider the following example.

■ **Example 3.2** We consider a deconvolution problem of images. The two dimensional convolution model with finite number of observations, corresponding to an $n \times n$ pixel image, can be written in the form

$$g(s_j) = \int_Q a(s_j - t) f(t) dt, \quad 1 \leq j \leq N = n^2,$$

where $Q = [0, 1] \times [0, 1] \in \mathbb{R}^2$ is the image area, the sampling points $s_j \in Q$ correspond to the centers of the pixels, and a is the convolution kernel. The data corresponds to observations of the values $g(s_j)$ corrupted by noise. Representing the unknown as a discrete image with pixel values $x_j = f(t_j)$, $1 \le j \le N$, and assuming that the observation noise is additive, the finite dimensional model becomes

$$b = \mathsf{A}x + e, \tag{3.2}$$

where $e \in \mathbb{R}^N$ denotes the additive noise and $\mathsf{A} \in \mathbb{R}^{N \times N}$ is the finite dimensional approximation of the kernel action, obtained by applying an appropriate quadrature rule.

Besides the choice of the discretization grid and the quadrature rule for arriving at the finite dimensional model (3.2), the likelihood model depends on the subjective probability assignment of the noise. Assume that an observer, agent A, has a reason to believe that the additive noise is a realization of a Gaussian white noise with variance σ^2. Defining the random variables X, B and E corresponding to the realizations x, b and e in the model (3.2), the agent thus assigns the probability model $E \sim \mathcal{N}(0, \sigma^2 I)$ and arrives at the likelihood model

$$\pi_{\text{likelihood}}(b \mid x) \propto \exp\left(-\frac{1}{2\sigma^2}\|b - \mathsf{A}x\|^2\right). \tag{3.3}$$

Another observer, agent B, believes that the noise contains not only the exogenous noise modeled by white noise, but also a contribution of a background radiation. Assuming that the background radiation is Gaussian with mean intensity $\mu > 0$ and variance η^2, agent B writes a likelihood model

$$\pi_{\text{likelihood}}(b \mid x) \propto \exp\left(-\frac{1}{2}(b - \mu\mathsf{A}e_0 - \mathsf{A}x)^{\mathsf{T}}\Gamma^{-1}(b - \mu\mathsf{A}e_0 - \mathsf{A}x)\right),$$

where

$$e_0 = \begin{bmatrix} 1 \\ \vdots \\ 1 \end{bmatrix}, \quad \Gamma = \sigma^2\mathsf{I} + \eta^2\mathsf{A}^{\mathsf{T}}\mathsf{A}.$$

Neither one of the likelihood models above account for the effect of discretization, and often it is tacitly assumed that the exogenous noise level is chosen high enough to mask the approximation error which clearly depends on the unknown to be estimated. If the quality of the data is good, this assumption may be risky: the approximation error has a particular probability distribution that may be estimated from the prior distribution, and masking it with unspecified additive noise may lead to an unnecessary loss of quality of the posterior estimate and thus to a waste of the often so precious high quality of the data; see Kaipio and Somersalo (2004, 2007).

Another concern about the use of Gaussian models is that they may not correspond to what is understood about the noise in imaging applications. Indeed, the normality assumption of the data can often be justified as an average net effect of microscopic events, thanks to the central limit theorem, while when the data consist, e.g. of photon counts, a more detailed model has to be employed. For deblurring with non-Gaussian noise model, see, e.g. Starck *et al.* (2002).

3.3.2 Adding Layers: Hypermodels

Both the likelihood and the prior model may contain parameters that are poorly known. They may be parameters of the probability densities such as mean or covariance, or they may correspond to physical quantities in the generative model of the data. An example of the latter is the convection velocity in dynamic impedance tomography applied to the estimation and control of fluid flows in pipelines, see Seppänen *et al.* (2007). In line with the Bayesian approach, each quantity that is incompletely known should be modeled as a random variable, the probability distribution indicating the level of ignorance.

Consider the random variable X with prior probability density depending on a parameter vector $\theta \in \mathbb{R}^k$. The dependence is denoted by using the notation of conditional densities, i.e. we write $X \sim \pi_{\text{prior}}(x \mid \theta)$. Similarly, assume that the likelihood density depends on a parameter vector $\sigma \in \mathbb{R}^\ell$, denote it by $\pi_{\text{likelihood}}(b \mid x, \sigma)$. With the parameter vectors θ and σ fixed, the joint probability density of (X, B) can be written as

$$\pi(x, b \mid \theta, \sigma) = \pi_{\text{prior}}(x \mid \theta)\pi_{\text{likelihood}}(b \mid x, \sigma).$$

When the values θ and σ are poorly known, i.e., the agent is not willing or able to specify values for these parameters, the solution is to model them as random variables Θ and Σ, respectively, and assign a probability density, a hyperprior,

$$(\Theta, \Sigma) \sim \pi_{\text{hyper}}(\theta, \sigma).$$

This leads us naturally to consider the joint probability density of all unknowns,

$$\begin{aligned}\pi(x, b, \theta, \sigma) &= \pi(x, b \mid \theta, \sigma)\pi_{\text{hyper}}(\theta, \sigma) \\ &= \pi_{\text{prior}}(x \mid \theta)\pi_{\text{likelihood}}(b \mid x, \sigma)\pi_{\text{hyper}}(\theta, \sigma).\end{aligned}$$

There are now various options on where to go from here. A natural way is to define, via Bayes' law, the posterior probability density

$$\pi(x, \theta, \sigma \mid b) \propto \pi_{\text{prior}}(x \mid \theta)\pi_{\text{likelihood}}(b \mid x, \sigma)\pi_{\text{hyper}}(\theta, \sigma), \quad b = b_{\text{measured}},$$

and infer on the distribution of the three unknowns X, Θ and Σ simultaneously. Another possibility is to treat the model parameters as nuisance parameters and integrate them out, i.e. define first the marginal posterior distribution

$$\pi(x, b) = \int_{\mathbb{R}^k} \int_{\mathbb{R}^\ell} \pi(x, b, \theta, \sigma) d\theta d\sigma,$$

and then infer on the posterior distribution of X,

$$\pi(x \mid b) \propto \pi(x, b), \quad b = b_{\text{observed}}.$$

Furthermore, one might integrate out first X and estimate the parameters Θ and Σ, and use the estimated values as fixed parameter values to infer on X. There is no single best solution, and the selection of the method of inference needs to be based on the particular needs and goals on one hand, and on computational issues such as complexity, stability, memory requirements and the computation time on the other. Some of these issues are discussed in the computed examples.

Hypermodels are deeply rooted to the concept of *exchangeability*, introduced and discussed by de Finetti (de Finetti 1930, 1974). Consider an experiment where a sequence of 'similar' observations are made, the classic example being white and black stones extracted from an urn one at a time and returned back. The random events are modeled by a sequence $X_1, X_2, \ldots$ of random variables. Let $\pi(x_1, x_2, \ldots, x_n)$ denote the joint probability density of n of these variables. The similarity of the events is expressed by exchangeability, i.e. for any n and any permutation σ of the index set $\{1, 2, \ldots, n\}$,

$$\pi(x_1, x_2, \ldots, x_n) = \pi(x_{\sigma(1)}, x_{\sigma(2)}, \ldots, x_{\sigma(n)}).$$

In de Finetti (1930) it was shown that if the random variables X_j are $\{0, 1\}$ valued, there exists a parametric model $\pi(x \mid \theta)$ and a probability density $\pi(\theta)$ of the parameter θ such that

$$\pi(x_1, x_2, \ldots, x_n) = \prod_{j=1}^{n} \int \pi(x_j \mid \theta) \pi(\theta) d\theta,$$

i.e. the variables X_j are conditionally independent, and any finite subset of observations is a random sample of a parametric model. In particular, if we assume the exchangeability of the prior, then it must arise from a hierarchical model! Generalizations of this result for more general random variables are found, e.g. in Diaconis and Freedman (1987); Hewitt and Savage (1955), and further discussion in Bernardo and Smith (1994).

■ **Example 3.3** Consider again a linear inverse problem with additive noise. Assume that observer's information about the additive white noise prior is: 'The signal-to-noise ratio (SNR) is 50', a very typical description in the engineering literature. How should the observer select the variance σ^2 in the model? It is not uncommon to interpret the observed signal

$b = b_{\text{observed}}$ as signal, the standard deviation σ as noise level and write something of the type $\sigma = \max(|b_j|)/\text{SNR}$. This probability assignment, in the absence of anything better, may give reasonable results, in particular when the SNR is high and the observed signal consists mostly of the useful signal; it is, however, easy to find situations where this assignment goes completely astray. For instance, if the data b, in a particular measurement event, consists of noise only with no meaningful signal contents, the assignment does not make much sense. Evidently, if the observer has no prior information about the signal, the information seems insufficient for reasonably determining the likelihood, emphasizing how the likelihood is no less subjective than the prior.

Hypermodels provide a consistent way of estimating the noise level. Assume that the likelihood model, conditional on the noise variance v, is

$$\pi_{\text{likelihood}}(b \mid x, v) \propto v^{-m/2} \exp\left(-\frac{1}{2v}\|b - \mathsf{A}x\|^2\right)$$
$$= \exp\left(-\frac{1}{2v}\|b - \mathsf{A}x\|^2 - \frac{m}{2}\log v\right),$$

where $x \in \mathbb{R}^n$, $b \in \mathbb{R}^m$, $\mathsf{A} \in \mathbb{R}^{m\times n}$. Assume further that the observer has an a priori estimate of the signal amplitude, $\|x\| \approx \mu$, and decides to use this information to set up a prior model,

$$\pi_{\text{prior}}(x) \propto \exp\left(-\frac{1}{2\mu^2}\|x\|^2\right).$$

To make use of the scarce SNR information, the observer sets up a hypermodel for v as an inverse gamma distribution,

$$\pi_{\text{hyper}}(v) \propto v^{-\alpha-1} \exp\left(-\frac{v_0}{v}\right), \quad \alpha > 2, \tag{3.4}$$

and fixes the scaling parameter v_0 and the shape parameter α so that the prior expectation $\overline{v} = v_0/(\alpha - 1)$ of v corresponds to the SNR, i.e. $\mu\overline{v}^{-1/2} = \text{SNR}$, or equivalently, $v_0 = (\alpha - 1)\,(\mu/\text{SNR})^2$. With these choices, the joint probability density of X, B and V is

$$\pi(x, b, v) \propto \exp\Bigg(-\frac{1}{2\mu^2}\|x\|^2 - \left(v_0 + \frac{1}{2}\|b - \mathsf{A}x\|^2\right) v^{-1} - \left(\frac{m}{2} + \alpha + 1\right)\log v\Bigg). \tag{3.5}$$

A strategy often advocated in the statistical literature is to marginalize the above expression with respect to x, condition the joint density of b and v

by setting $b = b_{\text{observed}}$ and estimate v from the one-dimensional posterior density $\pi(v \mid b)$, e.g. by the posterior mean,

$$\widehat{v} = \int v\pi(v \mid b)dv,$$

and then use the value $\widehat{v}$ for v to estimate x. This approach may seem particularly attractive in the present example because the joint density (3.5) is Gaussian with respect to x and has a marginal density in closed form. Indeed, by defining

$$\mathsf{B}_v = \frac{v}{\mu^2}\mathsf{I} + \mathsf{A}^\mathsf{T}\mathsf{A},$$

we have, after some tedious calculations,

$$\int_{\mathbb{R}^n} \pi(x, b, v)dx \propto \exp\Bigg(- \left(v_0 + \frac{1}{2} b^\mathsf{T} \left(\mathsf{I} - \mathsf{A}\mathsf{B}_v^{-1}\mathsf{A}^\mathsf{T}\right) b \right) v^{-1} \\ - \left(\alpha + 1 + \frac{1}{2}(m - n) \right) \log v - \frac{1}{2} \log\left(\det(\mathsf{B}_v)\right) \Bigg),$$

The computational problem with this formula is the evaluation of the determinant of B_v because in several applications the matrix A is not explicitly known, but there is an algorithm that computes the matrix-vector product $x \mapsto \mathsf{A}x$ with any given input vector x. The question of calculating determinants of large matrices appears, e.g., in statistical physics, and the computational problem is addressed in Bai and Golub (1998). It turns out that it may be easier to explore the conditional density $\pi(x, v \mid b)$ directly, e.g. with Markov Chain Monte Carlo methods (Calvetti and Somersalo 2004; Gilks *et al.* 1996; Kaipio and Somersalo 2004). We return to this point later on.

3.4 Computed Examples Involving Hypermodels

We consider computed examples that illustrate the use of Bayesian methods in inverse problems. The forward model in these examples is a simple linear model, the emphasis being on the introduction of information complementary to the data.

■ **Example 3.4** To demonstrate the use of hierarchical models related to the likelihood, consider the Bayesian version of the classical problem of fitting a smooth curve to observed data.

Let $f : I = [0, 1] \to \mathbb{R}$ be a smooth function that is measured with noise at few sampling points, i.e. the data are

$$y_j = f(s_j) + e_j, \quad 1 \leq j \leq m, \quad 0 < s_1 < s_2 < \cdots < s_m < 1.$$

The noise components e_j are assumed to be mutually independent, of unknown level and distribution. Although a rough estimate of the noise variance may be available, it is likely that at some observation points, the noise variance may be significantly larger than estimated. Such observations are often called outliers, and it is assumed here that no information about the number, location or magnitude of these outliers is available.

In setting up a model, the interval I is first discretized by defining $t_k = k/(n+1)$, $0 \leq k \leq n+1$, and the notation $x_k = f(t_k)$. The values x_k are unknown and therefore modeled as random variables. Let X_k denote the corresponding random variables. For simplicity, we assume that the observation points s_j coincide with discretization points, $s_j = t_{k_j}$ for some t_{k_j}, $1 \leq k \leq m$.

The next step consists of assigning a prior for X. A priori, the function f is believed to be smooth, although it is up to us to decide what this means. In this example, we encode the information in the prior by writing a second order autoregressive Markov process for the variables X_k,

$$X_k = \frac{1}{2}(X_{k-1} + X_{k+1}) + \gamma W_k, \quad 1 \leq k \leq n,$$

where W_k's, known as innovations, are independent Gaussian white noise variables, $W_k \sim \mathcal{N}(0,1)$, and $\gamma > 0$. By doing so we convey the idea that the value X_k does not differ significantly from the mean value of its neighbors, with this deviation expressed by the magnitude of the parameter γ. This Markov process is naturally related to the smoothness of the function f, since

$$f''(t_k) \approx (n+1)^2(x_{k-1} - 2x_k + x_{k+1}),$$

so the standard deviation of the finite difference approximation of the second derivative of $f(t_k)$ is $2\gamma(n+1)^2$.

The Markov process above described is naturally suited to describe the distribution of the interior points, but a modification is in order at the boundary of the domain of f. Assuming for the moment that the boundary values are fixed,

$$X_0 = f(0) = \theta_1, \quad X_{n+1} = f(1) = \theta_2,$$

the Markov process can be written in matrix form as

$$\mathsf{L}X = \gamma W + \frac{1}{2}\begin{bmatrix} \theta_1 \\ 0 \\ \vdots \\ 0 \\ \theta_2 \end{bmatrix} = \gamma W + \mathsf{P}\theta,$$

where $X = [X_1\, X_2, \cdots X_n]^\mathsf{T}$, $\theta = [\theta_1\, \theta_2]^\mathsf{T}$, and

$$\mathsf{L} = \frac{1}{2}\begin{bmatrix} 2 & -1 & & \\ -1 & 2 & & \\ & & \ddots & -1 \\ & & -1 & 2 \end{bmatrix} \in \mathbb{R}^{n\times n}, \quad \mathsf{P} = \frac{1}{2}\begin{bmatrix} 1 & 0 \\ 0 & 0 \\ \vdots & \vdots \\ 0 & 0 \\ 0 & 1 \end{bmatrix} \in \mathbb{R}^{n\times 2}. \tag{3.6}$$

Therefore we may write the prior model for X conditioned on knowing the boundary values $X_0 = \theta_1$, $X_{n+1} = \theta_2$ as

$$\pi_{\text{prior}}(x \mid \theta) \propto \exp\left(-\frac{\gamma^2}{2}\|\mathsf{L}x - \mathsf{P}\theta\|^2\right).$$

If the boundary values are given, this model can be used indeed as a prior. If, on the other hand, the boundary values θ_j are unknowns or poorly known, they need to be defined as random variables by a stochastic extension of the model. In this example, we assume that the boundary values are close to zero with an error bar $\pm 2\Delta$. To implement such belief, we define the random variable

$$\Theta = \begin{bmatrix} \Theta_1 \\ \Theta_2 \end{bmatrix} \sim \mathcal{N}(0, \Delta^2 \mathsf{I}),$$

that is, the boundary values are within the error bars with a probability of approximately 98 %. We can then define a proper probability density,

$$\pi(x,\theta) = \pi_{\text{prior}}(x \mid \theta)\pi_{\text{hyper}}(\theta) \propto \exp\left(-\frac{\gamma^2}{2}\|\mathsf{L}X - \mathsf{P}\theta\|^2 - \frac{1}{2\Delta^2}\|\theta\|^2\right)$$

$$= \exp\left(-\frac{1}{2}\left\|\begin{bmatrix} \gamma\mathsf{L} & -\gamma\mathsf{P} \\ 0 & \Delta^{-1}\mathsf{I} \end{bmatrix}\begin{bmatrix} x \\ \theta \end{bmatrix}\right\|^2\right).$$

We now consider the likelihood model. The observation points s_j coincide with some of the interior discretization points, so we may write an observation model

$$Y = \mathsf{S}X + E,$$

where S is a sampling matrix with elements $\mathsf{S}_{jk} = \delta_{k_j,k}$, $1 \le j \le m$, $1 \le k \le n$. Since the noise variance is poorly known, we use a hierarchical Gaussian model for the additive noise. Assuming at first that the variances v_j of the noise components e_j are known, we can write a noise model of the form

$$\pi_{\text{noise}}(e \mid v) \propto \left(\frac{1}{v_1 v_2 \cdots v_m}\right)^{1/2} \exp\left(-\frac{1}{2}\sum_{j=1}^m \frac{e_j^2}{v_j}\right)$$

$$= \exp\left(-\frac{1}{2}\sum_{j=1}^m \frac{e_j^2}{v_j} - \frac{1}{2}\sum_{j=1}^m \log v_j\right).$$

The likelihood model conditioned on the noise variance is then

$$\pi_{\text{likelihood}}(y \mid x, v) \propto \pi_{\text{noise}}(y - \mathsf{S}x \mid v).$$

To reflect the lack of information about the level of the noise in the observations, we consider the variance vector v as a random variable V whose probability distribution needs to be specified. Since we assume that the noise level remains for the most part close to a prescribed level while rare outliers should be allowed, we choose a probability distribution with a heavy tail. The choice for distributions of this type is wide. Here, we postulate that the components V_j are independent and their distribution is the inverse gamma distribution,

$$V_j \sim \pi_{\text{hyper}}(v_j) \propto v_j^{-\beta-1} \exp(-v_0/v_j),$$

where β and v_0 are called the shape and the scaling parameters, respectively. The inverse gamma distributions are related to the conjugate analysis of probability distributions, see Bernardo and Smith (1994).

We may now write the full joint probability density as

$$\begin{aligned}\pi(x,\theta,v,y) &\propto \pi_{\text{noise}}(y - \mathsf{S}x \mid v)\pi_{\text{hyper}}(v)\pi_{\text{prior}}(x \mid \theta)\pi_{\text{hyper}}(\theta) \\ &\propto \exp\Bigg(-\frac{\gamma^2}{2}\|\mathsf{L}x - \mathsf{P}\theta\|^2 - \frac{1}{2\Delta^2}\|\theta\|^2 - \frac{1}{2}\big\|\mathsf{V}^{-1/2}(y - \mathsf{S}x)\big\|^2 \\ &\qquad - \sum_{j=1}^{m}\left(\left(\beta + \frac{3}{2}\right)\log v_j + \frac{v_0}{v_j}\right)\Bigg),\end{aligned}$$

where $\mathsf{V} = \text{diag}(v_1, v_2, \ldots, v_m)$, and the posterior probability density, up to a normalizing constant, is obtained by conditioning, that is, by substituting the measured values of y in this expression.

Having the posterior probability density of the triple (X, Θ, V), several possibilities and strategies to find useful estimates are available. We follow here an approach that leads to a computationally efficient algorithm of approximating the Maximum A Posteriori (MAP) estimate of the unknowns. The iterative sequential algorithm for approximating the MAP estimate $(x_{\text{MAP}}, \theta_{\text{MAP}}, v_{\text{MAP}})$ proceeds as follows:

1. Set $k = 0$ and initialize the variables, $v = v^0$, $x = x^0$, $\theta = \theta^0$.

2. Increase $k \to k+1$ and update $(x^{k-1}, \theta^{k-1}) \to (x^k, \theta^k)$ by maximizing

$$(x, \theta) \mapsto \pi(x, \theta, v^{k-1} \mid y);$$

3. Update $v^{k-1} \to v^k$ by maximizing

$$v \mapsto \pi(x^k, \theta^k, v \mid y);$$

4. If the current estimate satisfies the stopping criterion, stop, otherwise return to 2.

The advantage of this algorithm is that, due to the conditionally Gaussian prior and likelihood models, the steps above are straightforward to implement. Indeed, consider the updating step of (x, θ). Holding the variance vector fixed, $v = v^{k-1}$, and denoting $\mathsf{V}^{k-1} = \mathrm{diag}(v^{k-1})$, we have

$$\pi(x, \theta \mid v^{k-1}, y)$$
$$\propto \exp\left(-\frac{1}{2}\left\|\begin{bmatrix}(\mathsf{V}^{k-1})^{-1/2}\mathsf{S} & 0\\ \gamma\mathsf{L} & -\gamma\mathsf{P}\\ 0 & \Delta^{-1}\mathsf{I}\end{bmatrix}\begin{bmatrix}x\\ \theta\end{bmatrix} - \begin{bmatrix}(\mathsf{V}^{k-1})^{-1/2}y\\ 0\\ 0\end{bmatrix}\right\|^2\right),$$

the maximizer being the least squares solution of the linear system

$$\begin{bmatrix}(\mathsf{V}^{k-1})^{-1/2}\mathsf{S} & 0\\ \gamma\mathsf{L} & -\gamma\mathsf{P}\\ 0 & \Delta^{-1}\mathsf{I}\end{bmatrix}\begin{bmatrix}x\\ \theta\end{bmatrix} = \begin{bmatrix}(\mathsf{V}^{k-1})^{-1/2}y\\ 0\\ 0\end{bmatrix}.$$

On the other hand, the probability density of v conditioned on $(x, \theta) = (x^k, \theta^k)$ is

$$\pi(v \mid y, x^k, \theta^k) \propto \exp\left(-\sum_{j=1}^{m}\left(\frac{(y - \mathsf{S}x^k)_j^2}{2v_j} + \left(\beta + \frac{1}{2}\right)\log v_j + \frac{v_0}{v_j}\right)\right).$$

The variables v_j can therefore be updated independently, since the maximum is attained at the point where the derivative of the expression in the exponent vanishes,

$$v_j^k = \frac{1}{\beta + 3/2}\left(\frac{1}{2}(y - \mathsf{S}x^k)_j^2 + v_0\right).$$

We illustrate the performance of the proposed algorithm with a computed example. Figure 3.1 shows a smooth curve and the results of noisy sampling of it. The data displayed here is generated by adding normally distributed random noise with fixed variance to each observations, with the exception of two data points where the variance of the noise is significantly larger. In Figure 3.2, we show four iterations of the suggested algorithm, with parameters $\gamma = 100$, $\Delta = 0.5$ and $v_0 = (0.08)^2$. After few iteration steps, the algorithm is able to identify the outlier data and rather than trying to fit the smooth solution to those values, finds a trade-off between the fidelity of the data and smoothness of the solution by letting the noise variance of the outliers grow.

In the discussion above, we settled for finding a single estimate for both x and v, by computing an approximation of the MAP estimate. The Bayesian methodology, however, is much more versatile, allowing to assess how much belief one should have in the estimates. This requires methods of exploring the posterior probability densities, which is often a non-trivial and computationally challenging task as the posterior densities are defined in high-dimensional

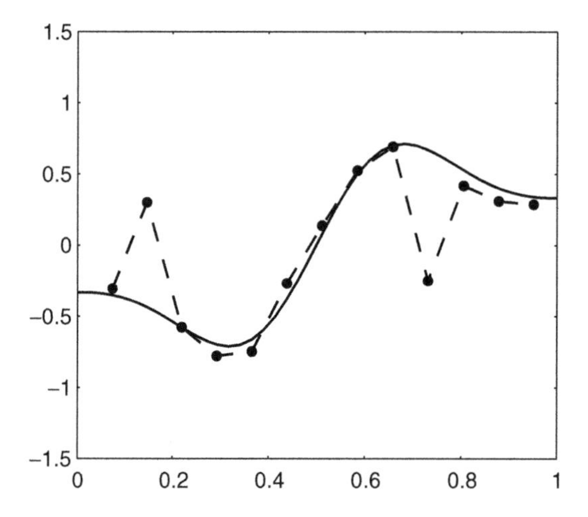

Figure 3.1 Noisy sampling of a smooth function. Two of the observations are corrupted by noise with significantly large standard deviation.

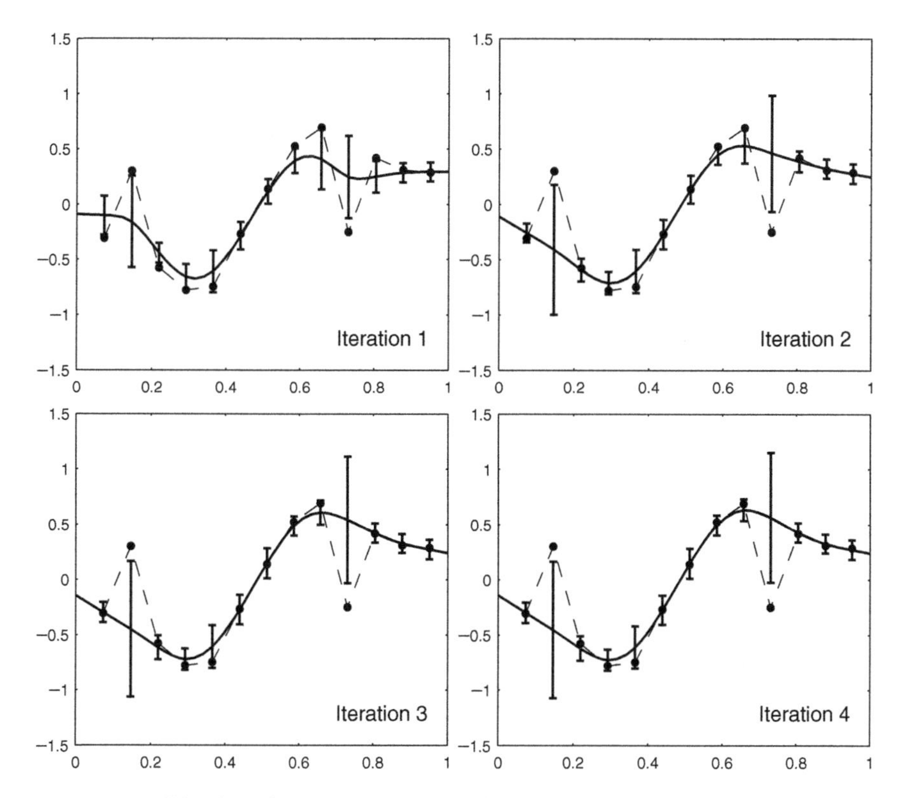

Figure 3.2 The first four iterations of the sequential updating algorithm. The estimated standard deviations are indicated as error bars.

spaces, and getting a handle of them may be complicated. As pointed out earlier, the Markov Chain Monte Carlo (MCMC) methods are developed to explore high dimensional densities. Given a probability density $\pi(x)$ in $\mathbb{R}^n$, MCMC methods seek a transition probability kernel $K(x, y)$ such that π is the invariant density of it, i.e.

$$\pi(x) = \int K(x, y)\pi(y)dy.$$

A sample $\{x^0, x^2, \ldots, x^N\}$ of vectors in $\mathbb{R}^n$ that are distributed according to the density π is then generated by recursively drawing x^k from the density $x \mapsto K(x, x^{k-1})$. For details and algorithms, we refer to Calvetti and Somersalo (2004); Gilks *et al.* (1996); Kaipio and Somersalo (2004); Liu (2004).

The updating algorithm for finding an approximation of the MAP estimate suggests a method for generating a sample of points (x^k, θ^k, v^k) via the following steps.

1. Set $k = 0$ and initialize the variables, $v = v^0$, $x = x^0$, $\theta = \theta^0$. Decide the sample size N.

2. Increase $k \to k + 1$ and draw (x^k, θ^k) from the Gaussian density

$$(x, \theta) \mapsto \pi(x, \theta, v^{k-1} \mid y);$$

3. Draw v^k component by component from the inverse gamma density

$$v \mapsto \pi(x^k, \theta^k, v \mid y);$$

4. If $k = N$, stop, otherwise continue from 2.

The above algorithm is a version of a block wise Gibbs sampler (Geman and Geman (1984)), see Calvetti and Somersalo (2008) for further details.

We run the algorithm starting from the initial point found by the MAP estimation algorithm, and compute a sample of size $N = 15\,000$. In a standard dual 1.6 GHz processor, 1 GB RAM, the sample generation requires about 15 seconds. Having the sample computed, we then plot pointwise percentile predictive envelopes: at each point t_j we seek an interval that contains a given percentage of the points x_j^n, $1 \le n \le N$. In Figure 3.3, the envelopes of 50 % and 90 % are plotted with different shades of gray. Notice that the envelopes do not get wider at the points of the outliers. However, in the same figure, we have also plotted the predictive output intervals, i.e., intervals that contain 90 % of the model predicted simulated measurements,

$$b_k^n = x_{s_k}^n + e_k^n, \quad e_k^n \sim \mathcal{N}(0, v_k^n), \quad 1 \le n \le N.$$

As expected, at two of the measurement locations the predicted intervals are considerably wider.

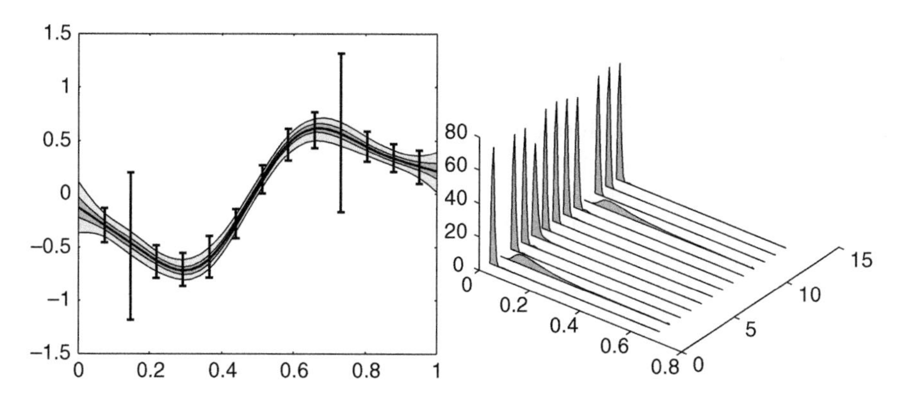

Figure 3.3 Pointwise predictive envelopes corresponding to 50 % and 90 % percentiles of the sample, and predictive output intervals (Left). The sample based probability distributions of the variances v_k (Right). The sample size is 15 000.

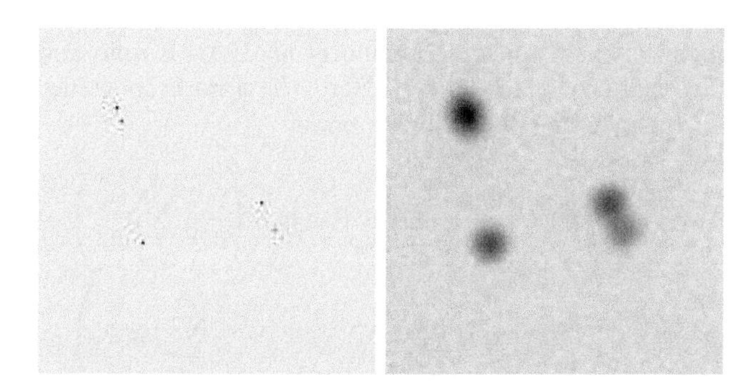

Figure 3.4 The original image (left) and the blurred noisy version of it (right). The FWHV of the Airy kernel is 10 pixels, the full image being 128 × 128.

Finally, in Figure 3.3, we plot the sample based computed posterior probability distribution functions of the noise variances v_k. The effect of outlier data is clearly visible in this plot.

The next example employs a similar algorithm, but this time the main uncertainty is in the parameters of the prior rather than those of the likelihood.

■ **Example 3.5** Consider a deconvolution problem of finding a pixel image $x \in \mathbb{R}^N$, $N = n^2$ from a blurred and noisy version of it, denoted by $b \in \mathbb{R}^N$. The convolution kernel is assumed to be known and defined by the Airy kernel,

$$a(t-s) = A_0 \left(\frac{J_1(\gamma|t-s|)}{\gamma|t-s|} \right)^2, \quad \gamma > 0.$$

The images x and b are supported in the unit square $Q = [0,1] \times [0,1]$, and the scaling parameter γ is related to the Full Width Half Value (FWHV), expressed in number of pixels, by $\gamma \approx (n/\mathrm{FWHV})\, 3.216$. We assume that the convolution model is discretized as in Example 3.2 using a piecewise constant approximation as a quadrature rule, $n = 128$, and the likelihood corresponds to additive Gaussian white noise. The signal to noise ratio SNR is estimated to be 50. The likelihood model based on this information is then (3.3), where the variance σ^2 is estimated by the user as $\sigma^2 = (\max(b_{\mathrm{observed}})/\mathrm{SNR})^2$, a choice which can be contested, as discussed in Example 3.3.

Suppose that in the present problem, the observer is convinced that the underlying image x represents an almost black object with few focal objects, such as stars. The exact location of the stars is not known, although the blurred image gives a pretty good hint. More importantly, some of the blurred objects in the data could actually represent double or multiple focal objects closer to one another than the Rayleigh limit. Hence, the real challenge is to find a prior that allows superresolution (see Figure 3.4). One suitable prior for this problem is the ℓ^1–prior (Donoho *et al.* 1992; Kaipio and Somersalo 2004) that favors almost black objects. Here, the prior is constructed by using hyperpriors. Consider therefore a prior model

$$\pi_{\mathrm{prior}}(x \mid \theta) \propto \left(\frac{1}{\theta_1\theta_2\cdots\theta_N}\right)^{1/2} \exp\left(-\frac{1}{2}\sum_{j=1}^{N}\frac{x_j^2}{\theta_j}\right)$$

$$= \exp\left(-\frac{1}{2}\|\mathsf{D}_\theta^{-1/2}x\|^2 - \frac{1}{2}\sum_{j=1}^{N}\log\theta_j\right),$$

where $\mathsf{D}_\theta = \mathrm{diag}(\theta_1,\ldots,\theta_N)$. Intuitively, the variance θ_j should be large for a pixel corresponding to a star, while it should be almost zero for the background pixels. The expected number of nonzero pixels is small, i.e. they could be interpreted as rare outliers. A probability density favoring few pronounced outliers is, e.g., the inverse gamma distribution (3.4), leading to the joint density

$$\pi(x,\theta,b) \propto \exp\Bigg(-\frac{1}{2\sigma^2}\|b - \mathsf{A}x\|^2 - \frac{1}{2}\|\mathsf{D}_\theta^{-1/2}x\|^2 - \sum_{j=0}^{N}\frac{\theta_0}{\theta_j} - \left(\alpha+\frac{3}{2}\right)\sum_{j=1}^{N}\log\theta_j\Bigg),$$

from which the posterior density $\pi(x,\theta \mid b)$ is obtained by conditioning, i.e. fixing $b = b_{\mathrm{observed}}$.

In other words, we have transformed the problem of describing the prior information in terms of the pixel values themselves into one of describing the prior information in terms of the pixel variances. What has been gained by doing this?

First, the new formulation is rather forgiving: if we assume that the variance of a pixel is large, it does not necessarily mean that the pixel value is large. Similarly, the pixel variance needs not to be precisely known to allow the pixel value to adjust itself to the data. This feature makes hypermodels attractive when the prior information is qualitative rather than quantitative. The second point in using the hypermodel is that the prior density is conditionally Gaussian, allowing to design fast and effective iterative solvers.

A fast iterative solver, an alternating sequential algorithm, for approximating the mode of the posterior density in a similar inverse problem was suggested in Calvetti and Somersalo (2008). Each iteration consists of two steps, as in the previous example. Given the current estimate (x^k, θ^k),

1. update $x^k \to x^{k+1}$,

$$x^{k+1} = \arg\max\left\{\pi(x, \theta^k \mid b)\right\},$$

2. update $\theta^k \to \theta^{k+1}$,

$$\theta^{k+1} = \arg\max\left\{\pi(x^{k+1}, \theta \mid b)\right\}.$$

As pointed out above, the conditionally Gaussian structure of the prior makes both steps computationally simple. Indeed, the updating of the pixel image requires that we solve the linear system

$$\begin{bmatrix} \sigma^{-1}\mathsf{A} \\ \mathsf{D}_{\theta^k}^{-1/2} \end{bmatrix} x = \begin{bmatrix} \sigma^{-1} b \\ 0 \end{bmatrix} \tag{3.7}$$

in the least squares sense, while the updating of θ, due to the mutual independency of the components, is obtained by a closed formula,

$$\theta_j^{k+1} = \frac{1}{\alpha + 3/2}\left(\frac{1}{2} x_j^{k+1} + \theta_0\right).$$

The effective numerical solution of the system (3.7) using iterative solvers as well as convergence questions have been discussed in the article (Calvetti and Somersalo (2008)), so we present here only the results of some computed examples. In Figure 3.5, the approximate MAP estimates of x and θ (iterations 3, 5 and 8) are shown. Each iteration requires about 0.2 seconds on a PC (dual processor, 1.6 GHz, 1 GB of RAM).

The iterates, in particular the double object on the upper left corner, show that the algorithm is capable of superresolution, i.e. to distinguish separate sources closer than the Rayleigh radius. Since the information to achieve this is not present in the data, the superresolution is due to prior information.

Our third example shows that the prior information can be used to discriminate mixed signals with different statistics.

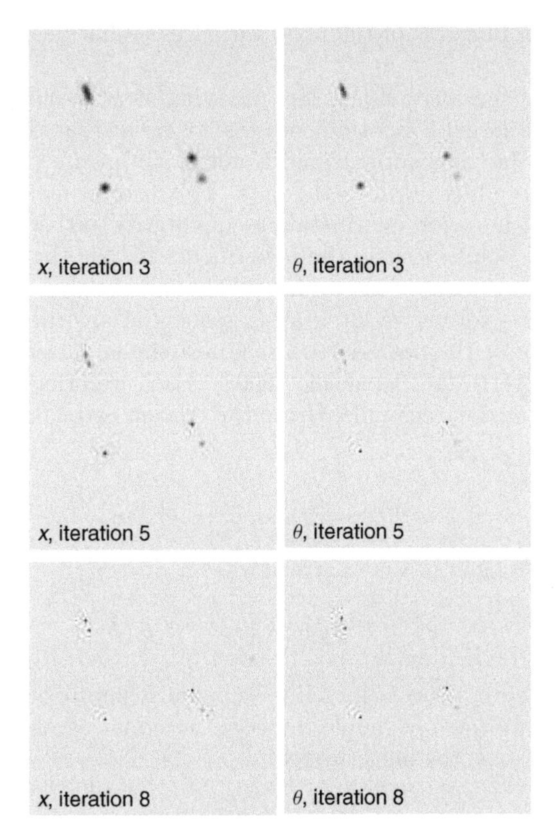

Figure 3.5 Iterates 3, 5 and 8 of the alternating sequential algorithm of the image x (left) and the variance θ (right).

■ **Example 3.6** The classical cocktail party problem is to discriminate two or more added signals from each other based on the different statistical nature of the signals. The method of choice for this problem would be the Independent Component Analysis (ICA) with its various implementations, see, e.g. Hyvärinen and Oja (2000). While the ICA assumes that the mixed signal is observed directly, in inverse problems it is not uncommon that the data consists of filtered or blurred version of the mixed signal. Markov Chain Monte Carlo methods can be employed to successfully discriminate the statistically different signals, see Kaipio and Somersalo (2004, 2002). Here we revisit the problem using hypermodels and fast iterative solvers.

Consider a deconvolution problem similar to that of the previous example, except that instead of a purely focal, or almost black object, the observer assumes that the underlying image is a sum of two qualitatively different images, one that is focal, the other distributed and blurry. An example of this kind could be an astronomical image consisting of bright stars and a nebular object.

The prior distribution that adequately describes the object is a mixture model. The observer sets up a stochastic model for the image X interpreted as a random variable of the form

$$X = U + V,$$

where the random variables U and V are mutually independent and follow a given statistics. If V represents the focal object, the hierarchical hypermodel of the previous example can be applied. The variable U, representing the blurred or fuzzy object, is postulated to follow statistics described by a Gaussian second order smoothness prior. To construct the prior distribution, let L be the second order finite difference matrix in (3.6) with n equal to the number of pixels per coordinate direction. The second order finite difference matrix in two dimensions can be defined using the matrix Kronecker product. Thus, if I denotes the $n \times n$ identity matrix, we define

$$\mathsf{L}_2 = \mathsf{I} \otimes \mathsf{L} + \mathsf{L} \otimes \mathsf{I} \in \mathbb{R}^{N \times N}, \quad N = n^2, \tag{3.8}$$

where the Kronecker products are given in block form as

$$\mathsf{I} \otimes \mathsf{L} = \begin{bmatrix} \mathsf{L} & & & \\ & \mathsf{L} & & \\ & & \ddots & \\ & & & \mathsf{L} \end{bmatrix}, \quad \mathsf{L} \otimes \mathsf{I} = \frac{1}{2} \begin{bmatrix} 2\mathsf{I} & -\mathsf{I} & & \\ -\mathsf{I} & 2\mathsf{I} & & \\ & & \ddots & -\mathsf{I} \\ & & -\mathsf{I} & 2\mathsf{I} \end{bmatrix}.$$

The prior probability distribution of the smooth object is now chosen as

$$U \sim \pi_{\rm prior}(u) \propto \exp\left(-\frac{\gamma^2}{2}\|\mathsf{L}_2 u\|^2\right).$$

The value of the parameter $\gamma > 0$ is related to what we a priori believe to be the amplitude of the image. If the amplitude is poorly known, the parameter could be modeled as a random variable in a hierarchical model. Here, it is assumed that the observer chooses the value of γ, e.g. based on an analysis of the intensity of the image. Implicitly, it is assumed here that the intensity of the fuzzy object vanishes at the boundaries of the field of view. To overcome this restriction, see Calvetti and Somersalo (2005); Calvetti *et al.* (2006); Lehikoinen *et al.* (2007).

The joint probability density of the random variables U, V, Θ and B is given by

$$\pi(u, v, \theta, b) \propto \exp\Bigg(- \frac{1}{2\sigma^2}\|b - \mathsf{A}u - \mathsf{A}v\|^2 - \frac{1}{2}\|\mathsf{D}_\theta^{-1/2} v\|^2 \\ - \frac{\gamma^2}{2}\|\mathsf{L}_2 u\|^2 - \sum_{j=0}^{N} \frac{\theta_0}{\theta_j} - \left(\alpha + \frac{3}{2}\right)\sum_{j=1}^{N} \log\theta_j \Bigg),$$

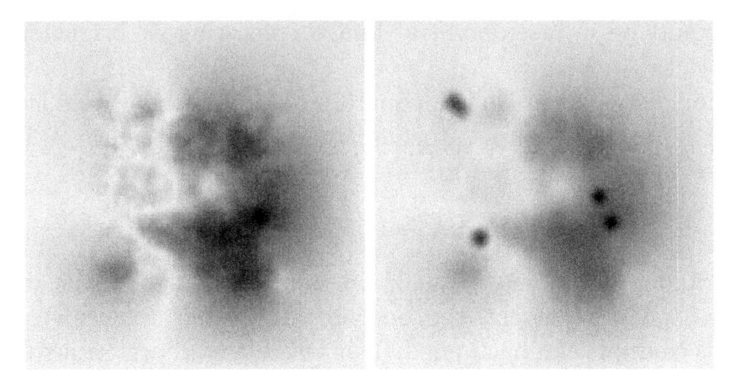

Figure 3.6 The blurry object (left) used to generate the data (right), where the focal object is as in the previous example, Figure 3.4.

which is then conditioned on the observation $b = b_{\text{observed}}$ to obtain the posterior.

As in the previous example, we apply the iterative sequential alternating algorithm to update

$$(u^k, v^k, \theta^k) \to (u^{k+1}, v^k, \theta^k) \to (u^{k+1}, v^{k+1}, \theta^k) \to (u^{k+1}, v^{k+1}, \theta^{k+1}),$$

requiring the solution of two linear systems in the least squares sense plus a componentwise updating of the variance vector.

Figure 3.6 shows the fuzzy object which is added to the focal image of the previous example and then blurred to generate the data, shown in the same figure. The FWHV of the Airy kernel is 5 pixels and the SNR is 1000.

In Figure 3.7 we show the iterates of the presumably smooth component and the focal component, respectively, after 1, 3 and 5 iterations of sequential algorithm. The computation time per iteration in this example varies from 8 to 14 seconds with a PC specified in the previous example. Note that although the focal object is visually reconstructed rather well, there is a slight visible cross-talk between the components.

3.5 Dynamic Updating of Beliefs

The Bayesian formalism and the subjective interpretation behind provide a flexible framework for updating dynamically one's beliefs by a process that resembles everyday reasoning. As new information arrives, old experience is not thrown away but instead is assimilated to the new data. Yesterday's posterior density is today's prior. This idea can be clarified by considering the following question: if the observed data comes as a stream of small bits of information rather than in a big chunk, how should one use it to dynamically update the information on the unknown? For simplicity, we consider first the case where the data arrives sequentially as single projections, and extend the

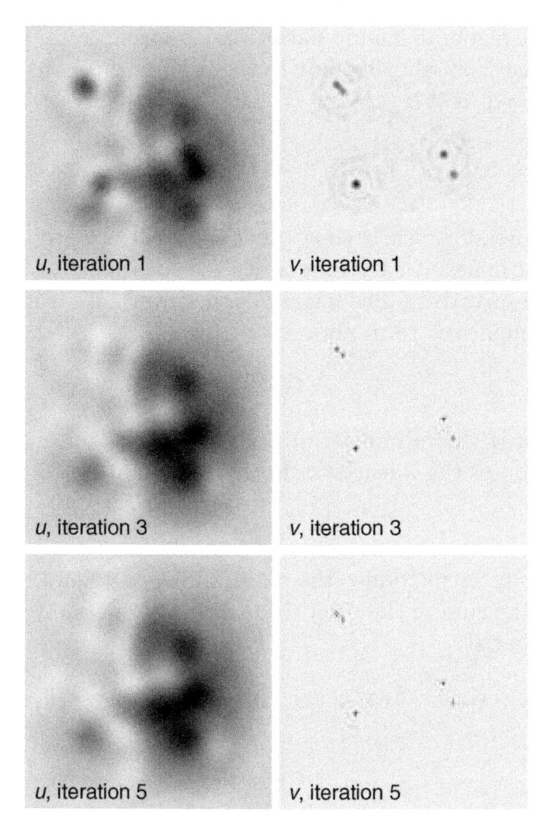

Figure 3.7 Iterates 1, 3 and 5 of the alternating sequential algorithm of the fuzzy component u (left) and the focal component v (right).

discussion to higher dimensional data. The discussion will be a convenient prelude for Kalman filtering (Brockwell and Davis 1991), and more generally, Bayesian filtering, a statistical methodology for data assimilation.

■ **Example 3.7** Let X denote a random variable representing the quantity of interest, which is a priori assumed to follow a Gaussian distribution, $X \sim \mathcal{N}(x_0, \mathsf{G}_0)$. Consider a one-dimensional observation model,

$$Y = v^{\mathsf{T}} X + E, \quad E \sim \mathcal{N}(0, \sigma^2).$$

We scale Y and v with the weight σ^{-1}, so that without loss of generality we may assume that $\sigma = 1$. By Bayes' formula, the posterior probability density is

$$\pi(x \mid y) \propto \exp\left(-\frac{1}{2}(v^{\mathsf{T}} x - y)^2 - \frac{1}{2}(x - x_0)^{\mathsf{T}} \mathsf{G}_0^{-1}(x - x_0) \right).$$

Upon a simple algebraic manipulation and ignoring terms independent of x in the exponent, contributing only to the normalizing constant, the density can be expressed as

$$\pi(x \mid y) \propto \exp\left(-\frac{1}{2}\left(x^{\mathsf{T}}(\mathsf{G}_0^{-1} + vv^{\mathsf{T}})x - 2x^{\mathsf{T}}(\mathsf{G}_0^{-1}x_0 + yv)\right)\right).$$

The latter expression clearly shows that the posterior probability density is Gaussian, and makes it easy to identify the covariance and mean, denoted by G and $\overline{x}$, respectively. The inverse of the covariance is in fact the matrix defining the quadratic term above,

$$\mathsf{G}^{-1} = \mathsf{G}_0^{-1} + vv^{\mathsf{T}},$$

and the mean is the minimizer of the quadratic form in the exponential, i.e. the solution of the linear system

$$\mathsf{G}^{-1}\overline{x} = \mathsf{G}_0^{-1}x_0 + vy.$$

For later use, let us introduce the notation $\mathsf{K} = \mathsf{G}^{-1}$ and $\mathsf{K}_0 = \mathsf{G}_0^{-1}$. If $\overline{x}$ is expressed as the sum of the prior mean and a correction term, $\overline{x} = x_0 + w$, then w must satisfy

$$\mathsf{K}w = \mathsf{K}\overline{x} - \mathsf{K}x_0 = \mathsf{K}_0x_0 + vy - (\mathsf{K}_0 + vv^{\mathsf{T}})x_0 = v(y - v^{\mathsf{T}}x_0),$$

implying that

$$\overline{x} = x_0 + \mathsf{G}v(y - v^{\mathsf{T}}x_0).$$

Consider now the case where the data consist of a sequence of observed noisy one-dimensional projections of X,

$$Y_j = v_j^{\mathsf{T}}X + E_j, \quad 1 \leq j \leq N.$$

Using the updating formulas derived above, we may now apply the sequential updating algorithm, where the density of X, which was initially $\mathcal{N}(x_0, \mathsf{G}_0)$, is updated sequentially based on one-dimensional observed realizations, $Y_j = y_j$. The algorithm for updating the density proceeds as follows:

1. Set $j = 0$, and $\mathsf{K}_j = \mathsf{G}_0^{-1}$, $\overline{x}_0 = x_0$.
2. Increase $j \to j + 1$, and let

   $$\mathsf{K}_j = \mathsf{K}_{j-1} + v_jv_j^{\mathsf{T}}.$$

   Update the mean by

   $$\overline{x}_j = \overline{x}_{j-1} + \mathsf{K}_j^{-1}v_j(y_j - v_j^{\mathsf{T}}\overline{x}_{j-1}).$$

3. If $j = N$, stop, else repeat from 2.

Observe that since the covariance matrix of X after j observations is

$$\mathsf{G}_j = \mathsf{K}_j^{-1},$$

the updating scheme for the covariance can be written, using the Sherman-Morrison-Woodbury formula (Golub and Van Loan 1996) as a rank-1 update

$$\begin{aligned} \mathsf{G}_j &= \left(\mathsf{G}_{j-1}^{-1} + v_j v_j^{\mathsf{T}}\right)^{-1} \\ &= \mathsf{G}_{j-1} - \frac{1}{1 + v_j^{\mathsf{T}} G_{j-1} v_j} \left(\mathsf{G}_{j-1} v_j\right)\left(\mathsf{G}_{j-1} v_j\right)^{\mathsf{T}}. \end{aligned} \tag{3.9}$$

Similarly, the update of the mean, in terms of the covariance matrix, can be written as

$$\overline{x}_j = \overline{x}_{j-1} + \mu_j \mathsf{G}_{j-1} v_j, \quad \mu_j = \frac{y_j - v_j^{\mathsf{T}} \overline{x}_{j-1}}{1 + v_j^{\mathsf{T}} \mathsf{G}_{j-1} v_j}. \tag{3.10}$$

From the point of view of computational cost, formulas (3.9)–(3.10) are enlightening: A complete update sweep $(\overline{x}_{j-1}, \mathsf{G}_{j-1}) \rightarrow (\overline{x}_j, \mathsf{G}_j)$ requires the computation of the matrix-vector product $\mathsf{G}_{j-1} v_j$. In large-scale problems, the memory allocation required for this operation may become an issue, since the covariance matrices are typically non-sparse. However, since G_j is a rank one update of G_{j-1}, if G_0 is sparse, the matrix-vector product can be computed effectively by storing only the vectors $\mathsf{G}_{k-1} v_k$, $1 \leq k \leq j$ without explicitly forming the matrix G_j. Eventually, the memory will fill up, and approximate formulas need to be introduced. This will be demonstrated in the following computed example.

Equation (3.9) shows clearly how the information increases at each iteration step. In fact, uncertainty can be measured in terms of the narrowness of the covariance matrix. As covariance matrices are positive definite symmetric matrices, the information content can be measured in terms of the quadratic form they define. We observe that

$$z^{\mathsf{T}} \mathsf{G}_j z \leq z^{\mathsf{T}} \mathsf{G}_{j-1} z$$

for all $z \in \mathbb{R}^N$, and the equality holds if and only if z is perpendicular to $\mathsf{G}_{j-1} v_j$. We can therefore say that a projection measurement of X in direction v_j increases information, or decreases the width of the covariance matrix, in direction $\mathsf{G}_{j-1} v_j$. This observation is important in design of experiments: measurements should be planned in a way that the information gain is maximized, and the information increases in directions that are meaningful from the point of view of the application.

■ **Example 3.8** In this example, we use the updating of the posterior density to two-dimensional X-ray tomography, when the data consist of a stream of parallel beam projection radiographs with a fixed illumination angle. Let $f : Q \to \mathbb{R}_+$ denote the mass absorption function of the object contained in the unit square $Q = [-1/2, 1/2] \times [-1/2, 1/2]$ around the origin. The continuous noiseless model for the radiograph data is written as a Radon transform,

$$\mathcal{R}f(s, \theta) = \int f(te + se^{\perp})dt,$$

where $e = \cos\theta e_1 + \sin\theta e_2$ is the direction vector of the illumination and $e^{\perp}$ is the unit vector perpendicular to it. To obtain a discrete model, Q is divided in $N = n \times n$ regular sub-squares, or pixels, the mass absorption function is approximated by a piecewise constant pixel image, and the integral is approximated by weighted sums of the pixel values, where the weights equal to the lengths of intersections of the rays and the pixels, see Kaipio and Somersalo (2004) for details. Let $x \in \mathbb{R}^N$ denote the vector containing the pixel values of f, and let s_k, $1 \le k \le K$, θ_j, $1 \le j \le J$, denote the discrete values of the parameters s and θ, respectively. For $\theta = \theta_j$ fixed, we denote the matrix approximation of the Radon transform as

$$\mathcal{R}f(s_k, \theta_j) \approx (\mathsf{A}_j x)_k, \quad 1 \le k \le K,$$

where $\mathsf{A}_j \in \mathbb{R}^{K \times N}$. The measurement configuration is presented schematically in Figure 3.8. The observation model for a noisy recording of a radiograph with fixed illumination angle θ_j is written as

$$Y_j = \mathsf{A}_j X + E_j, \quad j \ge 0,$$

where E_j is an $\mathbb{R}^K$ valued noise with covariance matrix $\Sigma_j \in \mathbb{R}^{K \times K}$.

As in the previous example, we assume that X is a priori a Gaussian random variable with mean x_0 and covariance G_0. The modifications needed to generalize the updating algorithm of the previous example are straightforward. By replacing the vector v_j^{T} by the scaled matrix $\Sigma_j^{-1/2}\mathsf{A}_j$ and the data y_j by $\Sigma_j^{-1/2} y_j$, the generalized form of the updating algorithm can be written as follows:

1. Set $j = 0$, and $\mathsf{K}_j = \mathsf{G}_0^{-1}$, $\overline{x}_0 = x_0$.

2. Increase $j \to j + 1$, let

$$\mathsf{K}_j = \mathsf{K}_{j-1} + \mathsf{A}_j^{\mathsf{T}} \Sigma_j^{-1} \mathsf{A}_j.$$

   Update the mean by

$$\overline{x}_j = \overline{x}_{j-1} + \mathsf{K}_j^{-1} \mathsf{A}_j^{\mathsf{T}} \Sigma_j^{-1} (y_j - \mathsf{A}_j \overline{x}_{j-1}).$$

3. If $j = J$, stop, else repeat from 2.

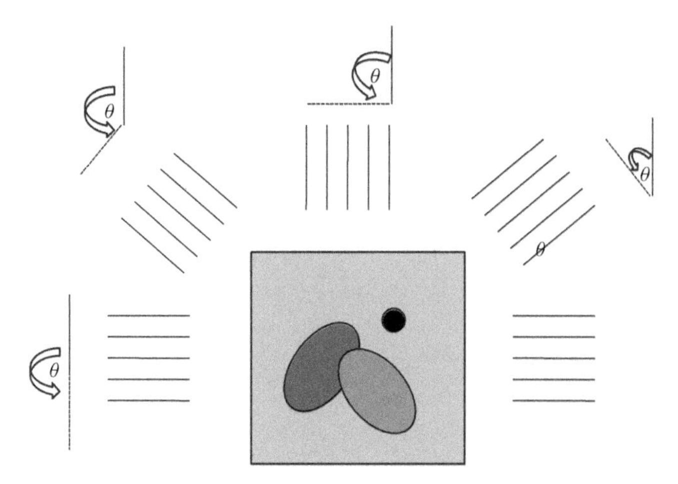

Figure 3.8 The geometric configuration of the tomography example.

If A_j is a matrix with few rows, i.e., $K \ll N$, the updating of the K–matrix is a low rank update. Therefore, the covariance matrix can conveniently be updated via the Sherman-Morrison-Woodbury formula,

$$\begin{aligned}\mathsf{G}_j &= \left(\mathsf{G}_{j-1}^{-1} + \mathsf{A}_j^\mathsf{T}\Sigma_j^{-1}\mathsf{A}_j\right)^{-1} \\ &= \mathsf{G}_{j-1} - \left(\mathsf{G}_{j-1}\mathsf{A}_j^\mathsf{T}\right)\left(\Sigma_j + \mathsf{A}_j\mathsf{G}_{j-1}\mathsf{A}_j^\mathsf{T}\right)^{-1}\left(\mathsf{G}_{j-1}\mathsf{A}_j^\mathsf{T}\right)^\mathsf{T}.\end{aligned}$$

The corresponding updating formula for the mean can be written as

$$\overline{x}_j = \overline{x}_{j-1} + \left(\mathsf{I} - \mathsf{G}_{j-1}\mathsf{A}_j^\mathsf{T}\left(\Sigma_j + \mathsf{A}_j\mathsf{G}_{j-1}\mathsf{A}_j^\mathsf{T}\right)^{-1}\mathsf{A}_j\right)\mathsf{G}_{j-1}\mathsf{A}_j^\mathsf{T}\Sigma_j^{-1}(y_j - \mathsf{A}_j\overline{x}_{j-1}).$$

Observe that the updating scheme is almost as simple as in the case of one-dimensional projection data: for the update $(x_{j-1}, \mathsf{G}_{j-1}) \to (x_j, \mathsf{G}_j)$, one needs to calculate the matrix-matrix product $\mathsf{G}_{j-1}\mathsf{A}_j^\mathsf{T}$ and in addition, the inverse of the matrix $\Sigma_j + \mathsf{A}_j\mathsf{G}_{j-1}\mathsf{A}_j^\mathsf{T} \in \mathbb{R}^{K\times K}$. Again, if K is small compared to N, the latter operation may be relatively uncomplicated from the point of view of computational complexity, while the memory requirements of updating the full matrix $\mathsf{G}_j \in \mathbb{R}^{N\times N}$ may become the limiting factor.

To design a memory-efficient approximate updating algorithm, let us define

$$\mathsf{R}_j = \mathsf{G}_{j-1}\mathsf{A}_j^\mathsf{T} \in \mathbb{R}^{N\times K}, \quad \mathsf{M}_j = \Sigma_j + \mathsf{A}_j\mathsf{G}_{j-1}\mathsf{A}_j^\mathsf{T} \in \mathbb{R}^{K\times K}.$$

After observing that

$$\mathsf{G}_j = \mathsf{G}_0 - \sum_{\ell=1}^{j} \mathsf{R}_\ell\mathsf{M}_\ell^{-1}\mathsf{R}_\ell^\mathsf{T},$$

we obtain the following approximate r-limited memory updating algorithm:

1. Set $j = 0$, and $\overline{x}_0 = x_0$. Define $\mathsf{R}_\ell = 0 \in \mathbb{R}^{N\times K}$ and $\mathsf{M}_\ell = 0 \in \mathbb{R}^{K\times K}$, $\ell = -r+1, \ldots, -1, 0$.

2. Increase $j \to j+1$, calculate the r-limited memory update

$$\mathsf{R}_j = \left(\mathsf{G}_0 - \sum_{\ell=j-r}^{j-1} \mathsf{R}_\ell \mathsf{M}_\ell^{-1} \mathsf{R}_\ell^\mathsf{T}\right) \mathsf{A}_j^\mathsf{T},$$

   and set

$$\mathsf{M}_j = \Sigma_j + \mathsf{A}_j \mathsf{R}_j.$$

3. Store R_j over R_{j-r} and M_j over M_{j-r}.

4. Update the posterior mean by the formula

$$\overline{x}_j = \overline{x}_{j-1} + \mathsf{R}_j \left(\mathsf{I} - \mathsf{M}_j^{-1} \mathsf{A}_j \mathsf{R}_j\right) \Sigma_j^{-1} \left(y_j - \mathsf{A}_j \overline{x}_{j-1}\right). \tag{3.11}$$

5. If $j = J$, end, otherwise repeat from 2.

We demonstrate the performance of the algorithm in the following example, where the data is generated by using a gray scale image of size 256×256 shown in Figure 3.9. The data consists of vectors $\{y_1, y_2, \ldots, y_J\}$, $J = 100$, $y_j \in \mathbb{R}^{100}$, corresponding to equally distributed projection angles θ_j between $\theta_1 = 0$ (illumination from the right) and $\theta_J = \pi$ (illumination from the left). The measurement geometry is illustrated graphically in Figure 3.8, and the data matrix $\mathsf{D} = [y_1, y_2, \cdots, y_J]$ is plotted as a sinogram in Figure 3.10.

We run the r-limited memory algorithm described above with $r = 1$, starting from a smoothness prior density, $x_0 = 0$, $\mathsf{K}_0 = \mathsf{G}_0^{-1} = \alpha \mathsf{L}_2^\mathsf{T} \mathsf{L}_2$, where L_2 is a five point finite difference approximation (3.8) of the Laplacian in $\mathbb{R}^2$ and $\alpha = 20$. Since G_0 is a full $10\,000 \times 10\,000$ matrix, while its inverse K_0 is sparse,

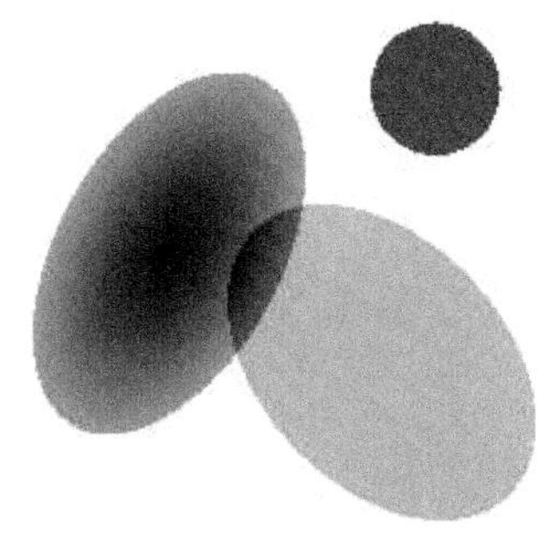

Figure 3.9 The test target used to generate the data.

instead of computing G_0, we define the matrix-vector product $z \mapsto \mathsf{G}_0 z = \zeta$ by solving the system $\mathsf{K}_0 \zeta = z$.

In Figure 3.11, a few snapshots of the updated posterior means are shown. In the computations, the data are generated with additive Gaussian noise with standard deviation $\sigma = 0.001$, which corresponds to a signal-to-noise ratio SNR$\approx$ 1000. The computation time per iteration is about two seconds.

To test how much the mean of the prior original probability density affects the results, we run the algorithm with the same data, but instead of $x_0 = 0$, we use a smooth initial guess shown in Figure 3.12. As the algorithm proceeds, the initial guess that is obviously in conflict with the sinogram data, is gradually erased and the final posterior mean is similar to the one using a flat initial guess. Notice that since the flat initial guess is, too, in conflict with the data, the result should not be surprising.

In the examples above, it is assumed that the quantity of interest, modeled by the random variable X, does not change from measurement to measurement. This situation, however, cannot always be guaranteed. If the estimated variable changes as a function of j, referred to as time, we need to model it by a stochastic process $\{X_j\}_{j\geq 0}$. Obviously, if the random variables X_j are mutually independent, information about X_k with $k < j$ is not helpful for the estimation of X_j. To obtain a recursive algorithm updating the current information, we therefore have to assume a model linking the variables X_j together. A popular model that can be often justified by evolutionary models

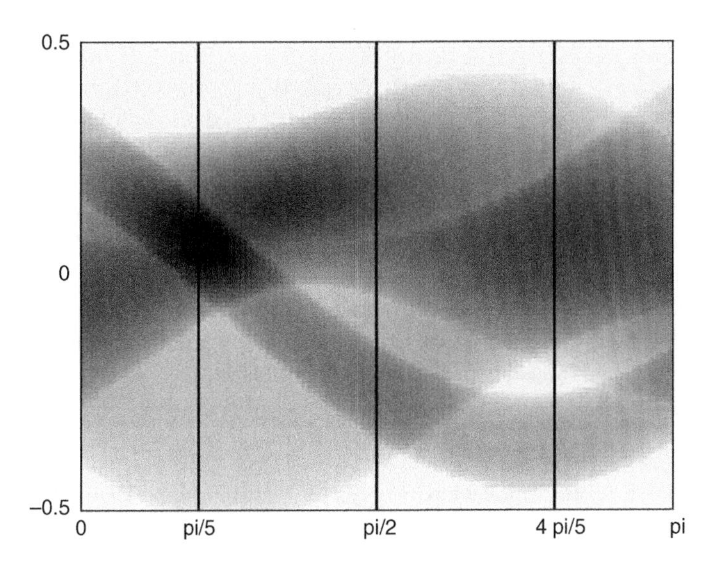

Figure 3.10 The sinogram data. The reconstructions shown in Figure 3.11 are computed using the data on the left of the vertical lines.

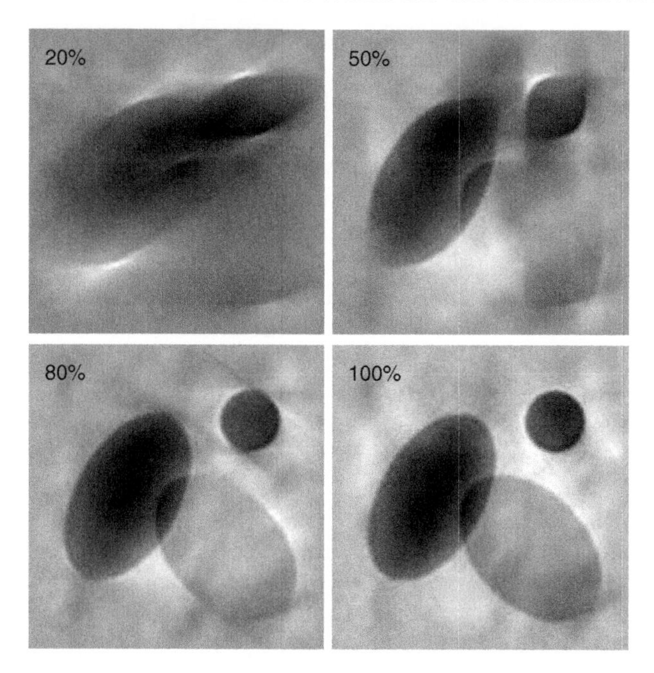

Figure 3.11 Posterior mean estimates using the approximate formula (3.11) with $r = 1$, based on 20 %, 50 %, 80 % and 100 % of the full sinogram data shown in Figure 3.10.

is a linear autoregressive Markov process,

$$X_j = \mathsf{B}_j X_{j-1} + W_j.$$

Here, $\mathsf{B}_j \in \mathbb{R}^{n\times n}$ is the Markov propagation matrix and W_j is the innovation process, expressing the uncertainty in predicting the successive value based on the previous one. It is often assumed that W_j is independent of X_j. If $\mathsf{B}_j = \mathsf{B}$ and the innovation processes are identically distributed, i.e. the propagation step is statistically time independent, the model is a stationary Markov process. In the discussion below, we will adhere to this paradigm.

Assume that at the time $t = j$, the current information concerning X_{j-1} is encoded in the Gaussian probability distribution,

$$X_{j-1} \sim \mathcal{N}(\overline{x}_{j-1}, \mathsf{G}_{j-1}),$$

and assume that the innovation process W_j is Gaussian, $W_j \sim \mathcal{N}(0, \mathsf{D})$ and independent of X_{j-1}. We use the Markov process to calculate the probability density of X_j. As a sum of two Gaussian variables, X_j is Gaussian, with the mean

$$\widehat{x}_j = \mathrm{E}\{X_j\} = \mathrm{E}\{\mathsf{B}X_{j-1} + W_j\} = \mathsf{B}\overline{x}_{j-1}, \tag{3.12}$$

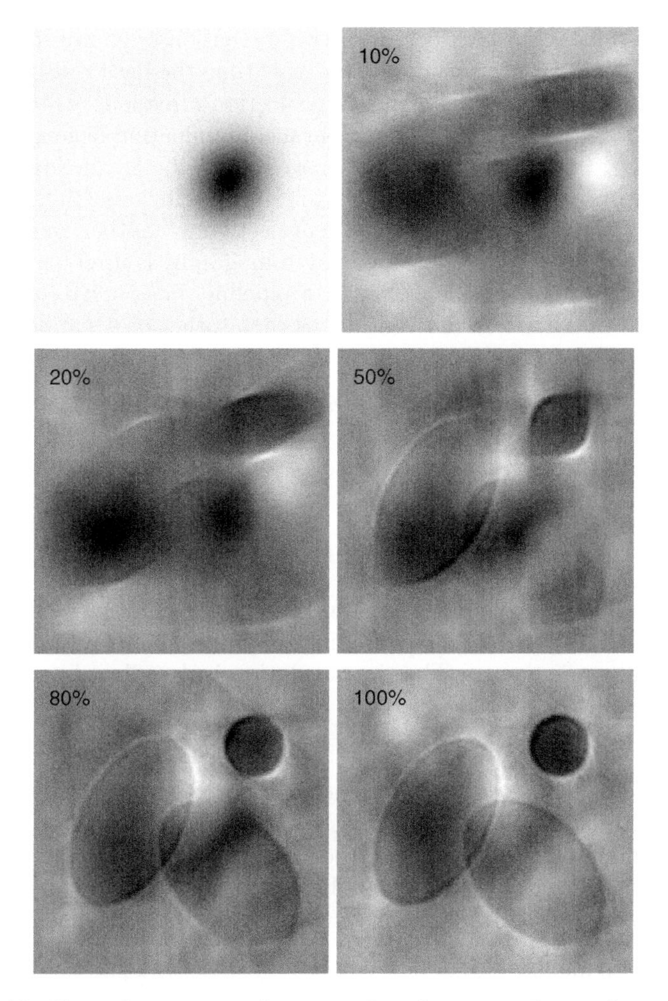

Figure 3.12 Posterior mean estimates using the approximate formula (3.11) with $r = 1$, starting from a prior mean shown in the top left figure. Approximate estimates based on 10 %, 20 %, 50 %, 80 % and 100 % of the full sinogram data shown in Figure 3.10.

and covariance

$$\begin{aligned}\widehat{\mathsf{G}}_j &= \mathrm{E}\{X_j X_j^{\mathsf{T}}\} - \widehat{x}_j \widehat{x}_j^{\mathsf{T}} = \mathsf{B}\big(\mathrm{E}\{X_{j-1}X_{j-1}^{\mathsf{T}}\} - \overline{x}_{j-1}\overline{x}_{j-1}^{\mathsf{T}}\big)\mathsf{B}^{\mathsf{T}} + \mathrm{E}\{W_j W_j^{\mathsf{T}}\} \\ &= \mathsf{B}\mathsf{G}_{j-1}\mathsf{B}^{\mathsf{T}} + \mathsf{D}.\end{aligned} \tag{3.13}$$

These formulas are the backbone of a dynamic updating algorithm that, together with the observation updating considered in previous examples, amounts to the classical Kalman filter algorithm. For the sake of concreteness, we discuss the algorithm in the light of a computed example.

■ **Example 3.9** We consider the two-dimensional X–ray tomography problem of the previous example, but this time the data consist of sparse projections in randomly chosen few projection directions. More precisely, at each $t = j$, ten randomly chosen parallel illumination beams are picked, and the object is illuminated in ten randomly chosen directions from 0 to π. The tomography matrix thus obtained is denoted by A_j. In this example, the target changes from one projection to another. Such situation could be encountered, e.g. when X–ray tomography is used for monitoring an industrial process, a fluid flow in a pipeline, for quality controlling of logs in sawmills, or for the continuous monitoring of a patient during a surgical intervention.

Let $X_j \in \mathbb{R}^N$ denote the random variable modeling the discretized mass absorption distribution in Q at $t = j$, $j \geq 0$. We assume a priori that X_0 is Gaussian with mean x_0 and covariance G_0. The observation model is, as in the previous example,

$$Y_j = \mathsf{A}_j X_j + E_j, \quad j \geq 0,$$

where E_j is an $\mathbb{R}^K$ valued noise with covariance matrix $\Sigma_j \in \mathbb{R}^{K\times K}$, $K = 100$, and A_j is the X–ray matrix corresponding to sparse parallel beam projections into the random directions. The time evolution is described by a stationary Markov process,

$$X_j = \mathsf{B} X_{j-1} + W_j,$$

where $\mathsf{B} \in \mathbb{R}^{N\times N}$ is fixed and the innovation process W_j is zero mean and Gaussian with covariance D, and W_j is independent of X_{j-1}.

By combining the formulas derived in the previous examples with the updating formulas (3.12)–(3.13), we obtain the classical Kalman filter algorithm (Brockwell and Davis (1991)):

1. Set $j = 0$ and $\overline{x}_0 = x_0$.
2. **Prediction step**: Increase $j \to j + 1$, and propagate the mean and covariance matrices according to the formulas
$$\begin{aligned} \widehat{x}_j &= \mathsf{B}\overline{x}_{j-1}, \\ \widehat{\mathsf{G}}_j &= \mathsf{B}\mathsf{G}_{j-1}\mathsf{B}^\mathsf{T} + \mathsf{D}. \end{aligned}$$
3. **Updating step**: On arrival of the observation y_j, update the mean and covariance according to the formulas
$$\begin{aligned} \mathsf{G}_j &= \widehat{\mathsf{G}}_j - \big(\widehat{\mathsf{G}}_j\mathsf{A}_j^\mathsf{T}\big)\big(\Sigma_j + \mathsf{A}_j\widehat{\mathsf{G}}_j\mathsf{A}_j^\mathsf{T}\big)^{-1}\big(\widehat{\mathsf{G}}_j\mathsf{A}_j^\mathsf{T}\big)^\mathsf{T}, \\ \overline{x}_j &= \widehat{x}_j + \mathsf{G}_j\mathsf{A}_j^\mathsf{T}\Sigma_j^{-1}\big(y_j - \mathsf{A}_j\widehat{x}_j\big). \end{aligned}$$
4. If $j = N$, stop, otherwise repeat from 2.

In the literature, the matrix $\mathsf{K}_j \in \mathbb{R}^{K\times K}$, defined as

$$\mathsf{K}_j = \mathsf{A}_j^\mathsf{T}\left(\Sigma_j + \mathsf{A}_j\widehat{\mathsf{G}}_j\mathsf{A}_j^\mathsf{T}\right)^{-1}\mathsf{A}_j$$

is often referred to as the *Kalman gain* matrix. The updating of the covariance can be written in terms of K_j as

$$\mathsf{G}_j = \left(\mathsf{I} - \widehat{\mathsf{G}}_j\mathsf{K}_j\right)\widehat{\mathsf{G}}_j.$$

The computational challenges of the Kalman filtering are similar to the ones encountered in the continuous updating of the posterior densities of Example 3.8. Luckily, it is possible to overcome these problems by means of a similar limited memory variant. Indeed, using the notations

$$\begin{aligned}\mathsf{R}_j &= \widehat{\mathsf{G}}_j\mathsf{A}_j^\mathsf{T},\\ \mathsf{M}_j &= \Sigma_j + \mathsf{A}_j\widehat{\mathsf{G}}_j\mathsf{A}_j^\mathsf{T},\end{aligned}$$

the updating step for the covariance matrix can be written as

$$\begin{aligned}\mathsf{G}_j &= \widehat{\mathsf{G}}_j - \mathsf{R}_j\mathsf{M}_j^{-1}\mathsf{R}_j^\mathsf{T}\\ &= \mathsf{B}\mathsf{G}_{j-1}\mathsf{B}^\mathsf{T} + \mathsf{D} - \mathsf{R}_j\mathsf{M}_j^{-1}\mathsf{R}_j^\mathsf{T},\end{aligned}$$

and by iterating, we have

$$\mathsf{G}_j = \mathsf{B}^j\mathsf{G}_0\left(\mathsf{B}^\mathsf{T}\right)^j + \sum_{\ell=1}^{j}\mathsf{B}^{j-\ell}\mathsf{D}\left(\mathsf{B}^\mathsf{T}\right)^{j-\ell} - \sum_{\ell=1}^{j}\mathsf{B}^{j-\ell}\mathsf{R}_\ell\mathsf{M}_\ell^{-1}\mathsf{R}_\ell\left(\mathsf{B}^\mathsf{T}\right)^{j-\ell}.$$

We may therefore define an r-limited memory Kalman filtering algorithm which is particularly attractive when the covariance matrix D of the innovation process is sparse and therefore does not require excessive memory allocation.

1. Set $j = 0$, and $\overline{x}_0 = x_0$. Define $\mathsf{R}_\ell = 0 \in \mathbb{R}^{N\times K}$ and $\mathsf{M}_\ell = 0 \in \mathbb{R}^{K\times K}$, $\ell = -r+1, \ldots, -1, 0$.

2. Increase $j \to j+1$, calculate the r-limited memory update

$$\mathsf{R}_j = \left(\mathsf{B}^j\mathsf{G}_0\left(\mathsf{B}^\mathsf{T}\right)^j + \sum_{\ell=1}^{j}\mathsf{B}^{j-\ell}\mathsf{D}\left(\mathsf{B}^\mathsf{T}\right)^{j-\ell} - \sum_{\ell=j-r}^{j-1}\mathsf{B}^{j-\ell}\mathsf{R}_\ell\mathsf{M}_\ell^{-1}\mathsf{R}_\ell\left(\mathsf{B}^\mathsf{T}\right)^{j-\ell}\right)\mathsf{A}_j^\mathsf{T},$$

   and

$$\mathsf{M}_j = \Sigma_j + \mathsf{A}_j\mathsf{R}_j.$$

3. Store R_j over R_{j-r} and M_j over M_{j-r}.

4. Update the posterior mean by the formula

$$\overline{x}_j = \overline{x}_{j-1} + \mathsf{R}_j \left(\mathsf{I} - \mathsf{M}_j^{-1}\mathsf{A}_j\mathsf{R}_j\right)\Sigma_j^{-1}\left(y_j - \mathsf{A}_j\overline{x}_{j-1}\right). \tag{3.14}$$

5. If $j = J$, end, otherwise repeat from 2.

The performance of this algorithm is demonstrated by an example. In Figure 3.13, the left column shows the true target used to calculate the simulated data, and the right column shows the dynamic posterior mean estimates. In this example, the prior is a second order Gaussian smoothness prior with zero mean as in the previous example, and the noise level is also the same. The propagation model is a Gaussian random walk model, $\mathsf{B} = \mathsf{I}$ and $\mathsf{D} = \delta\mathsf{I}$ with $\delta = 0.1$. The full data consists of $J = 100$ snapshots of projection data. We observe that the repeated measurements lead to improvement of the static object, while the reconstruction quality of the moving object is about the same in the beginning and at the end of the scan.

Finally, observe that the idea of Bayesian updating the posterior density does not depend on the simplifying assumptions such as normality of the distributions, linearity of the observation model, stationarity and linearity of the propagation model. However, the updating formulas are usually no longer explicitly known, and one has to resort to local linearization and iterative updating, or Markov Chain Monte Carlo methods. The former alternative gives rise to extended Kalman filtering algorithms, the latter to particle filtering.

3.6 Discussion

A famous quotation of Cornelius Lanczos states that 'no mathematical trickery can remedy lack of information'. The use of priors has been unjustly accused of attempting to circumvent the obstacle of the paucity of information. This is plainly and simply wrong. It is true that a useful prior provides information that is not present in the data, but the same is true of any useful regularization scheme as well, e.g. by providing a rule of selecting a particular element from the nontrivial numerical null space of a matrix. The Bayesian approach expresses the prior precisely in terms of the information, or the lack thereof, while, for instance, classical regularization schemes may be rather vague on what their information contents is. In fact, regularization is often described as a 'method' of solving an inverse problem, giving the wrong impression that the data and model themselves contain enough information to find a solution for the problem. A good example is the classical Tikhonov regularization scheme: formally, it penalizes the solution for growing in norm, which is in itself a rather strong condition. Moreover, when interpreted from the statistical viewpoint, it implicitly introduces an a priori covariance structure implying a mutual independence of the components of

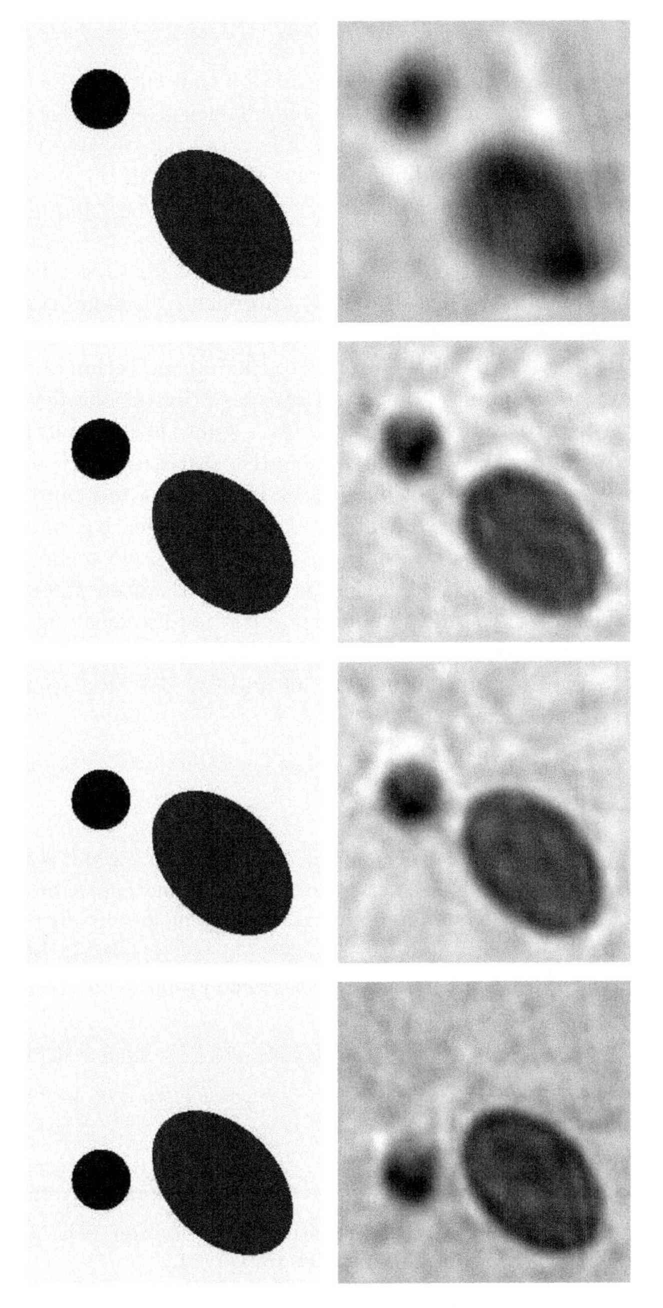

Figure 3.13 Dynamic updating of the posterior mean by formula 3.14 with $r = 1$. The left column is the object used to generate the data, the right column is the corresponding reconstruction at $t = 1, t = 21, t = 41$ and $t = 81$.

the unknown which may not be desirable and has generally a visible effect on the solution.

The prior is an expression of the information that the subject has at hand, and the implementation of this information in terms of probability assignment is a nontrivial and non-unique process. The same information can be interpreted by different agents in different ways that may all be justified. This is especially true when the information is qualitative rather than quantitative. The same is true for the process of setting up a reasonable likelihood model. Thus, the concept of probability is by its own nature subjective. Contrary to what the adversaries of the Bayesian approach sometimes claim, subjectivity is not the same as arbitrary. The attribute of subjective probability attached to the Bayesian approach refers to the interpretation of the meaning of probability, and as pointed out, the Bayesian prior probability assignment is no more arbitrary than the selection of regularization functional in the nonstatistical framework. Classical frequentist statistical methods also rely on probability assignments that need to be interpreted and can be contested (D'Agostini 2003). The part which is incontestably objective, namely the use of the laws of probability and analysis, remain the same in both strains of statistics, although the methodologies and final goals often differ.

Our emphasis on the subjective interpretation of the meaning of probability fits into a much wider discussion about the role of subjectivity in science. The relation between scientific subjectivity and the Bayesian approach is discussed in Press and Tanur (2001).

References

Arridge, S.R., Kaipio, J.P., Kolehmainen, V., Schweiger, M., Somersalo, E., Tarvainen, T. and Vauhkonen, M., 2006 Approximation errors and model reduction with an application in optical diffusion tomography. *Inverse Problems* **22**, 175–195.

Bai, Z. and Golub, G. 1998 Some unusual eigevalue problems. In *Lecture Notes In Computer Science* **1573** Springer.

Bernardo, J.M. and Smith, A.F.M. 1994 *Bayesian Theory*. John Wiley& Sons, Ltd.

Brockwell, P.J. and Davis, R.A. 1991 *Time Series: Theory and Methods*. Springer Verlag.

Calvetti, D. and Somersalo, E. 2004 *Introduction to Bayesian Scientific Computing – Ten Lectures on Subjective Computing*, Springer Verlag.

Calvetti, D. and Somersalo, E. 2005 Statistical compensation of boundary clutter in image deblurring. *Inverse Problems* **21** 1697–1714.

Calvetti, D. and Somersalo, E. 2008 Hypermodels in the Bayesian imaging framework. *Inverse Problems* **24** 034013.

Calvetti, D., Kaipio, J.P. and Somersalo, E. 2006 Aristotelian prior boundary conditions. *Int. J. Math. Comp. Sci.* **1** 63–81.

Caves, C.M., Fuchs, C.A. and Schack, R. 2002 Quantum probabilities as Bayesian probabilities. *Phys. Rev. A* **65** 022305.

Caves, C.M., Fuchs, C.A. and Schack, R. 2007 Subjective probability and quantum certainty. *Studies in History and Philosophy of modern Physics* **38** 255–274.

D'Agostini, G. 2003 *Bayesian Reasoning in Data Analysis: A Critical Introduction.* World Scientific.

D'Agostini, G. 2005 From observations to hypotheses. Probabilistic reasoning versus falsificationism and its statistical variations. *arXiv:physics/0412148*.

de Finetti, B. 1930 Funzione caratteristica di un fenomeno aleatorio. *Mem. Acad. Naz. Lincei* **4** 86–133

de Finetti, B. 1974 *Theory of Probability, Vol. 1*, John Wiley & Sons, Ltd.

Diaconis, P. and Freedman, D. 1987 A dozen de Finetti-style results in search of a theory. Ann. Inst. H. Poincaré **23** 397–423.

Donoho, D., Johnstone, I.M., Hoch, J.C. and Stern, A.S. 1992 Maximum entropy and the nearly black object. *J. R. stat. Soc. Ser B* **54** 41.

Geman, S. and Geman, D. 1984 Stochastic relaxation, Gibbs distributions and the Bayesian restoration of images. *IEEE Trans. Pattern Anal. Machine Intell.* **6** 721–741.

Gilks, W.R., Richardson, S. and Spiegelhalter, D.J. 1996 *Markov Chain Monte Carlo in Practice*. CRC Press.

Golub, G. and Van Loan, C. 1996 *Matrix Computations* (3rd ed.), Johns Hopkins.

Hewitt, E. and Savage, L.J. 1955 Symmetric measures on Cartesian products. *Trans. Amer. Math. Soc.* **80** 470–501.

Hyvärinen, A. and Oja, E. 2000 *Independent Component Analysis: Algorithms and Applications*, John Wiley & Sons, Ltd.

Jeffrey, R. 2004 *Subjective Probability: The Real Thing*. Cambridge University Press.

Kaipio, J.P. and Somersalo, E. 2002 Estimating anomalies from indirect observations. *J. Comp. Phys.* **181** 398–406.

Kaipio, J.P. and Somersalo, E. 2004 *Statistical and Computational Inverse Problems*, Springer Verlag.

Kaipio, J.P. and Somersalo, E. 2007 Statistical inverse problems: discretization, model reduction and inverse crimes. *J. Comp. Appl. Math.* **198** 493–504.

Laplace, P.S. 1774 Mémoire sur la probabilité des causes par les évènemens. In *Laplace's 1774 Memoir in Inverse Probability, Statistical Science* **1** 1986 (Stigler SM), pp. 359–363.

Lehikoinen, A., Finsterle, S., Voutilainen, A., Heikkinen, L.M., Vauhkonen, M. and Kaipio, J.P. 2007 Approximation errors and truncation of computational domains with application to geophysical tomography. *Inverse Problems and Imaging* **1** 371–389.

Liu, J.S. 2004 *Monte Carlo Strategies in Scientific Computing*. Springer Verlag.

O'Hagan, A. and Forster, J. 2004 *Kendall's Advanced Theory of Statistics, Volume 2B: Bayesian Inference*. Arnold

Press, S.J. and Tanur, J.M. 2001 *The Subjectivity of Scientists and the Bayesian Approach*. John Wiley & Sons, Ltd.

Seppänen, A., Heikkinen, L., Savolainen, T., Voutilainen, A., Somersalo, E. and Kaipio, J.P. 2007 An experimental evaluation of state estimation with fluid dynamical models in process tomography. *Chem. Eng. Journal* **127** 23–30.

Starck, J.L., Pantin, E. and Murtagh, F. 2002 Deconvolution in astronomy: A review. *Publ. Astron. Soc. of the Pacific* **114** 1051–1069.

Tarantola, A. 2005 *Inverse Problem Theory (2nd ed.)*. SIAM.

Tarantola, A. 2006 Inverse problems, Popper and Bayes. *Nature Physics* **2** 492–494.

Chapter 4

Bayesian and Geostatistical Approaches to Inverse Problems

P.K. Kitanidis

Stanford University, USA

We examine key ideas related to the rationale for using stochastic Bayesian and Geostatistical methods to solve inverse problems. We point out that the frequentist approach employed in classical statistics is not well suited for inverse problems that are characterized by insufficient data. The Bayesian approach differs in substantial ways from the classical approach, both in terms of methodology and in terms of the meaning of the results one obtains. We discuss certain misconceptions about the so-called prior probability distribution. This distribution can be used to express known facts about the structure of the unknown. It does not have to be preceding in time to the data and can be described through expressions that involve parameters that can be estimated from the data. Finally, we review a geostatistical approach to the inverse problem in light of the discussion of the Bayesian methodology.

4.1 Introduction

The parameters that govern flow and transport in geologic formations are the result of complex geologic and physicochemical processes acting over long

---

Large-Scale Inverse Problems and Quantification of Uncertainty Edited by L. Biegler, G. Biros, O. Ghattas, M. Heinkenschloss, D. Keyes, B. Mallick, Y. Marzouk, L. Tenorio, B. van Bloemen Waanders and K. Willcox

periods of time. Similarly, variables that describe atmospheric process vary constantly, governed by complex thermo- and aerodynamics. The parameters vary at all scales and in ways that are impossible to predict without a plethora of measurements. The parameters of geologic formations are also hard to measure directly. Thus, a crucial component of modeling in the applied Earth Sciences is using sparse data to evaluate model parameters, i.e. the solution of inverse problems. In a typical practical situation, one is faced with an inverse problem that is ill-posed because, at least at first appearance, the primary data do not suffice to determine a unique solution. We are concerned with distributed parameters, for example: hydraulic conductivity, which is a function of space; pollutant release history, which is a function of time or of both time and space; and moisture content, in unsaturated flow, which is a function of matric potential. Thus, when we refer to model parameters, we actually mean one or more unknown functions.

The root of the nonuniqueness is usually in the relation between the data and the parameters, which makes the results be overly sensitive to the data, such as small fluctuations in the data, to the extent that one cannot determine a solution without some form of suppression of noise or regularization of the solution. An additional, and even more common, cause of nonuniqueness is gaps in the data. In the discretized form in which $\boldsymbol{y}$ is the $n \times 1$ array of the data and $\boldsymbol{s}$ is the $m \times 1$ array of the model parameters, one may start by formulating the problem through the relation

$$\boldsymbol{y} = \mathbf{h}(\boldsymbol{s}) \tag{4.1}$$

Even if the mapping $\boldsymbol{s} \rightarrow \boldsymbol{y}$ were one-to-one, the inverse mapping $\boldsymbol{y} \rightarrow \boldsymbol{s}$ is not. Typically, the problem is algebraically underdetermined ($m > n$). Adding to the complexity of the problem, the mapping $\boldsymbol{s} \rightarrow \boldsymbol{y}$ is not really one-to-one because either the mathematical model is an imperfect representation of the actual system or the data are corrupted by error due to measurement methods and their implementation. Thus,

$$\boldsymbol{y} = \mathbf{h}(\boldsymbol{s}) + \boldsymbol{\varepsilon} \tag{4.2}$$

The discrepancy is represented through term $\boldsymbol{\varepsilon}$ which is known as the *observation* error (or noise). It is usually called observation error even though much of it is due to model limitations. In the author's experience, in practical applications this measurement error is significant, which means that it is meaningless and even counterproductive to strive to select model parameters that reproduce observations exactly.

Let us consider an example. Assume we are faced with a deconvolution problem,

$$y(t) = \int g(t-\tau)\, s(\tau)\, d\tau \tag{4.3}$$

where g is a known smooth kernel, such as $g(t) = \exp\left(-\frac{t^2}{2T^2}\right)$, s is the unknown function, and the output y has been measured at n points in time,

$$y(t_i) = \int g(t_i - \tau)\, s(\tau)\, d\tau + \varepsilon_i, \quad i = 1, \ldots n \tag{4.4}$$

It is obvious that one cannot uniquely identify the function $s(t)$ from sparse and noise-affected observations of y. The observations are not sensitive to fine-scale characteristics of $s(t)$, i.e. fluctuations at scales much shorter than T; however, they may be quite informative about features at scales larger than T. The challenge is to try to resolve $s(t)$ as finely as possible while also staying aware of the limitations in estimation. It would be great if one could obtain: (a) An estimate that essentially shows only features that are supported by data. This means that the estimate should not show details that are not justified by the data or other credible information. (b) A credibility interval that contains the actual solution with some degree of reliability and gives us an idea of the kind of error we should expect. (c) A collection of plausible s functions.

Summarizing, the three main causes of nonuniqueness are the functional form of the relation between data and parameters, gaps in the data (insufficient observations), and observational error (noise in the available data and limitations in the model). Essentially, these problems constitute lack of information. Many approaches used in practice add information, in the form of additional restrictions on the solution, in an effort to formulate a problem with a well-posed problem. For example, in Tikhonov-type methods (Tikhonov *et al.* 1997; Tikhonov and Arsenin 1977), an objective of smoothness or flatness of the solution is added. This is known as regularization. However, this approach, although sufficient to produce a mathematically unique solution, does not really eliminate uncertainty. In fact, in the case of geophysical parameters, I believe that it is now widely understood that the solution thus obtained is not the actual parameter function. Instead, the solution is viewed as an estimate and there is interest in evaluating how close it may be to the actual solution and how it can be improved.

One general alternative approach is to recognize that the data are not sufficient to fix a single solution, even with some form of regularization. Then, to generate in a systematic way multiple solutions that are in some sense equally valid candidates to represent the actual solution (or, equivalently, to somehow evaluate the degree of plausibility of all possible solutions). This is the approach adopted by a number of methods, including Bayesian and geostatistical approaches. Although these methods have enthusiastic proponents, they are not widely understood and have no lack of detractors. In my view, some skepticism echoes concern about Bayesian methods as used in completely different applications. Others can only be attributed to inadequate understanding of Bayesian statistical methods and to confusing them with classical statistical methods. Questions that are often raised include:

- Is not the objective of statistical methods to extract information through analysis of voluminous data sets? If so, how can statistical methods be used when the very source of the difficulty is inadequate data?
- How can probabilities be assigned to the multiple possible solutions? What do these probabilities mean?
- In particular, Bayesian methods make use of prior probabilities. How are prior probabilities selected and how do they affect the results? Are Bayesian methods sufficiently objective?

My original motivation for writing this chapter was to discuss the logical underpinnings of the stochastic method of Kitanidis and Vomvoris (1983) and Kitanidis (1995) for solving inverse problems. It soon became clear, however, that it would be better to discuss some aspects of the Bayesian approach forsolving inverse problems and, in particular, the role and meaning of prior information. I believe that, at the end, any approach should be judged by its usefulness so I will try not to wax philosophical. Nevertheless, my experience indicates that when one does not understand the rationale of an approach, one is more likely to avoid it, be discouraged at the first difficulty or, perhaps worse of all, misuse it. Words are surprisingly powerful and through their emotional appeal often determine the outcome of a debate. By that, I mean words such as 'probability', 'subjective', 'prior', and 'Gaussian'. Unfortunately, many terms that are used in discussing Bayesian methods are either misunderstood or carry excessive baggage.

4.2 The Bayesian and Frequentist Approaches

Perhaps a good starting point in understanding the Bayesian approach (BA) is to realize that it should not be confused with the classical statistical methodology (frequentist approach or FA) that is usually taught in undergraduate classes! This is not to say that they do not share common features. They do, which unfortunately makes it easier to confuse them. Let us review important similarities and differences.

4.2.1 Frequentist Approach

Both BA and FA utilize the calculus of probability, i.e. deal with variables that are described with probability distributions. These variables are commonly known as random, a prevalent though unfortunate (at least in the context of inverse methods) choice of a term. However, in BA, the variables that have distributions are the unknowns; in FA the variables that have distributions are the data and the estimates whereas the unknowns are treated as fixed. This is an important point that is often misunderstood by practitioners. So, the terms distribution and uncertainty refer to the unknown $\boldsymbol{s}$ in BA methods and to the estimate $\widehat{\boldsymbol{s}}$ in FA methods. For example, the 95 % confidence interval in FA is random and includes the unknown with probability 0.95 and should not necessarily be interpreted as *the* interval that contains the unknown with 95 % probability, since in FA the estimator is random but the unknown is fixed.

A practical person may dismiss this difference as nitpicking since we often are interested in describing the distribution of the estimation error, $\widehat{\boldsymbol{s}} - \boldsymbol{s}$. If both could evaluate the distribution of the error, then BA and FA would simply offer two different approaches, both involving a conceptual model with a high degree of abstraction, to solve the same problem: quantify the error.

However, FA has some disadvantages when it comes to its application to inverse problems. FA requires the analyst to somehow come up with an estimator, which is a function of the data. The estimate is produced by plugging the actual data in the estimator. One can develop an estimator, for example, through minimization of a fitting criterion, maximizing a likelihood function, or matching of some averages or moments. The estimator should have some 'good properties' in the sense that the estimation error should have zero mean and be small in some sense.

One difficulty, however, is that in inverse problems, especially nonlinear ones, it is hard to say which estimator one should use, unlike for some simpler problems studied in introductory statistics. For example, in estimating the mean of the population of independent and identically distributed variables, the sample mean is a reasonable estimator meeting most criteria for 'good estimators'. The asymptotic argument is sometimes used in FA, i.e. that the estimator will be optimal or distributed in a certain way (e.g. distribution asymptotically Gaussian by virtue of the central limit theorem) independently of the distribution of the data as the sample size increases. Unfortunately, asymptotic arguments are meaningless when the main issue is that there are no even enough data to even come up with a reasonable estimator.

For example, consider the linear version of (4.2),

$$\boldsymbol{y} = \boldsymbol{H}\boldsymbol{s} + \boldsymbol{\varepsilon} \tag{4.5}$$

where $\boldsymbol{H}$ is $n \times m$ and of rank n (algebraically underdetermined case) and let us further postulate that $\boldsymbol{\varepsilon}$ is random with mean $\mathbf{0}$ and covariance matrix $\boldsymbol{R}$. Consider a linear estimator

$$\widehat{\boldsymbol{s}} = \boldsymbol{G}\boldsymbol{y}$$

with associated error

$$\begin{aligned} \widehat{\boldsymbol{s}} - \boldsymbol{s} &= \boldsymbol{G}\boldsymbol{y} - \boldsymbol{s} \\ &= (\mathbf{G}\boldsymbol{H} - \boldsymbol{I})\,\boldsymbol{s} + \mathbf{G}\boldsymbol{\varepsilon} \end{aligned} \tag{4.6}$$

If one could find a $\boldsymbol{G}$ that satisfies $\mathbf{G}\boldsymbol{H} = \boldsymbol{I}$, then the estimator would be unbiased for any unknown $\boldsymbol{s}$ and the covariance matrix of the estimation error would be

$$E\left[(\widehat{\boldsymbol{s}} - \boldsymbol{s})(\widehat{\boldsymbol{s}} - \boldsymbol{s})^T\right] = \boldsymbol{GRG}^T \tag{4.7}$$

Notice that this approach would quantify the distribution of the error but not really of the unknown.

However, for underdetermined problems the rank of $\mathbf{G}\boldsymbol{H}$ cannot be larger than n which is less than m, the rank of $\boldsymbol{I}$. So, one cannot really find an unbiased estimator or evaluate the estimation-error covariance matrix of the estimation error.

The bottom line is that the classical approach is not suited to deal with algebraically underdetermined inverse problems. Perhaps even more importantly, this approach does not even consider it reasonable to speculate about the probability of a unique situation, such as the hydraulic conductivity of a geologic formation. This is because the approach goes at great lengths to use only data and distrusts anything that could be viewed as opinion or subjective judgement. This feature of the classical approach makes it desirable in, say, evaluating the efficacy of a new drug in double-blind controlled trials, but narrows its scope too much to address inverse problems.

4.2.2 Bayesian Approach

A fundamental difference between the Frequentist Approach and the Bayesian Approach appears to be philosophical. While in the FA one can only postulate the probability of events in the sample space associated with data, the Bayesian Approach considers it meaningful to make claims such as 'There is a 30 % probability that the maximum concentration in a specific contaminant plume exceeds 10 *ppm*.' In my view, the main difference is that BA has a broader scope of applications than FA. Jaynes' book (Jaynes 2003), appropriately titled *Probability Theory: The Logic of Science*, discusses the broad perspective of Bayesian-type methods. Independently of whether on agrees with Jaynes' perspective, there is much food for thought for anyone who is interested in the role of probabilistic methods in inverse problems. Other books that explain the Bayesian viewpoint include Jeffreys (1957).

BA utilizes Bayes theorem, which makes the application of the approach much more straightforward than the search of estimators in FA. The key modeling steps involve introducing the so-called prior probability distribution of the unknowns and the distribution of the observation errors. Specifically, if $p'(\boldsymbol{s})$ is the prior probability density function of the unknowns and $p(\boldsymbol{y} \mid \boldsymbol{s})$ is the likelihood function (the pdf of $\boldsymbol{y}$ conditional on $\boldsymbol{s}$), then the posterior pdf is given by Bayes theorem:

$$p''(\boldsymbol{s}) = Cp(\boldsymbol{y} \mid \boldsymbol{s})\, p'(\boldsymbol{s}) \tag{4.8}$$

It is important to emphasize that BA quantifies the distribution of unknowns so that a 95 % credible interval (or credibility interval) is supposed to contain the unknown with probability 95 %.

Gelman (2008) has listed objections to Bayesian inference made by an imaginary anti-Bayesian statistician. Although one may find the style in the article to be occasionally sardonic, the article summarizes very well most objections that have been heard over the years. Gelman does not dismiss them but states that '...these are strong arguments to be taken seriously – and ultimately accepted in some settings and refuted in others.' According to Gelman,

(a) The first major objection is that 'Bayesian methods are presented as an automatic inference engine' that, in effect, is too easy to use and presumably abuse.

(b) The second major objection 'addresses the subjective strand of Bayesian inference: the idea that prior and posterior distributions represent subjective states of knowledge. Here the concern from outsiders is, first, that as scientists we should be concerned with objective knowledge rather than subjective belief, and second, that it's not clear how to assess subjective knowledge in any case.'

To counter the first point one only needs to point out the excellent expositions of the Bayesian methodology and tools (e.g. Berger 1985; Box and Tiao 1973) that make it clear that Bayesian methods are much more than the mechanical application of Bayes' theorem. The second point is admittedly tougher to address because one would need to use terms that either are hard to define or are widely misconstrued. To begin with, the concept of probability is primitive and thus hard to explain in terms of simpler concepts (one notable exception being when probability is the average or limit of a frequency of occurrence). Dictionary definitions, such as the chance or likelihood that something will happen, are not helpful since they resort to synonyms. Also, to say that probabilities in Bayesian inference express state of knowledge, state of information, or someone's degree of belief is of limited assistance because these statements are fuzzy and use other difficult concepts (like 'information'). I believe that probabilistic modeling should be evaluated by its usefulness, since I doubt that we will ever agree on what 'probability' really means. Nevertheless, in the next section, I will try to discuss some misunderstandings related to the prior probability distribution.

4.3 Prior Distribution

One occasionally hears the old objection, 'We do not want to use the Bayesian approach because it relies on a prior distribution that is subjective.' Though there have been many interesting discussions in the general statistical literature of the use of priors, I think that the issue has not been discussed adequately in the context of inverse problems. So, let us think this issue through starting with the terms 'prior' and 'posterior'.

Traditionally, the distributions $p'(\boldsymbol{s})$ and $p''(\boldsymbol{s})$ that appear in Bayes theorem, Equation (4.8), have been known as prior and posterior, respectively. It is perhaps a testament to the power of words that some of the criticisms to Bayesian methods stem from the perception that these terms should be taken literally, prior denoting preceding and posterior meaning coming after, in time. That is wrong. Instead, we should be thinking of $p'(\boldsymbol{s})$ and $p(\boldsymbol{y} \mid \boldsymbol{s})$ as expressing two types of known facts and there is no reason that one should come earlier in time. There is considerable confusion regarding the very meaning of the term prior. According to Jaynes (2003, pg. 87), 'The term "a priori" was introduced by Immanuel Kant to denote a proposition whose truth can be known independent of experience; which is most emphatically what we do not mean here.' Indeed a so-called prior may very well be the product of theory, experience, *and* new data.

Let us break with tradition and adopt the more descriptive terms (in the context of many problems we are interested in) *structural* instead of prior and *combined* instead of posterior. Then, we can interpret Bayes' theorem as the engine that allows us to combine known facts from two sources: knowledge of the structure of the variable we want to estimate and the numerical values of specific measurements. It would be absurd to state that knowledge about the structure precedes chronologically the acquisition of observations! In fact, we know most about the structure after all the measurements are accounted for. Thus, it is imperative to adjust the structural distribution on the basis of data. This is exactly what methods known as empirical and hierarchical Bayes do (e.g. Carlin and Louis 2000). I consider that these methods are among the most practical and most effective Bayesian tools available for the solution of inverse problems so I will later try to explain the rationale for this approach. But first, let us hear a common criticism in the words of Gelman's dreamed-up anti-Bayesian: '... don't these empirical and hierarchical Bayes methods use the data twice? If you're going to be Bayesian, then be Bayesian: it seems like a cop-out and contradictory to the Bayesian philosophy to estimate the prior from the data.'

It is definitely neither a cop-out nor 'contradictory to Bayesian philosophy' to use data to obtain a reasonable $p'(\boldsymbol{s})$; in fact, it would be foolish not to. Double counting of data is an altogether different issue. It would be incorrect of a method to double count data–that would be the inferential equivalent of an engine with a coefficient of efficiency higher than 1 and would result in trusting one's results much more that one should. However, common sense indicates that data should be useful in two ways: by informing about the structure and by providing specific constraints that must be honored by the unknown function.

Using data for two purposes is not the same as double counting – just think that the data affect the results through two channels. The mathematics of how double counting can be avoided will be explained in the next section.

Returning to the question of subjectivity of the prior, again in Gelman's words: 'Subjective prior distributions don't transfer well from person to person, and there's no good objective principle for choosing a noninformative prior (even if that concept were mathematically defined, which it's not). Where do prior distributions come from, anyway?' These are interesting questions that, although already debated elsewhere, are worth revisiting in the context of inverse problems in geophysical applications.

Let us start by thinking through the meaning of the word subjective. It seems that some researchers effectively consider subjective any statement that is not directly based on hard data. However, there are many statements that express obvious or well accepted facts that could be useful when taken into account in prediction and decision making (e.g. that hydraulic or electric conductivity is nonnegative, that porosity takes values between zero and one, that the mean square value of solute concentration is finite). Such statements are not subjective, because they cannot be dismissed as someone's opinions with which a reasonable and knowledgeable person might disagree. Of course, there remains the issue of how such knowledge can be translated into a mathematical object that we can use to perform quantitative analysis. According to Jaynes (2003, pg. 88) '... the conversion of verbal prior information into numerical probabilities is an open-ended problem of logical analysis' but I do not see any reason to use the word subjective in this context any more than in the context of modeling using regression or ANOVA techniques.

As Berger (2006, pg. 386) points out '... the process of data analysis typically involves a host of subjective choices, such as model building, and these choices will typically have a much greater effect on the answer than will such things as choice of prior distributions for model parameters... Model-building is not typically part of the objective/subjective debate, however - in part because of the historical success of using models, in part because all the major philosophical approaches to statistics use models and, in part, because models are viewed as testable, and hence subject to objective scrutiny.' If this is the case, it should be emphasized that the model underlying structural distribution $p'(\boldsymbol{s})$, is also subject to criticism and revision in the light of new information (see Kitanidis 1988, 1991, in the context of geostatistical methods).

The loaded term 'subjective probability' is actually used rather indiscriminately by many (including some Bayesians) for any probability that does not correspond to a well-defined ratio in repeated trials (i.e. a frequency). In my view, this is tiresome and also unfair because it undermines the credibility of valuable methods. Every time an actual problem is translated into a mathematical one, some modeling decisions and approximations must be made. To refer to them constantly as subjective is to belittle the thought and analysis that underlies modeling. Furthermore, there is considerable diversity within the Bayesian school of thought, which is hardly surprising given the broad scope of applications. Methods used in practice often are in the spirit of what

Berger calls 'objective Bayesian analysis', see Berger (2006); this is certainly true about the method of Kitanidis (1995).

Let us now review two approaches for representing structure within the geostatistical approach.

1. One approach is to utilize simple geostatistical models and tools (see Kitanidis 1997a; Olea 1999, for review), such as semivariograms or generalized covariance functions (Kitanidis 1993; Matheron 1973). These methods produce Gaussian priors that capture the fact that the variance of the function is finite but without enforcing any continuity (when using the nugget semivariogram); or the function is continuous (like when using the linear, exponential, and spherical semivariograms), etc. We will see an example in the next section.

2. Another approach attempts to use more geological information. One way is through training images which are used to calibrate a model of structure. For a review of such methods, see Hu and Chugunova (2008).

It is interesting that a popular topic in discussions of Bayesian methods and comparisons with frequentist methods is the topic of noninformative priors. What is called a noninformative prior may be the one that is uniform so that the posterior is the same as the likelihood function; then, the maximum-a-posteriori value (i.e. the one that maximizes the posterior pdf) is the same as the frequentist's maximum likelihood estimate (which maximizes the likelihood function). However, it is now better understood that there is no such thing as a truly noninformative prior distribution (see, for example, Tarantola 2005, p. 11). For example, the uniform distribution may appear to assign equal weight a-priori to all values until one realizes that this prior distribution is not uniform anymore if we think in terms of a transformation of the same parameter (such as consider $\ln K$ instead of conductivity K). Incidentally, trying to solve an inverse problem using a minimum of prior information, in the name of 'objectivity', is not necessarily a good idea (Kitanidis 1997b).

A related idea that I consider very important and relevant is that of encoding limited information in terms of a probability distribution. For example, let us consider the example problem introduced beginning with Equation (4.3). Assume that we have decided that a reasonable description of the structure is that the values are uncorrelated and fluctuate about the mean value β with variance θ, where β and θ are two structural parameters (*i.e.*, parameters that describe the structural distribution of $\boldsymbol{s}$) that are either known from past experience or need to be estimated from fresh data. What probability should we assign to describe these statements? To be sure, one of the first things one learns in probability theory is that there are many probability distributions consistent with given values for the mean and the variance. E. T. Jaynes has championed the principle of maximum entropy for the assignment of probabilities based on incomplete information (Jaynes 1968). The idea is that of all distributions that have, say, the same mean and the same variance, the most reasonable choice is the Gaussian because it is the one with the highest Shannon entropy (Tribus 1969); this means that this distribution allows the highest number of possible solutions consistent with available constraints. To put it in more casual terms, the maximum-entropy assignment is the least restrictive consistent with some facts.

It is well beyond the scope of this work to discuss the justification, pros, and cons of the maxentropic assignment of distributions. However, a few words are in order because I think that this point is widely misunderstood and misrepresented. This idea is neither a cure-all nor a sleight of hand. It is a rational way to deal with an irksome part of the problem formulation but the probabilities one obtains are just a transcription in a mathematically convenient form of the information on the table. For example, knowing nothing but the mean and the variance does justify the use of Gaussian more than any other distribution; however, one should not think that the probability of an extremely rare event can be computed with confidence from this Gaussian. In other words, the fact that one can assign a value to the probability of any event does not mean that this number will end up looking good at the proverbial end of the day. In modeling, whether deterministic or probabilistic, there is a difference between what can be computed and what can be computed reliably. In data analysis, if one positively needs to know the probability of an extreme event, one better get hold of the kind of data that contain this knowledge.

4.4 A Geostatistical Approach

We will illustrate many of the ideas we have discussed by reviewing the key elements of a 'geostatistical' methodology that has already found several applications. Such an approach has been applied to both Gaussian (Snodgrass and Kitanidis 1997) and non-Gaussian prior distributions (Michalak and Kitanidis 2004). Geostatistics in a Bayesian framework have been discussed in Kitanidis (1986) and an approach to inverse problems along the same lines can be found in Kitanidis (1995) where more details are presented.

We will review the main steps of the methodology. We will be referring to the discretized form in which $\boldsymbol{y}$ is the $n \times 1$ array of the data and $\boldsymbol{s}$ is the $m \times 1$ array of the model parameters. Double-primed probability density functions are conditional on $\boldsymbol{y}$; i.e. $p''(\boldsymbol{s})$ is the same as $p(\boldsymbol{s} \mid \boldsymbol{y})$ whereas $p'(\boldsymbol{s})$ is unconditional. Consider that the structural (aka prior) and likelihood functions are parameterized, with parameter vector $\boldsymbol{\Theta}$. For example, the structural pdf $p'(\boldsymbol{s})$ is Gaussian, with mean with elements $\mu_i = \beta$ and covariance matrix with elements $Q_{ij} = \theta_1 \delta_{ij}$, where $i, j = 1, \ldots, m$ and δ_{ij} is Kronecker's delta. The likelihood is Gaussian with covariance matrix with elements $R_{ij} = \theta_2 \delta_{ij}$, $i, j = 1, \ldots, n$. Then, $\boldsymbol{\Theta} = [\beta, \theta_1, \theta_2]$.

The 'prior' joint pdf of $\boldsymbol{s}$ and $\boldsymbol{\Theta}$ is

$$p'(\boldsymbol{s}, \boldsymbol{\Theta}) = p'(\boldsymbol{s} \mid \boldsymbol{\Theta})\, p'(\boldsymbol{\Theta}) \tag{4.9}$$

where the prior pdf for given parameters is denoted $p'(\boldsymbol{s} \mid \boldsymbol{\Theta})$ and the prior pdf of $\boldsymbol{\Theta}$ is denoted $p'(\boldsymbol{\Theta})$. The latter can be taken as uniform, $\propto 1$, but other choices are possible and may be preferable in specific cases.

The joint posterior (i.e., conditional on $\boldsymbol{y}$) is

$$p''(\boldsymbol{s}, \boldsymbol{\Theta}) \propto p(\boldsymbol{y} \mid \boldsymbol{s}, \boldsymbol{\Theta})\, p'(\boldsymbol{s} \mid \boldsymbol{\Theta})\, p'(\boldsymbol{\Theta}) \tag{4.10}$$

From this, one can derive the posterior marginal pdf of model parameters

$$p''(\boldsymbol{\Theta}) = \int p''(\boldsymbol{s}, \boldsymbol{\Theta})\, d\boldsymbol{s} \tag{4.11}$$

or

$$p''(\boldsymbol{\Theta}) \propto \int p(\boldsymbol{y} \mid \boldsymbol{s}, \boldsymbol{\Theta})\, p'(\boldsymbol{s} \mid \boldsymbol{\Theta})\, p'(\boldsymbol{\Theta})\, d\boldsymbol{s} \tag{4.12}$$

This posterior pdf is the key to making inferences about the $\boldsymbol{\Theta}$ parameters. It describes our state of knowledge about $\boldsymbol{\Theta}$, for a given model, the data, and other information. A common approach is to obtain a representative value of the parameters by maximizing its posterior pdf, Equation (4.12). This estimate is known as maximum a posteriori or MAP estimate and denoted $\hat{\boldsymbol{\Theta}}$. Note that the problem of estimation of $\boldsymbol{\Theta}$ could be addressed within the framework of frequentist methods, since it is an overdetermined problem.

We return to the central problem, which is to estimate $\boldsymbol{s}$. We are interested in the marginal posterior pdf

$$p''(\boldsymbol{s}) = \int p''(\boldsymbol{s}, \boldsymbol{\Theta})\, d\boldsymbol{\Theta} \tag{4.13}$$

or

$$p''(\boldsymbol{s}) = \int p''(\boldsymbol{s} \mid \boldsymbol{\Theta})\, p''(\boldsymbol{\Theta})\, d\boldsymbol{\Theta} \tag{4.14}$$

where $p''(\boldsymbol{s} \mid \boldsymbol{\Theta})$ is the pdf of $\boldsymbol{s}$ given a $\boldsymbol{\Theta}$ and the data $\boldsymbol{y}$.

This key relationship, Equation (4.14) gives the posterior pdf of the unknown given structural information and data. Notice the important difference between $p''(\boldsymbol{s})$ and $p''(\boldsymbol{s} \mid \boldsymbol{\Theta})$. The former accounts for the effect on prediction of uncertainty about the model parameters, whereas the latter is for a given value of $\boldsymbol{\Theta}$. To distinguish between the two, $p''(\boldsymbol{s})$ is referred to as the *compound* or Bayesian distribution.

One cannot overemphasize that the distribution of interest is the compound one. However, to reduce computational effort, in run-of-the mill applications, it may be adequate to simplify by approximating the posterior pdf of $\boldsymbol{\Theta}$ with a Dirac delta function

$$p''(\boldsymbol{\Theta}) = \delta\left(\boldsymbol{\Theta} - \hat{\boldsymbol{\Theta}}\right) \tag{4.15}$$

where $\hat{\boldsymbol{\Theta}}$ is MAP estimate of $\boldsymbol{\Theta}$; then Equation (4.14) simplifies to

$$p(\boldsymbol{s}) = p\left(\boldsymbol{s} \mid \hat{\boldsymbol{\Theta}}\right) \tag{4.16}$$

This simplification generally leads to (usually slight) underestimation of measures of error.

A few clarifications:

1. The method of first focusing on the marginal pdf, Equation (4.12), and then on the compound pdf, Equation (4.14), is not the same as maximizing the joint posterior pdf, Equation (4.10). See discussion in Kitanidis (1996).
2. The method for determining the structural parameters $\boldsymbol{\Theta}$ can be interpreted as a cross-validation method (see Kitanidis 1991).

4.5 Conclusion

Geophysical processes in the lithosphere or the atmosphere are hard to characterize because they are highly heterogeneous in complex ways. New measurement techniques, such as hydrogeophysical and remote-sensing tools, can produce large sets, but the inversion problem is still ill-posed because effectively there are not enough measurements to single out a unique solution. The challenge is to develop practical methods that help us make progress towards the following objectives: (a) Arrive at better predictions; (b) Quantify the uncertainty in the predictions; (c) Provide input for better decisions in the face of imperfect information. These challenges can be met within the framework of stochastic methods, particularly Bayesian and geostatistical methods. It is hoped that the discussion in this chapter will motivate others to try such methods in the solution of important and challenging interpolation and inverse problems.

References

Berger, J.O. 1985 *Statistical Decision Theory and Bayesian Analysis*, second edition. Springer-Verlag, New York.

Berger, J.O. 2006 The case for objective Bayesian analysis. *Bayesian Analysis* **1**(3), 385–402.

Box, G.E.P. and Tiao, G.C. 1973 *Bayesian Inference in Statistical Analysis*. Addison-Wesley Reading, MA.

Carlin, B.P. and Louis, T.A. 2000 *Bayes and Empirical Bayes Methods for Data Analysis*, 2nd edition ed. Chapman & Hall CRC, Boca Raton.

Gelman, A. 2008 Objections to Bayesian statistics. *Bayesian Analysis*. International Society for Bayesian Analysis (online publication).

Hu, L.Y. and Chugunova, T. 2008 Multiple-point geostatistics for modeling subsurface heterogeneity: A comprehensive review. *Water Resour. Res.* **44**, doi:10.1029/2008WR006993.

Jaynes, E.T. 1968 Prior probabilities. *IEEE Trans. on Systems Science and Cybernetics* **4**, 227–241.

Jaynes, E.T. 2003 *Probability Theory: The Logic of Science*. Cambridge Univ. Press, Cambridge, UK.

Jeffreys, H. 1957 *Scientific Inference*, 2nd ed. Cambridge Univ. Press, Oxford, England.

Kitanidis, P.K. 1986 Parameter uncertainty in estimation of spatial functions: Bayesian analysis. *Water Resources Research* **22**(4), 499–507.

Kitanidis, P.K. 1988 The concept of predictive probability and a simple test for geostatistical model validation, in *Consequences of Spatial Variability in Aquifer Properties and Data Limitations for Groundwater Modelling Practice*. vol. IAHS Publication No 175, edited by A. Peck, S. Gorelick, G. DeMarsily, S. Foster, and V. Kovalevsky, pp. 178–190, International Association of Hydrological Sciences.

Kitanidis, P.K. 1991 Orthonormal residuals in geostatistics: model criticism and parameter estimation. *Mathematical Geology* **23**(5), 741–758.

Kitanidis, P.K. 1993 Generalized covariance functions in estimation. *Mathematical Geology* **25**(5), 525–540.

Kitanidis, P.K. 1995 Quasilinear geostatistical theory for inversing. *Water Resour. Res.* **31**(10), 2411–2419.

Kitanidis, P.K. 1996 On the geostatistical approach to the inverse problem. *Adv. Water Resour.* **19**(6), 333–342.

Kitanidis, P.K. 1997a *Introduction to Geostatistics*. Cambridge University Press.

Kitanidis, P.K. 1997b The minimum-structure solution to the inverse problem. *Water Resour. Res.* **33**(10), 2263–2272.

Kitanidis, P.K. and Vomvoris, E.G. 1983 A geostatistical approach to the inverse problem in groundwater modeling (steady state) and one-dimensional simulations. *Water Resour. Res.* **19**(3), 677–690.

Matheron, G. 1973 The intrinsic random functions and their applications. *Applied Probability* **5**, 439–468.

Michalak, A.M. and Kitanidis, P.K. 2004 Application of geostatistical inverse modeling to contaminant source identification at Dover AFB, Delaware. *IAHR Journal of Hydraulic Research* **42**(extra issue), 9–18.

Olea, R. 1999 *Geostatistics for Engineers and Earth Scientists*. Kluwer Academic Publishers Hingham, MA.

Snodgrass, M.F. and Kitanidis, P.K. 1997 A geostatistical approach to contaminant source identification. *Water Resour. Res.* **33**(4), 537–546.

Tarantola, A. 2005 *Inverse Problem Theory and Methods for Model Parameter Estimation.* SIAM Philadelphia, PA.

Tikhonov, A. Leonov, I. Tikhonov, A.N. Yagola, A.G. and Leonov, A.S. 1997 *Non-Linear Ill-Posed Problems.* CRC Press, Boca Raton, FL.

Tikhonov, A.N. and Arsenin, V.Y. 1977 *Solutions of Ill-Posed Problems.* Halsted Press/Wiley, New York.

Tribus, M. 1969 *Rational Descriptions, Decisions, and Designs.* Pergamon Press, NY.

Chapter 5

Using the Bayesian Framework to Combine Simulations and Physical Observations for Statistical Inference

D. Higdon, K. Heitmann, E. Lawrence and S. Habib

Los Alamos National Laboratory, USA

5.1 Introduction

This chapter develops a Bayesian statistical approach for characterizing uncertainty in predictions that are made with the aid of a computer simulation model. The computer simulation code – or simulator – models a physical system and requires a set of inputs, some or all of which are unknown. A limited number of observations of the true physical system are available to estimate the unknown inputs. The approach yields calibrated predictions while accounting for uncertainty regarding model inputs, uncertainty due to limitations on the available number of simulations, and noise in the observations of the true physical system. We utilize Gaussian

Large-Scale Inverse Problems and Quantification of Uncertainty Edited by L. Biegler, G. Biros, O. Ghattas, M. Heinkenschloss, D. Keyes, B. Mallick, Y. Marzouk, L. Tenorio, B. van Bloemen Waanders and K. Willcox

process models to model the simulator at untried settings. Parameter estimation is performed using Markov chain Monte Carlo.

The latter part of the paper considers an illustrative example using observations of the cosmic microwave background. In this case, we have a single functional observation of the temperature spectrum of the cosmic microwave background. We also have the ability to simulate a limited number of cosmologies and record the spectrum for each simulation. The goal is to rigorously compare the simulated data with the physical system in order to determine the simulation inputs that best reproduce reality.

5.2 Bayesian Model Formulation

5.2.1 General Formulation

We take $\eta(x, t)$ to denote the simulator output at input vector (x, t), where x contains initial conditions for the system, and t holds unknown parameter settings; note that t does not denote time. The latter will be calibrated – or constrained – to be consistent with the physical data. For the purposes of this chapter, we assume that the initial conditions x are known. For an appropriate choice of $t = \theta$, we also assume that $\eta(x, \theta)$ simulates the physical system $\zeta(x)$. Note the actual physical system $\zeta(x)$ does not depend on θ.

At various settings for x, observations $y(x)$ are made of the physical system

$$y(x_i) = \zeta(x_i) + \varepsilon_i,\ i = 1, \ldots, n,$$

where $\zeta(x_i)$ denotes the physical system at initial conditions x_i, and ε_i denotes the error for the i-th observation. Often the size and nature of the ε_i are sufficiently well characterized that their distribution can be treated as known. We take the n-vector $y = (y(x_1), \ldots, y(x_n))^T$ to denote the physical observations. For now we assume the observations are univariate. Section 5.2.4 extends this approach to multivariate output. The observed data are modeled statistically using the simulator $\eta(x, \theta)$ at the true calibration value θ according to

$$y(x_i) = \eta(x_i, \theta) + e_i,\ i = 1, \ldots, n,$$

where the error term e_i accounts for observation error as well as any discrepancy between the simulator $\eta(x_i, \theta)$ and reality $\zeta(x_i)$. While this systematic discrepancy can play an important role in many analyses, we assume it is negligible in this chapter.

Depending on the application, one may wish to treat a fixed set of m simulation runs

$$\eta(x_j^*, t_j^*), j = 1, \ldots, m,$$

as additional data to be used in the analysis. This is typically the case when the computational demands of the simulation code are so large that only a limited number of runs can be produced. In this case, a statistical model for $\eta(x, t)$ will be required for input combinations (x, t) at which the simulator has not been run. This will be discussed in more detail in Sections 5.2.3 and 5.2.4 of this chapter. Note that we use t to denote an input setting for the calibration parameters here. We reserve θ to denote the 'best' or 'true' value of the calibration parameters – a quantity which we wish to infer. The following subsections will step through increasingly complicated statistical analyses based on simulation runs and observed field data.

5.2.2 Unlimited Simulation Runs

This section considers a simulator that is fast enough for new outputs to be obtained nearly instantly. Because we can query the simulator at will, we can use it in the overall estimation scheme.

We now consider a simple example for which both x and t are one-dimensional and $\eta(x, t)$ simulates a physical system $\zeta(x)$ when t is set to the true, but unknown, value of $t = \theta$. We have $n = 5$ physical observations at five different values of x as shown in the first plot of Figure 5.1. We assume that the physical observations are noisy measurements of the true system $\zeta(x)$, and this observation noise is known to be normal, with a standard deviation of 0.25.

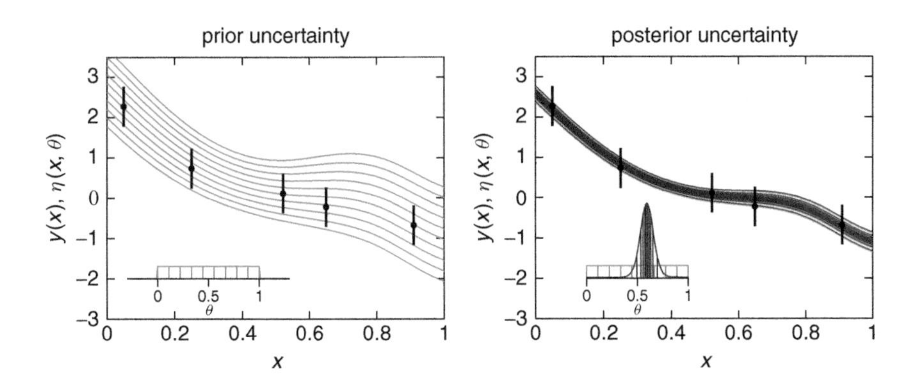

Figure 5.1 Prior and posterior simulator predictions. Left: Observed data (black dots) and uncertainties (vertical, black lines) are shown along with simulator output $\eta(x, \theta)$ at equally spaced percentiles $(5, 15, \ldots, 95)$ of the prior distribution for θ. Here the prior distribution $\pi(\theta)$ is uniform, as shown in the inset figure. Right: the posterior distribution for θ is computed using Bayes rule and is given by the dark blue lines on the inset figure. The resulting posterior distribution for prediction $\eta(x, \theta)$ is obtained by propagating the posterior uncertainty about θ through the simulation model $\eta(x, \theta)$.

For now we assume that the simulator sufficiently represents the physical system so that the model

$$y(x_i) = \eta(x_i, \theta) + \varepsilon_i,\ i = 1, \ldots, n, \tag{5.1}$$

is appropriate. Because the distribution of the ε_i is assumed to be i.i.d. $N(0, .25^2)$, the sampling model for y is

$$L(y|\eta(\theta)) \propto \exp\{-\frac{1}{2}(y - \eta(\theta))^T \Sigma_y^{-1} (y - \eta(\theta))\},$$

where $y = (y(x_1), \ldots, y(x_n))^T$, $\eta(\theta) = (\eta(x_1, \theta), \ldots, \eta(x_n, \theta))^T$, and the observation covariance matrix $\Sigma_y = I_n.25^2$. The sampling model density is called the likelihood when considered as a function of θ given y.

The Bayesian formulation is completed by specifying a prior distribution $\pi(\theta)$ for the unknown calibration parameter θ. For this example we specify that θ is a priori uniform over the interval $[0, 1]$, as shown in the first plot of Figure 5.1. This prior uncertainty about θ induces prior uncertainty about the simulator output $\eta(x, \theta)$. The lines in the left hand plot of Figure 5.1 show simulations of $\eta(x, \theta)$ for $x \in [0, 1]$ and the prior quantiles for θ.

The resulting posterior distribution for θ is determined by Bayes' rule and is given by

$$\begin{aligned} \pi(\theta|y) &\propto L(y|\eta(\theta)) \times \pi(\theta) \\ &\propto \exp\left\{-\frac{1}{2}(y - \eta(\theta))^T \Sigma_y^{-1} (y - \eta(\theta))\right\} \times I[0 \leq \theta \leq 1]. \end{aligned} \tag{5.2}$$

Bayes' rule is shown graphically in Figure 5.2. The posterior density (5.2) is typically a difficult expression to manipulate since it is often nonlinear and high dimensional. In some cases, the posterior can be approximated with a more tractable density (Evans and Swartz 1995). Monte Carlo methods

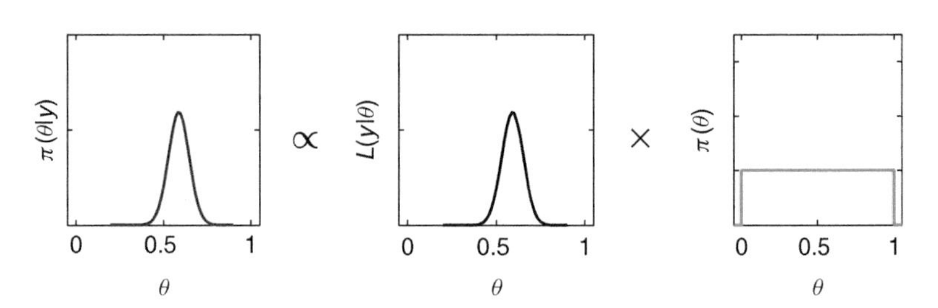

Figure 5.2 A graphical representation of Bayes' rule $\pi(\theta|y) \propto L(y|\theta) \times \pi(\theta)$. The posterior density $\pi(\theta|y)$ for the unknown parameter θ given the observed data y is obtained by taking the pointwise product of the likelihood function $L(y|\theta)$ and the prior density $\pi(\theta)$. This product must then be normalized so that $\pi(\theta|y)$ integrates to one.

are a very general approach to sampling from this complicated density. In particular, we use Markov chain Monte Carlo (MCMC) (Gilks *et al.* 1998; Robert and Casella 1999) to produce a sequence of dependent draws from the posterior distribution using a simple Markov chain construction.

A very simple MCMC implementation uses the Metropolis algorithm (Metropolis *et al.* 1953) as follows:

1. initialize θ^1 at some value;

2. given the current realization θ^t, generate θ^* from a symmetric distribution (i.e. the chance of generating θ^* given θ^t is the same as generating θ^t given θ^*);

3. compute the Metropolis acceptance probability

$$\alpha = \min\left\{1, \frac{\pi(\theta^*|y)}{\pi(\theta^t|y)}\right\};$$

4. set

$$\theta^{t+1} = \begin{cases} \theta^* & \text{with probability } \alpha. \\ \theta^t & \text{with probability } 1-\alpha; \end{cases}$$

5. iterate over steps 2–4.

This simple but general recipe is shown graphically in Figure 5.3; it has a number of features that are worth mentioning. This approach is applicable

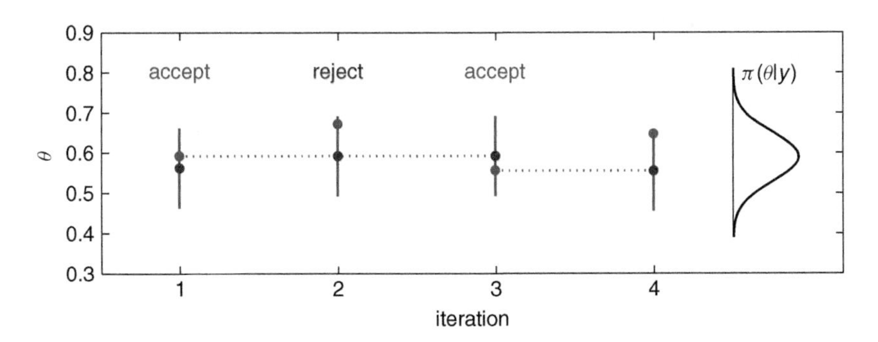

Figure 5.3 The first four Metropolis updates for the simple chain described above. At iteration 1 the chain is initialized at θ^1 given by the blue dot. A candidate θ^* (red dot at iteration 1) is generated from a uniform distribution centered at θ^1. Here the proposal is accepted so that θ^* becomes θ^2. The process is repeated for iterations 2–4. At iteration 2, θ^* is rejected; at iteration 3, θ^* is accepted. Also shown on the right of the figure is the posterior density $\pi(\theta|y)$ produced by this sample problem.

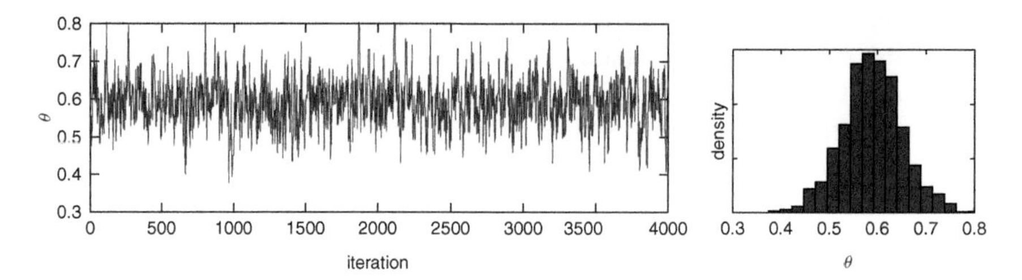

Figure 5.4 The trace for θ produced by the Metropolis scheme described in Section 5.2.2. The sequence $\{\theta^1, \ldots, \theta^{4000}\}$ are draws from a Markov chain whose stationary distribution is $\pi(\theta|y)$. A histogram of the first 4000 draws is shown on the right.

even when θ is a high dimensional vector. In this case components of θ can be updated individually (single site MCMC) or in groups. The above MCMC references give details regarding such implementations. In step 3, the acceptance probability requires only ratios of the posterior density. Hence $\pi(\theta|y)$ need only be specified up to a constant of integration. Typically, many thousands of MCMC steps are required to sufficiently sample the posterior distribution. Again, the previously mentioned references give guidance on selecting the number of MCMC steps. Since the simulator must be run to compute the acceptance probability, this recipe is only feasible if $\eta(x, \theta)$ can be computed very quickly. The first 4000 draws produced by this approach are shown in Figure 5.4.

This direct MCMC-based approach has a number of advantages when it is possible. It can readily handle a very large dimensional θ – see Rue (2001) or Kaipio and Somersalo (2004) for examples. It can also deal with large numbers of nuisance parameters as well as highly multivariate output. For such problems it may well be worth reducing the fidelity in the simulator in return for gains in simulation speed. The point being that a thorough exploration of an approximate posterior (which uses a low-fidelity simulator) may give more information than a very limited exploration of a more accurate posterior (which uses a high-fidelity simulator).

The second plot in Figure 5.1 shows the posterior distribution for θ which was computed from 4000 MCMC draws using the Metropolis scheme outlined above. Conditioning on the five noisy field observations leads to reduced uncertainty about θ. The induced posterior distribution for $\eta(x, \theta)$ also shown in the figure is produced by propagating the posterior draws of θ through the simulator $\eta(x, \theta)$.

5.2.3 Limited Simulation Runs

In many applications, the computational demands of the simulator make it impractical as a direct tool for exploring the posterior using MCMC. In such cases, it may be useful to build an *emulator* of the simulation model using a limited number of simulation runs. This emulator predicts the simulator output at untried input settings (x, t). In this section we describe a Bayesian analysis in which a Gaussian process (GP) prior is used to emulate $\eta(x, t)$. The additional uncertainty due to estimating the simulator at untried settings is reflected in the resulting posterior distribution. Using a GP to emulate the simulation model output has proven quite effective when the output changes smoothly as a function of the inputs (x, t), and is largely controlled by a small subset of the input vector (x, t) (Sacks *et al.* 1989; Welch *et al.* 1992).

We assume we have run the simulator at m different input settings

$$\eta(x_j^*, t_j^*),\ j = 1, \ldots, m.$$

The actual choice of input settings $(x_j^*, t_j^*),\ j = 1, \ldots, m$ is an important question, but is not discussed in this chapter. We point the interested reader to Santner *et al.* (2003) and the references therein.

The Bayesian approach requires that we treat $\eta(x, t)$ as unknown for inputs (x, t) that are not included in the original set of m simulator runs. We assume that the x and t variables have been transformed so that x is a vector in $[0, 1]^p$ and t a vector in $[0, 1]^q$. The function $\eta(\cdot, \cdot)$ maps $[0, 1]^{p+q}$ to R. A standard prior model for an unknown function is a Gaussian process (GP) (O'Hagan 1978; Sacks *et al.* 1989). Mean and covariance functions $\mu(x, t)$ and $\text{Cov}((x, t), (x', t'))$ are required to fully specify the GP prior model for $\eta(x, t)$. For simplicity, we assume the mean of the GP is zero for all x and t.

Following Sacks *et al.* (1989) and Kennedy and O'Hagan (2001) we specify the covariance function using a product correlation form

$$\begin{aligned} &\text{Cov}((x, t), (x', t')) = \\ &\frac{1}{\lambda_\eta} \exp\left\{ -\sum_{k=1}^{p} \beta_k |x_k - x'_k|^2 - \sum_{k=1}^{q} \beta_{p+k} |t_k - t'_k|^2 \right\} \end{aligned} \tag{5.3}$$

where the parameter λ_η controls the reciprocal of the marginal variance of $\eta(\cdot, \cdot)$, and the $p + q$-vector β controls the dependence strength in each of the component directions of x and t. This particular form of the covariance function ensures an infinitely differentiable representation for $\eta(\cdot, \cdot)$. While this covariance function can be too smooth to adequately model many physical processes, it works well in practice for emulating simulation output as a function of input settings (Sacks *et al.* 1989). We note that it is often useful to add a small amount to the diagonal of the covariance model (5.3) for numerical stability and/or to account for small numerical fluctuations in the simulation. For models with random outputs, such as epidemiological or agent based models, an additional independent error term will be required in (5.3) above.

As before, we assume the model (5.1) is appropriate and define the field observation vector $y = (y(x_1), \ldots, y(x_n))^T$. To hold the simulation outcomes, we define $\eta = (\eta(x_1^*, t_1^*), \ldots, \eta(x_m^*, t_m^*))^T$. We also define the joint vector $z = (y^T, \eta^T)^T$ with associated simulation input settings $(x_1, \theta), \ldots, (x_n, \theta)$ for its first n components and $(x_1^*, t_1^*), \ldots, (x_m^*, t_m^*)$ for its final m components. The sampling model, or likelihood, for the observed data z is then

$$L(z|\theta, \lambda_\eta, \beta, \Sigma_y) \propto |\Sigma_z|^{-\frac{1}{2}} \exp\left\{-\frac{1}{2} z^T \Sigma_z^{-1} z\right\} \tag{5.4}$$

where

$$\Sigma_z = \Sigma_\eta + \begin{pmatrix} \Sigma_y & 0 \\ 0 & 0 \end{pmatrix},$$

and Σ_y is the known, $n \times n$ observation covariance matrix, and elements of Σ_η are obtained by applying (5.3) to each pair of the $n + m$ simulation input points above.

Figure 5.5 shows an adaptation of the previous example in which we utilize only $m = 20$ simulations over the two-dimensional input space $[0, 1]^2$. Simulation output is obtained from $m = 20$ different input pairs $(x_1^*, t_1^*), \ldots, (x_m^*, t_m^*)$, shown in the top left frame of the figure. The simulation output and the $n = 5$ physical observations $y(x_1), \ldots, y(x_n)$, are shown in the top right plot; here the simulation outputs are the filled in circles. For this example, the simulation output is consistent with a mean 0, variance 1 GP. If this were not the case, we could standardize the simulation output so that this were appropriate. We also need to apply this same transformation to the physical observations. We usually take a simple shifting and scaling of the output: $\eta = (\eta_{\text{raw}} - \bar{\eta})/s$, where $\bar{\eta}$ and s are the sample mean and standard deviation of the raw simulation output.

We complete this formulation by specifying prior distributions for the remaining unknown parameters: θ, λ_η, and the $p + q$-vector β.

$$\begin{aligned}
\pi(\theta) &\propto I[0 \leq \theta \leq 1] \\
\pi(\lambda_\eta) &\propto \lambda_\eta^{a_\eta - 1} \exp\{-b_\eta \lambda_\eta\}, \; \lambda_\eta > 0 \\
\pi(\beta) &\propto \prod_{k=1}^{p+q} e^{-\beta_k}, \; \beta_k > 0.
\end{aligned}$$

Because of the standardization we use an informative prior for the marginal precision λ_η by taking $a_\eta = b_\eta = 5$. The prior specification for the spatial dependence parameter vector β is influenced by the scaling of the (x, θ) input space to $[0, 1]^{p+q}$.

Conditioning on the augmented observation vector $z = (y^T, \eta^T)^T$ results in the posterior distribution

$$\pi(\theta, \lambda_\eta, \beta^\eta | z) \propto L(z|\theta, \lambda_\eta, \beta)\pi(\theta)\pi(\lambda_\eta)\pi(\beta)$$

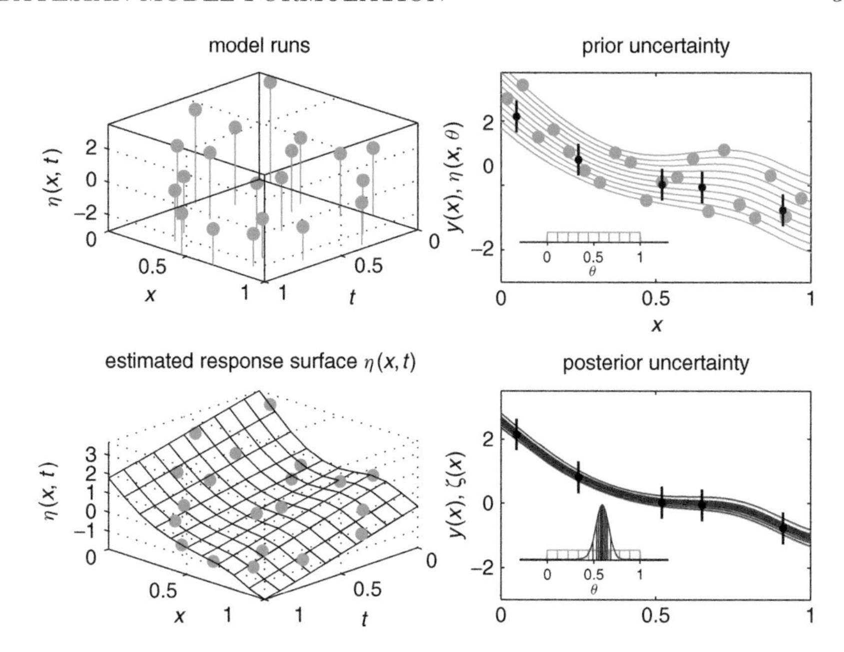

Figure 5.5 Prior and posterior uncertainty in model predictions when the number of model runs is limited. Top left: Model runs have been carried out only at the $m = 20$ (x^*, t^*) pairs. The simulation output is given by the circle plotting symbols. Top right: As with Figure 5.1, the prior density for θ and the implied simulations $\eta(x, \theta)$ at equally spaced values of the prior distribution are given by the light lines. The black dots show field observations and the corresponding black lines give 95% uncertainty bounds on the observations. Inference is carried out by conditioning on the $n = 5$ data points as well as the $m = 20$ model runs. Bottom left: The posterior mean of $\eta(\cdot, \cdot)$; the actual model runs are marked by the circle plotting symbols. Bottom right: The darker lines show the updated posterior density for θ and the resulting predictions are shown for percintiles $(5, 15, \ldots, 95)$ of the posterior distribution for θ. Uncertainty in the model prediction is due to the spread in the posterior distribution for θ, as well as uncertainty in $\eta(x, \theta)$.

which we explore using standard, single site MCMC which updates each parameter, in turn, conditional on the value of the remaining parameters.

Once the MCMC has been carried out on the $(\theta, \lambda_\eta, \beta)$ parameter space, posterior draws for $\eta(x, t)$ can be constructed using standard multivariate normal theory (Bernardo and Smith 2001; Wackernagel 1995) which states

$$\text{if } \begin{pmatrix} z_1 \\ z_2 \end{pmatrix} \sim N\left(\begin{pmatrix} 0 \\ 0 \end{pmatrix}, \begin{pmatrix} V_{11} & V_{12} \\ V_{21} & V_{22} \end{pmatrix}\right) \tag{5.5}$$

$$\text{then } z_2|z_1 \sim N(V_{21}V_{11}^{-1}z_1, V_{22} - V_{21}V_{11}^{-1}V_{12}).$$

Note that here $V_{21} = V_{12}^T$. Given a realization of the parameters $(\theta, \lambda_\eta, \beta)$, a posterior realization of $\eta(x^\star, t^\star)$ at arbitrary inputs $(x^\star, t^\star)$ (where both $x^\star$ and $t^\star$ are vectors of length $m^\star$) can be drawn with straightforward computation. This follows from the fact that conditional on $(\theta, \lambda_\eta, \beta)$

$$\begin{pmatrix} y \\ \eta \\ \eta(x^\star, t^\star) \end{pmatrix} \sim N_{n+m+m^\star}\left(\begin{pmatrix} 0 \\ 0 \\ 0 \end{pmatrix}, \Sigma_{\eta,\eta^\star} + \begin{pmatrix} \Sigma_y & 0 & 0 \\ 0 & 0 & 0 \\ 0 & 0 & 0 \end{pmatrix} \right),$$

where elements of $\Sigma_{\eta,\eta^\star}$ are obtained by applying (5.3) to each pair of the $n + m + m^\star$ simulation input points above. Hence the conditional distribution of $\eta(x^\star, t^\star)$ follows from (5.5) above. The bottom left plot of Figure 5.5 shows the posterior mean for $\eta(\cdot, \cdot)$ over a grid on $[0, 1]^2$. Uncertainty about the function $\eta(\cdot, \cdot)$ is determined by the location of simulations and is larger in regions where no simulation points (x^*, t^*) are nearby.

The resulting posterior inference for θ and $\eta(x, \theta)$ is summarized in the bottom right plot of Figure 5.5. For this simple example, there is very little additional uncertainty that can be ascribed to the restriction to $m = 20$ simulation runs – compare Figures 5.1 and 5.5. This uncertainty due to limited evaluations of $\eta(x, t)$ plays a larger role as the dimensions of x and t increase relative to the number of simulations m.

5.2.4 Limited Simulations Runs with Multivariate Output

In many applications, the simulation produces an array of outputs: a displacement trace over time; grayscale values over a rectangular lattice; or a power spectrum, as in the application of Section 5.3. One could use the formulation of the previous section and use one (or more) of the x dimensions to index the multivariate output. However, if the output is highly multivariate, the size of the covariance matrices in that formulation quickly becomes too large to carry out the inversions required in this GP formulation.

A more efficient alternative is to use a basis representation to model the multivariate output as described in Higdon *et al.* (2008). Now, rather than using the raw data and simulation output, the analysis is based on a reduced form of the data. Here we describe a simple version of this approach in the context of a simple example. Later in Section 5.3 this approach is used to infer cosmological parameters from observations on the cosmic microwave background.

Assume now that the simulator produces a vector response over a regular grid of n_η different τ values for a given input setting t. The lines in left plot of Figure 5.6 show the vector output for $\eta(t)$ at $m = 10$ different values of t. Note that we can also include the x-type input from previous sections, but we have dropped it here to simplify presentation.

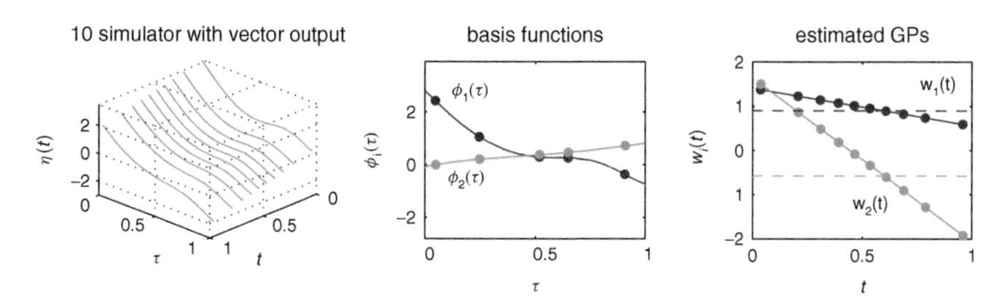

Figure 5.6 Elements of the basis approach for dealing with multivariate simulation output. Left: output vectors from 10 simulations at equally spaced values of t between 0 and 1. Middle: basis functions computed from the $m = 10$ simulations using principal components. The dots show the restriction of the basis functions that are used to connect the basis functions to the observed data y via $\Phi_y(\tau)$. Here we use $p = 2$ basis functions to represent the multivariate simulation output. Right: the solid lines give the posterior mean estimates for the two GPs $w_1(t)$ and $w_2(t)$. The dots show the transformed simulation output at the input settings $t_1^*, \ldots, t_{10}^*$. For each input t_i^*, two transformed outputs are produced $\hat{w}_{1i}$ and $\hat{w}_{2i}$. Finally, the two transformed data values $\hat{w}_{y1}$ and $\hat{w}_{y2}$ are given by the upper (blue) and lower (green) dashed horizontal lines.

With this basis approach, we model the simulation output using a p-dimensional basis representation

$$\eta(t) = \sum_{i=1}^{p} \varphi_i(\tau) w_i(t) + e, \; t \in [0,1]^q, \tag{5.6}$$

where $\{\varphi_1(\tau), \ldots, \varphi_p(\tau)\}$ is a collection of orthogonal, n_η-dimensional basis vectors, the $w_i(t)$ are GPs over the input space, and e is an n_η-dimensional error term. For the simple example, the $p = 2$ basis functions are shown in the middle plot of Figure 5.6. This type of formulation reduces the problem of building an emulator that maps $[0,1]^q$ to R^{n_η} to building p independent, univariate GP models for each $w_i(t)$. Each $w_i(t)$ is a mean 0 GP with correlation function controlled by $\beta_i = (\beta_{i1}, \ldots, \beta_{iq})$

$$\text{Cor}(w_i(t), w_i(t')) = \prod_{l=1}^{q} \exp\{-\beta_{il}|t_l - t_l'|^2\}. \tag{5.7}$$

Restricting to the m input design settings

$$t^* = \begin{pmatrix} t_1^* \\ \vdots \\ t_m^* \end{pmatrix} = \begin{pmatrix} t_{11}^* & \cdots & t_{1q}^* \\ \vdots & \ddots & \vdots \\ t_{m1}^* & \cdots & t_{mq}^* \end{pmatrix}$$

we define the m-vector w_i to be the restriction of the process $w_i(\cdot)$ to the input design settings

$$w_i = (w_i(t_1^*), \dots, w_i(t_m^*))^T, \;\; i = 1, \dots, q. \tag{5.8}$$

In addition we define $R(t^*; \beta_i)$ to be the $m \times m$ correlation matrix resulting from applying (5.7) to each pair of input settings in the design. The q-vector β_i gives the correlation distances for each of the input dimensions.

At the m simulation input settings, the mp-vector $w = (w_1^T, \dots, w_p^T)^T$ then has prior distribution

$$\begin{pmatrix} w_1 \\ \vdots \\ w_p \end{pmatrix} \sim N\left(\begin{pmatrix} 0 \\ \vdots \\ 0 \end{pmatrix}, \begin{pmatrix} \Lambda_{w_1} & 0 & 0 \\ 0 & \ddots & 0 \\ 0 & 0 & \Lambda_{w_p} \end{pmatrix} \right), \tag{5.9}$$

where

$$\Lambda_{w_i} \equiv \lambda_{w_i}^{-1} R(t^*; \beta_i),$$

which is controlled by p precision parameters held in λ_w and $p \cdot q$ spatial correlation parameters held in β. The prior above can be written more compactly as

$$w \sim N(0, \Sigma_w),$$

where Σ_w is covariance matrix given in (5.9).

We now describe how output from the m simulation runs are used to the prior GP models for the $w_i(\cdot)$'s. The simulation output over the m run input design given by t^* is given by $\eta(t_1^*), \dots, \eta(t_m^*)$. Each run produces a p-vector as output. For this basis formulation, we need only consider the transformed simulation output which is produced by projecting each simulation output vector onto the basis functions. Thus, the transformed simulation output is given by

$$\hat{w} = (\Phi^T \Phi)^{-1} \Phi^T \eta$$

where

$$\begin{aligned} \eta &= \text{vec}(\eta(t_1^*), \dots, \eta(t_m^*)), \text{ and} \\ \Phi &= [I_m \otimes \varphi_1; \cdots; I_m \otimes \varphi_p], \end{aligned}$$

so that $\hat{w}$ is modeled as a draw from (5.9).

Similarly, we produce transformed observational data $\hat{w}_y$ by projecting the data vector y onto the basis functions. Since the τ support of the observed data does not typically match that of the simulations, we must construct an analog to Φ above which is restricted to the τ values for which the observed data are taken. We define

$$\hat{w}_y = (\Phi_y^T \Sigma_y^{-1} \Phi_y)^{-1} \Phi_y^T \Sigma_y^{-1} y$$

where Φ_y is the $n \times p$ matrix obtained by column binding data supported basis vectors $[\varphi_1^y, \ldots, \varphi_p^y]$, and the data supported basis vectors $\varphi_i^y(\tau)$ are obtained by restricting the original basis vectors $\varphi_i(\tau)$ to the τ values for which the n data points are observed. The dots in the middle plot of Figure 5.6 give $\varphi_1^y(\tau)$ and $\varphi_2^y(\tau)$ for the simple example. The dashed lines in the right hand plot of Figure 5.6 give the values for the 2-vector $\hat{w}_y$. Φ_y is also defined by the data model

$$y = \Phi_y w(\theta) + \varepsilon,$$

where $w(\theta)$ is the p-vector $(w_1(\theta), \ldots, w_p(\theta))^T$. For additional details regarding the basis formulation, see Higdon *et al.* (2008).

The (marginal) distribution for the combined, reduced observational data $\hat{w}_y$ and the reduced simulations given the covariance parameters $\hat{w}$, has the form

$$\begin{pmatrix} \hat{w}_y \\ \hat{w} \end{pmatrix} \sim \mathrm{N}\left(\begin{pmatrix} 0 \\ 0 \end{pmatrix}, \begin{pmatrix} \Lambda_y^{-1} & 0 \\ 0 & 0 \end{pmatrix} + \begin{pmatrix} I_p & \Sigma_{w_y w} \\ \Sigma_{w_y w}^T & \Sigma_w \end{pmatrix} \right), \tag{5.10}$$

where Σ_w is defined in (5.9),

$$\begin{aligned} \Lambda_y &= \lambda_y \Phi_y^T \Sigma_y^{-1} \Phi_y, \\ I_p &= p \times p \text{ identity matrix}, \\ \Sigma_{w_y w} &= \begin{pmatrix} \lambda_{w1}^{-1} R(\theta, t^*; \beta_1) & 0 & 0 \\ 0 & \ddots & 0 \\ 0 & 0 & \lambda_{wp}^{-1} R(\theta, t^*; \beta_p) \end{pmatrix}. \end{aligned}$$

Above, $R(\theta, t^*; \beta_i)$ denotes the $1 \times m$ correlation submatrix for the GP modeling the simulator output obtained by applying (5.7) to the observational setting θ crossed with the m simulator input settings $t_1^*, \ldots, t_m^*$.

If we take $\hat{z}$ to denote the reduced data $(\hat{w}_y^T, \hat{w}^T)^T$, and $\Sigma_{\hat{z}}$ to be the covariance matrix given in (5.10), the posterior distribution has the form

$$\pi(\lambda_w, \beta, \theta | \hat{z}) \propto |\Sigma_{\hat{z}}|^{-\frac{1}{2}} \exp\left\{ -\frac{1}{2} \hat{z}^T \Sigma_{\hat{z}}^{-1} \hat{z} \right\} \times \tag{5.11}$$

$$\prod_{i=1}^{p} \lambda_{wi}^{a_w - 1} e^{-b_w \lambda_{wi}} \times \prod_{i=1}^{p} \prod_{l=1}^{p_\theta} e^{-\beta_{il}} \times I[\theta \in C],$$

where C denotes the constraint region for θ, which is typically a q-dimensional rectangle. In other applications C can also incorporate constraints between the components of θ. As with the previous formulations, this posterior can be sampled using standard MCMC techniques.

5.3 Application: Cosmic Microwave Background

Observations from several different cosmological probes have revealed a highly unexpected result: roughly 70 % of the Universe is made up of a mysterious dark energy which is responsible for a recent epoch of accelerated expansion. Understanding the nature of dark energy is the foremost challenge in cosmology today. This application uses cosmic microwave background (CMB) measurements from the Wilkinson Microwave Anisotropy Probe (WMAP) and a series of simulations (Figure 5.7) to infer a set of cosmological parameters that describe the makeup and evolution of the universe.

Figure 5.7 shows a reconstruction of the temperature field for CMB. A reconstruction is required since the observations from WMAP (and other sources) do not give a complete picture of the cosmic sky. From these incomplete measurements, estimates of the temperature spectrum (called the TT spectrum) are made at multipole moments whose order is indexed by ℓ (ℓ is related to the angular scale; the larger ℓ, the smaller the angular scale). Here we utilize TT measurements from the WMAP five year data archive. We use a Gaussian approximation to the likelihood of binned data on the natural log scale; the error bars correspond to ± 2 standard deviations. We will restrict our analysis to a seven-dimensional ΛCDM model. The ΛCDM – read as the lambda cold dark matter – model is the leading cosmological model to date; it describes a universe dominated by dark energy. Different groups analyzing different data sets (Spergel *et al.* 2007; Tegmark *et al.* 2006) found that the model specified by these seven parameters consistently fits all currently available data. The parameters and their ranges for this analysis are summarized

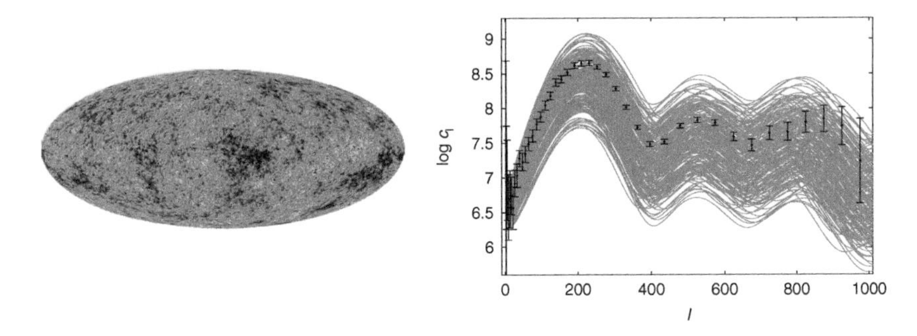

Figure 5.7 The TT temperature spectrum of the CMB as measured by WMAP. Left: a reconstruction of the spherical CMB temperature field from WMAP observations (Credit: NASA/WMAP Science Team). Right: simulated and measured temperature spectra for the CMB (on the natural log scale). The black vertical lines show 2 standard deviation uncertainties. (Credit: Legacy Archive for Microwave Background Data Analysis.)

Table 5.1 Parameters and their ranges used as input for CAMB simulations.

Parameter	Min	Max	Description
ω_b	0.018	0.025	baryonic density
ω_c	0.08	0.15	dark matter density
τ	0.01	0.18	optical depth
n	0.075	1.10	spectral index
w	-1.3	-0.7	dark energy equation of state
H	55	90	Hubble constant
A	$1.64 \cdot 10^{-9}$	$2.71 \cdot 10^{-9}$	amplitude fluctuation

in Table 5.1. The TT spectrum data y along with a local estimate for Σ_y are taken from the WMAP five year data release (Spergel *et al.* 2007).

We run the CAMB simulator (Lewis *et al.* 2000) at $m = 200$ input settings using a space-filling Latin hypercube design (Ye *et al.* 2000) over the $q = 7$ dimensional rectangular parameter space described in Table 5.1. The gray lines in the right hand plot of Figure 5.7 show the resulting simulated TT spectra given by c_ℓ for $\ell = 1, \ldots, 1000$. From these 200 simulations, we use $p = 6$ principal component bases to represent the multivariate simulation output. These six components account for over 99 % of the variation in the output. Tests on holdout simulations reveal that the GP-based emulator predicts the TT spectrum to within a few percent.

Figure 5.8 shows posterior means for the estimated simulator response $\eta(t)$ where each of the parameter inputs is varied over its prior (standardized) range of $[0, 1]$ while the remaining six inputs are held at their midpoint setting of 0.5. The posterior mean response conveys an idea of how the different parameters affect the highly multivariate simulation output. Other marginal functionals of the simulation response can be calculated such as sensitivity

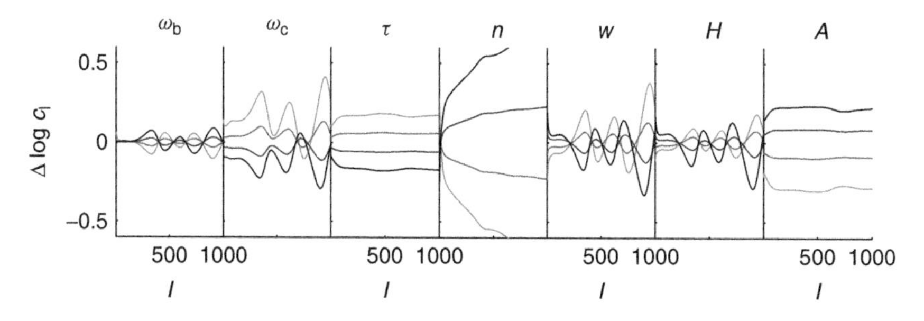

Figure 5.8 Changes to the posterior mean simulator predictions obtained by varying one parameter input, while holding others at their central values, i.e. at the midpoint of their range. The light to dark lines correspond to the smallest parameter setting to the biggest, for each parameter.

indicies or estimates of the Sobol decomposition (Oakley and O'Hagan 2004; Sacks *et al.* 1989).

Using the Bayesian approach described in this chapter, we produce 50,000 draws from the posterior distribution using MCMC. The resulting posterior distribution for the cosmological parameters is summarized in Figure 5.9. The data are rather informative about n and A. There is also a fair amount of posterior correlation between the parameters w and H. The resulting uncertainty for the predicted TT spectrum is shown in Figure 5.10. The uncertainty in the fitted spectrum is quite small, despite a moderate amount of uncertainty in the posterior for the parameters. Hence, it will likely take additional sources of information to further reduce uncertainties in cosmological parameters.

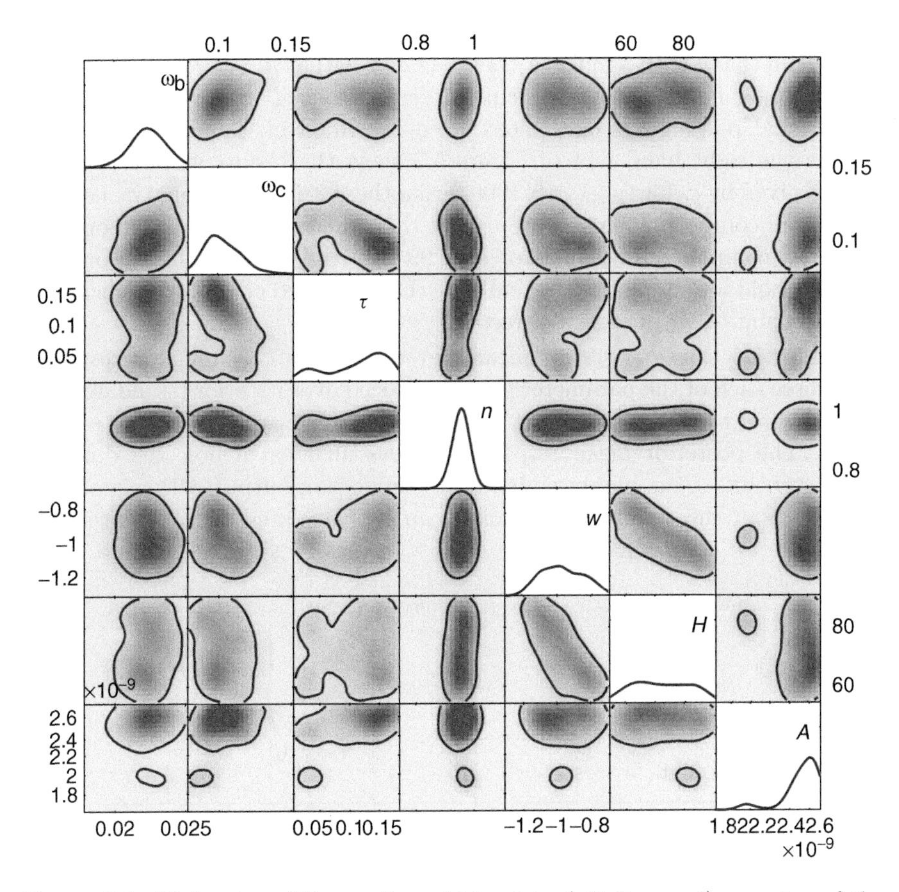

Figure 5.9 Univariate (diagonal) and bivariate (off diagonal) margins of the 7-dimensional posterior distribution for the cosmological parameters using the data from the WMAP TT power spectrum. The lines show estimates of 90% posterior probability regions for each 2-d margin.

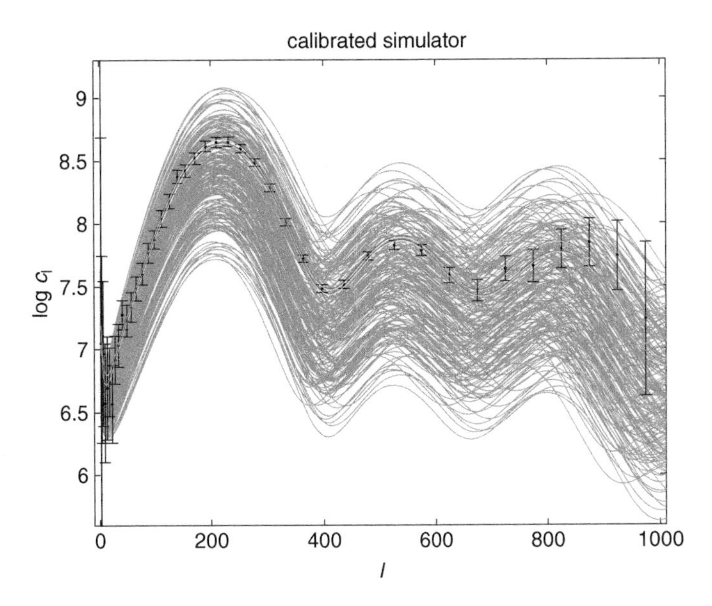

Figure 5.10 Posterior distribution for the CMB TT spectrum $\eta(x, \theta)$. Here the green lines give pointwise 80% intervals for the TT spectrum. This uncertainty is primarily due to the uncertainty in the cosmological parameters θ which is shown in Figure 5.9. Also shown are the 200 CAMB simulations along with the WMAP five year data.

5.4 Discussion

This chapter gives an overview of Bayesian methods for combining observed data and simulations. In order to minimize the complexity in the modeling details, we made a fair number of simplifications and assumptions. The aim here is to provide a gentle introduction to GPs and Bayesian methods. We caution that many real world analyses will likely call for more intricate modeling.

The presentation of the problem formulation is divided into categories based on the availability of simulation runs. This is really a division that is based on the overall computational cost of solving the problem. Estimation with a Gaussian process emulator (Sections 5.2.3 and 5.2.4) incurs two penalties when compared with simple MCMC (Section 5.2.2): greater uncertainty due to approximating the simulator based on limited runs and increased computational complexity due to a larger number of parameter and matrix inversion requirements. In the scenario considered in Section 5.2.2, these penalties would only hinder the estimation. Since the simulator is fast, there is no need to complicate the estimation. In the scenarios considered in Sections 5.2.3 and 5.2.4, the computational penalty from using the GP approach is considerably smaller than the one imposed by running an extremely complex simulator at every iteration and the uncertainty penalty is just the price that must be paid to obtain results in a reasonable amount of time. The result is a net gain in computational efficiency at the expense of some accuracy. These considerations need to be balanced when choosing one approach over the other.

We also note that we have provided a very rudimentary overview of MCMC. The references within this chapter should give the interested reader a starting point for learning more about MCMC methods. We also note that MCMC is certainly not the only way to approach Bayesian estimation. We use it here because it is a general approach for understanding a complicated posterior distribution. See Craig *et al.* (1996) or Kennedy and O'Hagan (2001) for Bayesian alternatives to MCMC in similar settings.

Finally, a key consideration in any inference involving a complex simulation model is the question of how well the model simulates reality. In many cases one must explicitly account for inadequacies in the simulation model or risk unrealistic predictions, with overly tight uncertainties. The discussion of simulation inadequacy in Kennedy and O'Hagan (2001) is a good place to start.

Acknowledgments

This research was supported, in part, by the DOE under contract W-7405-ENG-36, and by the Laboratory Directed Research and Development program at Los Alamos National Laboratory.

References

Bernardo, J. and Smith, A. 2001 Bayesian Theory. *Measurement Science and Technology* **12**, 221–222.

Craig, P.S., Goldstein, M., Seheult, A.H. and Smith, J.A. 1996 Bayes linear strategies for history matching of hydrocarbon reservoirs. In *Bayesian Statistics 5* (eds. Bernardo, J.M., Berger, J.O., Dawid, A.P., and Smith, A.F.M.) pp. 69–95. Oxford, Clarendon Press.

Evans, M. and Swartz, T. 1995 Methods for approximating integrals in statistics with special emphasis on Bayesian integration problems (with discussion). *Statistical Science* **10**, 254–272.

Gilks, W.R., Richardson, S. and Spiegelhalter, D.J. 1998 *Markov Chain Monte Carlo in Practice*. Chapman & Hall Ltd.

Higdon, D., Gattiker, J.R. and Williams, B.J. 2008 Computer model calibration using high dimensional output. *Journal of the American Statistical Association.*

Kaipio, J.P. and Somersalo, E. 2004 *Statistical and Computational Inverse Problems*. New York, Springer.

Kennedy, M. and O'Hagan, A. 2001 Bayesian calibration of computer models (with discussion). *Journal of the Royal Statistical Society (Series B)* **68**, 425–464.

Lewis, A., Challinor, A., and Lasenby, A. 2000 Efficient computation of cosmic microwave background anisotropies in closed Friedmann-Robertson-Walker models. *The Astrophysical Journal* **538**, 473–476.

Metropolis, N., Rosenbluth, A., Rosenbluth, M., Teller, A. and Teller, E. 1953 Equations of state calculations by fast computing machines. *Journal of Chemical Physics* **21**, 1087–1091.

Oakley, J. and O'Hagan, A. 2004 Probabilistic sensitivity analysis of complex models. *Journal of the Royal Statistical Society (Series B)* **66**, 751–769.

O'Hagan, A. 1978 Curve fitting and optimal design for prediction. *Journal of the Royal Statistical Society (Series B)* **40**, 1–24.

Robert, C.P. and Casella, G. 1999 *Monte Carlo Statistical Methods*. Springer-Verlag Inc.

Rue, H. 2001 Fast sampling of Gaussian Markov random fields. *Journal of the Royal Statistical Society: Series B* **63**, 325–338.

Sacks, J., Welch, W.J., Mitchell, T.J. and Wynn, H.P. 1989 Design and analysis of computer experiments (with discussion). *Statistical Science* **4**, 409–423.

Santner, T.J., Williams, B.J. and Notz, W.I. 2003 *Design and Analysis of Computer Experiments*. New York, Springer.

Spergel, D., Bean, R., Dore, O., Nolta, M.R., Bennett, C.L., Dunkley, J., Hinshaw, G. *et al.* 2007 Three-year Wilkinson Microwave Anisotropy Probe (WMAP) observations: implications for cosmology. *The Astrophysical Journal Supplement Series* **170**, 377–408.

Tegmark, M., Eisenstein, D., Strauss, M., Weinberg, D., Blanton, M., Frieman, J., Fukugita, M. *et al.* 2006 Cosmological constraints from the SDSS luminous red galaxies. *Physical Review D* **74**, 123507.

Wackernagel, H. 1995 *Multivariate Geostatistics. An Introduction With Applications*. Springer-Verlag.

Welch, W.J., Buck, R.J., Sacks, J., Wynn, H.P., Mitchell, T.J., and Morris, M.D. 1992 Screening, predicting, and computer experiments. *Technometrics* **34**, 15–25.

Ye, K.Q., Li, W. and Sudjianto, A. 2000 Algorithmic construction of optimal symmetric Latin hypercube designs. *Journal of Statistical Planning and Inference* **90**, 145–159.

Chapter 6

Bayesian Partition Models for Subsurface Characterization

Y. Efendiev, A. Datta-Gupta, K. Hwang, X. Ma and B. Mallick

Texas A&M University, USA

6.1 Introduction

Uncertainties on the detailed description of reservoir lithofacies, porosity, and permeability are major contributors to uncertainty in reservoir performance forecasting. Reducing this uncertainty can be achieved by integrating additional data in subsurface modeling . Integration of dynamic data is a critical aspect of subsurface characterization. Dynamic data such as transient pressure response, tracer or production history can particularly be effective in identifying preferential flow path or barriers of the flow. There are significant developments in the area of dynamic data integration. Several methods have been introduced which incorporate data with a primary focus on integrating seismic and well data. These methods include conventional techniques such as cokriging and its variations (Deutsch *et al.* (1996); Frykman and Deutsch

Large-Scale Inverse Problems and Quantification of Uncertainty Edited by L. Biegler, G. Biros, O. Ghattas, M. Heinkenschloss, D. Keyes, B. Mallick, Y. Marzouk, L. Tenorio, B. van Bloemen Waanders and K. Willcox

(1999); Xu *et al.* (1992); Xue and Datta-Gupta (1996)), Sequential Gaussian Simulation with Block Kriging (Behrens & Tran (1998)) and Bayesian updating of point Kriging (Doyen *et al.* (1997)).

Integration of production data into reservoir models is generally carried out in a Bayesian framework (e.g. Glimm and Sharp (1999)). This allows for rigorous quantification of uncertainty by defining a posterior distribution of reservoir properties such as permeability based on our prior geologic knowledge and the field production response. In general, it is difficult to calculate this posterior probability distribution explicitly because the process of predicting flow and transport in petroleum reservoirs is nonlinear. Moreover, inverse problems in the subsurface characterization are usually ill-conditioned and have multiple solutions. Instead, we use simulation based Markov chain Monte Carlo (MCMC method) to generate samples from this probability distribution. It is essential that these realizations adequately reflect the uncertainty of the parameters reflecting the reservoir properties, i.e. we need to correctly sample from this posterior probability distribution.

In this chapter we present a Bayesian Partition Model (BPM) (Denison *et al.* (2002)) for the integration of dynamic data in subsurface characterization and uncertainty quantification. Our main goal is to develop rapid and rigorous sampling algorithms by using surrogate surface obtained from the BPM. The primary idea of Bayesian Partition Models is to split parameter space into the regions and assume that, within each region, the data arise from the same simple probability distribution. We use Voronoi tessellation to define distinct regions because it provides a flexible method of partitioning. In each region, we assume that the probability distribution is a linear function. The BPM procedure used in the chapter allows the flexibility in the number of Voronoi tessellation by assuming the number of Voronoi tessellation is unknown. As a result, this procedure falls into category of Reversible Jump Markov chain Monte Carlo (RJ-MCMC) methods (Green (1995)) which are currently under active investigation. We note that the unknown number of Voronoi tessellation is important for our applications, since the multi-scale structure of the response surface (misfit surface) is, in general, unknown. This flexibility of the BPM provides robust approximation of the response surface. We would like to note that in previous findings the BPM is used for predictive computations, and our current work, to our best knowledge, is the first application of the BPM to inverse problems. The proposed strategy involving the BPM shares some similarities with recently proposed Neighborhood Approximation (NA) algorithms (see e.g. Subbey *et al.* (2004)); however, there are important differences. NA algorithms generate realizations using previous samples and Voronoi tessellation of the parameter space with the goal to identify the regions of lower misfit and preferential samples from these regions. The BPM algorithm used in the chapter allows us to consider unknown number of tessellation and provide a rigorous unified Bayesian framework.

Due to sparseness of the data, the obtained response surface fit from BPM is not an accurate representation of true unknown response surface. We suggest the use of two-stage Markov chain Monte Carlo method. In the first stage, a sample from the BPM is tested using Metropolis-Hasting criteria. If the proposal is accepted (conditionally) at the first-stage, then direct simulation is performed at the second stage to determine the acceptance probability of the proposal. The first stage of MCMC method modifies the proposal distribution. We can show that the modified Markov chain satisfies the detailed balance condition for the correct distribution. We would like to note that two-stage MCMC algorithms have been used previously (e.g. Christen and Fox (1992); Efendiev *et al.* (2005a), (2005b); Liu (2001)) in different situations.

We present numerical results to demonstrate the efficiency of the proposed approach. We assume that the permeability field is prescribed using two-point geostatistics. Using the Karhunen-Loève expansion, we represent the high dimensional permeability field by fewer numbers of parameters. We note that Karhunen-Loève expansion is used only for parameterization and helps to reduce the dimension of the parameter space. Furthermore, static data (the values of permeability field at some sparse locations) are incorporated into the Karhunen-Loève expansion to further reduce the dimension of the parameter space. Imposing the values of the permeability at some locations restricts the parameter space to a subspace (hyperplane). As for instrumental proposal distribution, we use independent sampler. Independent sampler typically has low acceptance rate, though independent realizations allow the efficient sampling of the response surface. Because the response surface is approximated by piece-wise linear functions, one can easily sample independent permeability fields. These permeability fields are used in second stage of MCMC for the final acceptance. Our numerical results show that the use of the BPM increases the acceptance rate by several times.

The chapter is organized as follows. In the next section, we present model equations and problem setting. Section 6.3 is devoted to approximation using the BPM and two-stage MCMC. In Section 6.4, we present numerical results.

6.2 Model Equations and Problem Setting

We briefly introduce a model used in the simulations. We consider two-phase flow in a reservoir (denoted by Ω) under the assumption that the displacement is dominated by viscous effects; i.e. we neglect the effects of gravity, compressibility, and capillary pressure. Porosity will be considered to be constant. The two phases will be referred to as water and oil, designated by subscripts w and o, respectively. We write for each phase as follows:

$$v_j = -\frac{k_{rj}(S)}{\mu_j} k \cdot \nabla p, \tag{6.1}$$

where v_j is the phase velocity, k is the permeability tensor, k_{rj} is the relative permeability to phase j ($j = o, w$), S is the water saturation (volume fraction) and p is pressure. In this work, a single set of relative permeability curves is used and k is taken to be a diagonal tensor. Combining Darcy's law with a statement of conservation of mass allows us to express the governing equations in terms of the so-called pressure and saturation equations:

$$\nabla \cdot (\lambda(S)k\nabla p) = h, \tag{6.2}$$

$$\frac{\partial S}{\partial t} + v \cdot \nabla f(S) = 0, \tag{6.3}$$

where λ is the total mobility, h is the source term, and v is the total velocity, which are respectively given by:

$$\lambda(S) = \frac{k_{rw}(S)}{\mu_w} + \frac{k_{ro}(S)}{\mu_o}, \tag{6.4}$$

$$f(S) = \frac{k_{rw}(S)/\mu_w}{k_{rw}(S)/\mu_w + k_{ro}(S)/\mu_o}, \tag{6.5}$$

$$v = v_w + v_o = -\lambda(S)k \cdot \nabla p. \tag{6.6}$$

The above descriptions are referred to as the model of the two-phase flow problem.

The problem under consideration consists of sampling permeability field given fractional flow measurements. Fractional flow, $F(t)$ (denoted simply by F in further discussion) is defined as the fraction of oil in the produced fluid and is given by q_o/q_t, where $q_t = q_o + q_w$, with q_o and q_w are the flow rates of oil and water at the production edge of the model,

$$F(t) = 1 - \frac{\int_{\partial\Omega^{out}} f(S)v_n dl}{\int_{\partial\Omega^{out}} v_n dl},$$

where $\partial\Omega^{out}$ are outflow boundaries and v_n is the normal velocity field. Pore volume injected (PVI) at time T, defined as $\frac{1}{V_p}\int_0^T q_t(\tau)d\tau$, with V_p the total pore volume of the system, provides the dimensionless time for the displacement. In further analysis, the notations q_o, q_w or q_t will not be used, and q will denote the proposal distribution.

Typically, the permeability is described on a finite grid that makes $k(x)$ finite dimensional. It is assumed that the permeability field has been exactly observed at some locations. This information can be incorporated into the prior models (distributions). Since the fractional flow is an integrated response, the map from the permeability field to the fractional flow is not one-to-one. Hence this problem is ill-posed in the sense that there exist many different permeability realizations for the given production data.

From the probabilistic point of view, this problem can be regarded as the conditioning of the permeability field to the fractional flow data with measurement errors. Consequently, our goal is to sample from the conditional distribution $P(k|F)$, where k is the permeability field defined on a grid and F is the fractional flow curve measured from production data. Using Bayes' theorem we can write

$$P(k|F) \propto P(F|k)P(k). \tag{6.7}$$

In the above formula, $P(k)$ is the unconditioned (prior) distribution of the permeability field, which is assumed to be log-normal or log-exponential. $P(F|k)$ denotes the conditional probability that the outcome of the measurement is F when the true permeability is k. In practice, the measured fractional flow contains measurement errors. In this paper, we assume that the measurement error follows a Gaussian distribution, thus, the likelihood function $P(F|k)$ takes the form

$$P(F|k) \propto \exp\Big(-\frac{\|F - F_k\|^2}{\sigma_f^2}\Big), \tag{6.8}$$

where F_k is the fractional flow computed by solving the nonlinear PDE system (6.1)–(6.3) on the numerical grid (typically large) for the given k, and σ_f is the measurement precision. Since both F and F_k are functions of t, we take the norm $\|F - F_k\|^2$ to be the L_2 norm, i.e.

$$\|F - F_k\|^2 = \int_0^T (F(t) - F_k(t))^2 dt.$$

We note that the methods discussed in this paper are not limited to Gaussian error functions, and any general covariance describing measurement errors can be used in the simulations.

6.3 Approximation of the Response Surface Using the Bayesian Partition Model and Two-Stage MCMC

In this chapter, we consider the sampling from the posterior distribution $P(k|F)$. For notational simplicity, we write it as $\pi(k)$ as a function of k. The direct MCMC simulations are generally very computationally expensive as each proposal requires solving a forward coupled nonlinear partial differential equations (F_k for each k) over a large time interval. The goal of this chapter is to employ BPM based method to develop more efficient MCMC algorithms. The posterior computations based on BPM are used to decide whether to run the expensive forward model in the next stage. This way we exploit BPM based method to increase the acceptance rate of the MCMC.

In the posterior computation of $\pi(k)$, the part say $G(k) = \|F - F(k)\|$ contains the expensive function F_k. Rather than computing F_k, we would like to fit the surface $G(k)$ based on some observed data using BPM. That way we approximate the posterior distribution π using this estimated function and denote it as π^*.

The fitting of a curve or surface through a set of observational data is a very frequent problem in different disciplines. There are several approaches for fitting surfaces. The new Bayesian approach, BPM, for surface fitting was originally suggested by Holmes *et al.* (Holmes *et al.* (2001, 2005)). BPM constructs arbitrarily complex regression surface by splitting the space into an unknown number of disjoint regions. Next, we present briefly BPM following Holmes *et al.* (2001, 2005).

The motivation behind the BPM is that points nearby in the parameter space have similar values in the response space. We construct a number of disjoint regions in parameter space using Voronoi tessellation and the data within regions are assumed to be exchangeable (conditionally independent) (De Finetti (1974)).

The Voronoi tessellation is determined by a number of sites $S = (s_1, \ldots, s_M), s_j \in R^\eta$, which split the parameter space into M disjoint regions such that points within R_i are closer to s_i than any other center $s_j, j \neq i$. i.e. $R_i = \{k : \|k - s_i\| < \|k - s_j\| \text{ for all } j \neq i\}$ where $\|(k_1, \ldots, k_\eta)\|^2 = \sum_{i=1}^{\eta} \omega_i^2 k_i^2$. The parameter $w = (\omega_1, \ldots, \omega_\eta)$ is a normalized weighting vector that places emphasis on different directions, although generally we could adopt any metric.

We now define some notation to ease the following exposition. We take the number of disjoint sets to be M and define g_{ij} to denote the jth observation of $G(k)$, which is in region R_i, $i = 1, \ldots, M, j = 1, \ldots, n_i$ where n_i is the number of points in R_i. Let $G_i = (g_{i,1}, \ldots, g_{i,n_i})'$ ($'$ denotes transpose) and K_i denote the $n_i \times \eta$ design matrix of points, including the intercept term, found in the ith partition. And let T denote number and location of the region centers and the weighting vector that define the tessellation, $T = \{S, w, M\}$, and $\delta = \{\beta_1, \ldots, \beta_M, \sigma^2\}$ denotes the set of all other model parameters. We assume that G_i follows a linear model in each region so that

$$\zeta(G_i \mid \delta) \sim \text{MVN}(G_i \mid K_i\beta_i, \sigma^2 \text{I}_{n_i}), \quad i = 1, \ldots, M,$$

where MVN refers to the multivariate normal distribution and β_i are a $\eta \times 1$ matrix of regression coefficients particular to the ith region, σ^2 is a global error variance, and I_k is the k-dimensional identity matrix. In the described notation, the likelihood of the complete dataset given the tessellation and model parameters is just

$$P(G \mid \delta, T) = \prod_{i=1}^{M} \zeta(G_i \mid \delta).$$

The ability to assign conjugate priors within the disjoint regions is central to this approach. This allows us to obtain the marginal likelihood of any proposed partition structure by integrating out the other parameters within the model

$$P(G \mid T) = \int_{\delta} p(G \mid \delta, T)\, p(\delta \mid T)\, d\delta.$$

and the above integral can be found explicitly.

We assign the standard conjugate normal prior distributions over β_i conditional on the noise variance which is allocated inverse gamma prior.

$$\begin{aligned} P(\beta_i \mid T, \sigma^2) &\sim \text{MVN}(\beta_i \mid 0, \sigma^2 \lambda_i^{-1});\ i = 1, \ldots, M \\ P(\sigma^{-2}) &\sim \text{Gamma}(\sigma^{-2} \mid \gamma_1/2, \gamma_2/2), \end{aligned}$$

where λ_i is the prior precision matrix of β_i and γ_1, γ_2 are hyperparameters. We then find that the marginal likelihood of the data given T is

$$P(G \mid T) = \frac{\gamma_2^{\gamma_1/2} \Gamma(\frac{1}{2}(\gamma_1 + n))}{\pi^{n/2} \Gamma(\gamma_1/2)} a^{(-\gamma_1 + n)/2} \prod_{i=1}^{M} |\lambda_i|^{1/2} |V_i|^{1/2}, \tag{6.9}$$

where $|\cdot|$ denotes the determinant of a matrix, $\Gamma(\cdot)$ is the usual Gamma function and

$$\begin{aligned} V_i &= (K_i' K_i + \lambda_i)^{-1} \\ a &= \gamma_2 + \sum_{i=1}^{M} G_i' (I_{n_i} - K_i V_i K_i') G_i. \end{aligned}$$

By assigning conjugate priors within regions, we are able to integrate out most of the parameters in the model and obtain analytic expressions for the marginal likelihood of any partition structure defined by the generating centers.

The parameter vector $T = \{S, w, M\}$ defines the partition or tessellation structure. We choose to assign flat priors on the space of tessellation structures

$$\begin{aligned} P(T) &= P(\omega) P(M) P(S \mid M) \\ &= Di(\omega \mid 1, \ldots, 1) U(M \mid 1, \ldots, M_{max}) \prod_{j=1}^{M} \prod_{l=1}^{\eta} U(s_{jl} \mid k_{1l}, \ldots, k_{nl}) \end{aligned}$$

where $Di(\cdot \mid 1, \ldots, 1)$ denotes the , the number of centers taken to be discrete uniform on $(1, M_{max})$ where M_{max} is the maximum number of regions set by the user, and the M sites, $S = \{s_1, \ldots, s_M\}$, $s_j = \{s_{j1}, \ldots, s_{j\eta}\}$, are uniformly distributed at the marginal predictor values $s_{jl} \sim U(k_{1l}, \ldots, k_{nl})$.

We resort to the reversible jump sampler to explore the parameter space of variable dimensions in MCMC. Within each dimension, the posterior distribution of T is analytically intractable and approximate sampling methods based on MCMC techniques are used. We propose one of the following moves, each with probability 1/4.

- Birth. Construct a new region by splitting one of the regions in two.
- Death. The exact opposite of the Birth step, remove a region by deleting one center.
- Update. Move a center by redrawing its position uniformly from the marginal predictor points.
- Alter the distance weighting w using proposal density tuned during a burn-in phase to give an acceptable rate of approximately 30 % .

The marginal likelihood is computed from (6.9) depending on the modeling task and the proposed change is accepted with probability

$$\alpha = \min\left(1, \frac{P(G \mid T_p)}{P(G \mid T_c)}\right)$$

where the subscripts c and p refer to the current and proposed models, respectively.

Posterior predictive distribution , $P(G \mid K)$ are approximated using the set of N retained MCMC tessellation samples $\{T^{(1)}, \ldots, T^{(N)}\}$, to form a mixture density

$$P(G \mid K) = \int_T \int_\delta P(G \mid T, \delta)\, P(T, \delta)\, dT\, d\delta \approx \frac{1}{N} \sum_{i=1}^{N} P(G \mid T^{(i)}, K) \quad (6.10)$$

where $P(G|T, K)$ is given by (6.9).

Now we have the MCMC samples of $G^*(k)$ which is fitted using BPM. First, we need to choose the best MCMC sample in the set of N retained MCMC tessellation samples after an initial burn-in period. To do this, we need to consider an error criterion that measures the distance between the true surface $G(k)$ and the fitted surface $G^*(k)$. One such error criterion is the relative error given by $\|G(k) - G^*(k)\|/\|G(k)\|$ where $G(k)$ is true surface, $G^*(k)$ is the surface fitted using the MCMC tessellation sample and $\|\cdot\|$ is L_2 norm. The best MCMC sample is obtained by choosing the MCMC sample with the minimum relative error.

We use this approximate response surface $G^*(k)$ for the approximate posterior calculation as

$$\pi^*(k) \propto \exp\left(-\frac{\{G^*(k)\}^2}{\sigma^2_{BPM}}\right), \quad (6.11)$$

where σ_{BPM} is chosen in our numerical simulations to be of the same order as σ_f. Next, we develop the two-stage MCMC procedure in the following way.

Algorithm Two-Stage MCMC

- Step 1. At k_n, generate k from distribution $q(k|k_n)$.
- Step 2. Accept k as a proposal with probability

$$a(k_n, k) = \min\left(1, \frac{q(k_n|k)\pi^*(k)}{q(k|k_n)\pi^*(k_n)}\right),$$

  i.e. pass k or k_n as a proposal to the next stage with probability $a(k_n, k)$ or $1 - a(k_n, k)$ respectively. Therefore, the final proposal to the fine-scale model is generated from

$$Q(k|k_n) = a(k_n, k)q(k|k_n) + \Big(1 - \int a(k_n, k)q(k|k_n)dk\Big)\delta_{k_n}(k).$$

- Step 3. Accept k as a sample with probability

$$\rho(k_n, k) = \min\left(1, \frac{Q(k_n|k)\pi(k)}{Q(k|k_n)\pi(k_n)}\right), \tag{6.12}$$

  i.e. $k_{n+1} = k$ with probability $\rho(k_n, k)$, and $k_{n+1} = k_n$ with probability $1 - \rho(k_n, k)$.

In the above algorithm, if the trial proposal k' is rejected at Step 2, k_n will be passed as the proposal. Since $\rho(k_n, k_n) \equiv 1$, no further computation is needed. Thus, the expensive direct computations can be avoided for those proposals which are unlikely to be accepted. In comparison, the regular MCMC method requires a direct simulation for every proposal k.

It is worth noting that there is no need to compute $Q(k|k_n)$ and $Q(k_n|k)$ in (6.12). The acceptance probability (6.12) can be simplified as

$$\rho(k_n, k) = \min\left(1, \frac{\pi(k)\pi^*(k_n)}{\pi(k_n)\pi^*(k)}\right). \tag{6.13}$$

In fact, (6.13) is obviously true for $k = k_n$ since $\rho(k_n, k_n) \equiv 1$. For $k \neq k_n$,

$$\begin{aligned} Q(k_n|k) = a(k, k_n)q(k_n|k) &= \frac{1}{\pi^*(k)} \min\Big(q(k_n|k)\pi^*(k),\ q(k|k_n)\pi^*(k_n)\Big) \\ &= \frac{q(k|k_n)\pi^*(k_n)}{\pi^*(k)} a(k_n, k) = \frac{\pi^*(k_n)}{\pi^*(k)} Q(k|k_n). \end{aligned}$$

Substituting the above formula into (6.12), we immediately get (6.13).

6.4 Numerical Results

For the model updating procedure, we apply an efficient parameterization of the permeability field using the Karhunen–Loève (K-L) expansion Loève

(1977); Wong (1971). This allows us to describe permeability field in terms of two-point statistics using fewer parameters. A key advantage of the K-L expansion is that these parameters can be varied continuously while maintaining the underlying geostatistical structure.

Using the K-L expansion, the permeability field can be expanded in terms of an optimal L^2 basis. By truncating the expansion we can represent the permeability matrix by a small number of random parameters. To impose the hard constraints (the values of the permeability at prescribed locations), we will find a linear subspace of our parameter space (a hyperplane) which yields the corresponding values of the permeability field. First, we briefly recall the facts of the K-L expansion. Denote $Y(x,\omega) = \log[k(x,\omega)]$, where ω denoting an elementary random event is included to remind us that k is a random field. Suppose $Y(x,\omega)$ is a second order stochastic process, that is, $Y(x,\omega) \in L^2(\Omega)$ with probability one. We will assume that $E[Y(x,\omega)] = 0$. Given an arbitrary orthonormal basis $\{\varphi_k\}$ in $L^2(\Omega)$, we can expand $Y(x,\omega)$ as $Y(x,\omega) = \sum_{j=1}^{\infty} Y_j(\omega)\varphi_j(x)$, where

$$Y_j(\omega) = \int_\Omega Y(x,\omega)\varphi_j(x)dx$$

are random variables. We are interested in the special L^2 basis $\{\varphi_k\}$ which makes Y_k uncorrelated, $E(Y_iY_l) = 0$ for all $i \neq l$. Denote the covariance function of Y as $R(x,y) = E\left[Y(x)Y(y)\right]$. Then such basis functions $\{\varphi_k\}$ satisfy

$$E[Y_iY_l] = \int_\Omega \varphi_i(x)dx \int_\Omega R(x,y)\varphi_l(y)dy = 0, \qquad i \neq l.$$

Since $\{\varphi_j\}$ is a complete basis in $L^2(\Omega)$, it follows that $\varphi_j(x)$ are eigenfunctions of $R(x,y)$:

$$\int_\Omega R(x,y)\varphi_j(y)dy = \lambda_j\varphi_j(x), \qquad j = 1, 2, \ldots, \tag{6.14}$$

where $\lambda_j = E[Y_j^2] > 0$. Furthermore, we have

$$R(x,y) = \sum_{j=1}^{\infty} \lambda_j\varphi_j(x)\varphi_j(y). \tag{6.15}$$

Denote $\theta_j = Y_j/\sqrt{\lambda_j}$, then θ_j satisfy $E(\theta_j) = 0$ and $E(\theta_i\theta_l) = \delta_{il}$. It follows that

$$Y(x,\omega) = \sum_{j=1}^{\infty} \sqrt{\lambda_j}\theta_j(\omega)\varphi_j(x), \tag{6.16}$$

where φ_j and λ_j satisfy (6.14). We assume that eigenvalues λ_j are ordered so that $\lambda_1 \geq \lambda_2 \geq \ldots$. The expansion (6.16) is called the K-L expansion. In

(6.16), the L^2 basis functions $\varphi_j(x)$ are deterministic and resolve the spatial dependence of the permeability field. The randomness is represented by the scalar random variables θ_j. Generally, we only need to keep the leading order terms (quantified by the magnitude of λ_k) and still capture most of the energy of the stochastic process $Y(x,\omega)$. For a $N-$term K-L approximation $Y_N = \sum_{j=1}^N \sqrt{\lambda_k}\theta_k\varphi_k$, we define the energy ratio of the approximation as

$$e(N) := \frac{E\|Y_N\|^2}{E\|Y\|^2} = \frac{\sum_{j=1}^N \lambda_j}{\sum_{j=1}^\infty \lambda_j}.$$

If $\lambda_j, j = 1, 2, \ldots,$ decay very fast, then the truncated K-L would be good approximations of the stochastic process in L^2 sense.

Suppose the permeability field $k(x,\omega)$ is a log normal homogeneous stochastic process, then $Y(x,\omega)$ is a Gaussian process and θ_k are independent standard Gaussian random variables. We assume that the covariance function of $Y(x,\omega)$ bears the form

$$R(x,y) = \sigma^2 \exp\Big(-\frac{|x_1-y_1|^2}{2L_1^2} - \frac{|x_2-y_2|^2}{2L_2^2}\Big). \tag{6.17}$$

In the above formula, L_1 and L_2 are the correlation lengths in each dimension, and $\sigma^2 = E(Y^2)$ is a constant.

The purpose of our first numerical example is to depict the response surface approximation obtained using the BPM. We set $L_1 = 0.4$, $L_2 = 0.4$ and $\sigma^2 = 2$. We choose the underlying grid to be 61×61 and first solve the eigenvalue problem (6.14) numerically and obtain the eigenpairs $\{\lambda_k, \varphi_k\}$. To propose permeability fields from the prior (unconditional) distribution, we maintain 11 terms in the K-L expansion. In all the examples we consider, there are eight producing wells in the corner cells and the cells that are adjacent to the middle of the edges of the global domain of size 1×1. Moreover, there is one injection well at the middle of the domain. This is known as the nine-spot pattern. The forward simulations are performed using 3D streamline simulator. We assume that the permeability is known at 9 well locations. This condition is imposed by setting

$$\sum_{j=1}^{11} \sqrt{\lambda_j}\theta_j\varphi_j(x_l) = \alpha_l, \tag{6.18}$$

where α_l $(l = 1, \ldots, 9)$ are prescribed constants. For simplicity, we set $\alpha_l = 0$ for all $l = 1, \ldots, 9$. In the simulations, we propose two θ_i and calculate the rest of θ_i by solving the linear system (6.18). These θs determine the unknown permeability field k on the finite grid. To plot the true response surface $G(k)$, we vary θ_i in the range $[-3, 3]$ and compute $G(k)$ using forward solutions. Next, we use 100 sample points and generate an approximation

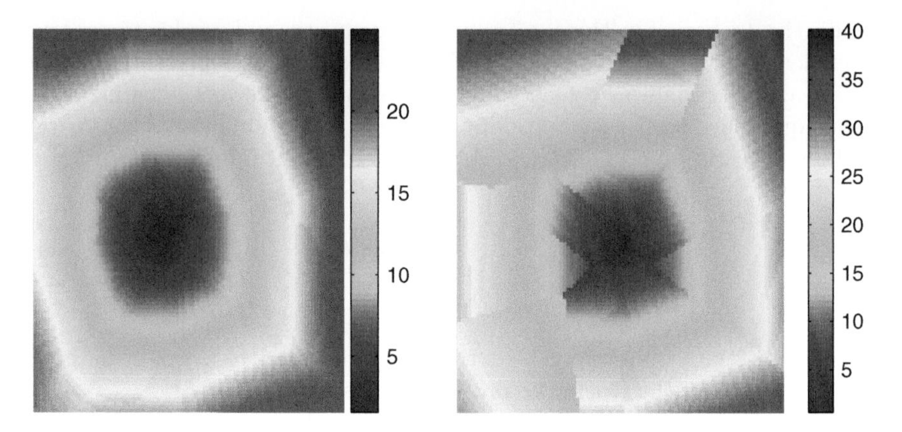

Figure 6.1 The true response surface $G(k)$ (left) and approximate response surface $G^*(k)$ obtained with the BPM (right).

of $G(k)$ with the BPM. The purpose of choosing lowdimensional parameter space (dimension is 2) is to depict the actual response surface $G(k)$ and the approximate response surface $G^*(k)$ obtained using the BPM. In Figure 6.1, we present the numerical results. On the left figure, the actual response surface is plotted. The approximation of $G(k)$ using 100 observation points is plotted on right figure. We observe from this figure that the BPM approximates the response surface fairly well in low dimensions. In particular, the approximation captures the lower values in the center and high values in the boundary of the true response surface.

In our next examples, we consider five dimensional parameter space. We choose the same variogram, however, we maintain total 14 terms in the K-L. The permeability values at well locations (total 9) are assumed to be 1. Thus, the dimension of parameter space is 5. We propose five θ_is and calculate the rest of θ_is by solving the linear system (6.18). In the simulation, we first generate a reference permeability field using all eigenvectors, and compute up to 2000 days of fractional flows for eight producing wells in the nine-spot pattern using 3D streamline simulator. To generate approximate response surface $G^*(k)$, 400 permeability fields are generated and corresponding L^2 norm data misfit is calculated. In our simulations, the initial realizations of permeability fields are generated uniformly, though one can design approaches for efficient initial sampling. After the marginal likelihood stabilized, 18 disjointed regions were generated. To perform sampling, we run two-stage MCMC, where at the first stage samples from the BPM are selected and if a sample passes the first stage, direct forwards simulation is performed. To sample from the approximate response surface, we use $\sigma_{BPM} = 0.2$. Typically, we have found that smaller values of σ_{BPM} increase the efficiency of two-stage MCMC. As for instrumental distribution $q(k|k_n)$,

Table 6.1 Acceptance rate.

σ_f	Independent MCMC	Two-stage MCMC
.10	.007	.05
.15	.010	.09
.20	.010	.12
.50	.050	.30

we use independent sampler. We compare the acceptance rates between direct MCMC and two-stage MCMC for different σ_f.

Table 6.1 shows that as the σ_f increases, the acceptance rate increases. This is because as we increase σ_f, the precision decreases and more samples are accepted. The acceptance rate is computed as the ratio between the total number of simulations and the total number of accepted realizations. We see from this table that even for large value of σ_f, such as $\sigma_f = 0.2$, the two-stage MCMC provides several times larger acceptance rate. Next we study the convergence of independent and two-stage MCMC methods. In Figure 6.2, the L^2 errors of water-cut (sum over all wells) for $\sigma_f = 0.1$ (left), $\sigma_f = 0.15$ (middle) and $\sigma_f = 0.2$ (right) are plotted as number of iterations. Here, water cut is $1 - F(t)$ and water cut error is the cumulative error over all 8 producing wells. We observe from this figure that two-stage MCMC has similar convergence properties, while it has higher acceptance rate.

Next, we plot the water cut curves in Figure 6.3. We have selected randomly several permeability samples after the chain reaches steady state. We see clearly from this figure that the samples of the permeability field closely reproduce the production history at the wells. The realizations of the permeability field are plotted in Figure 6.4. As one can observe from this figure that, the realizations closely match the reference permeability field. In particular, the main features of the reference permeability field are present in the samples derived through integration of water-cut data.

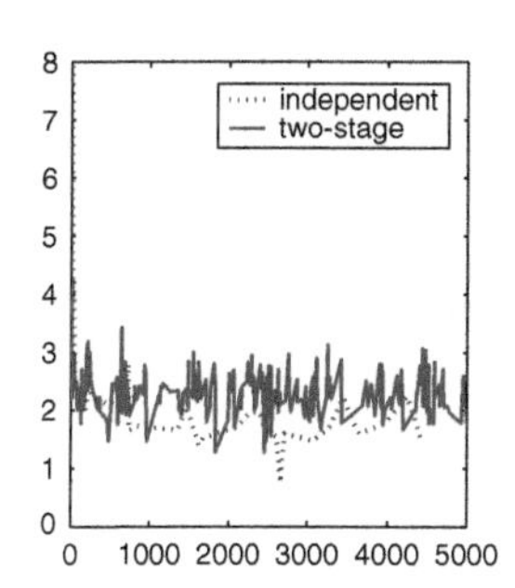

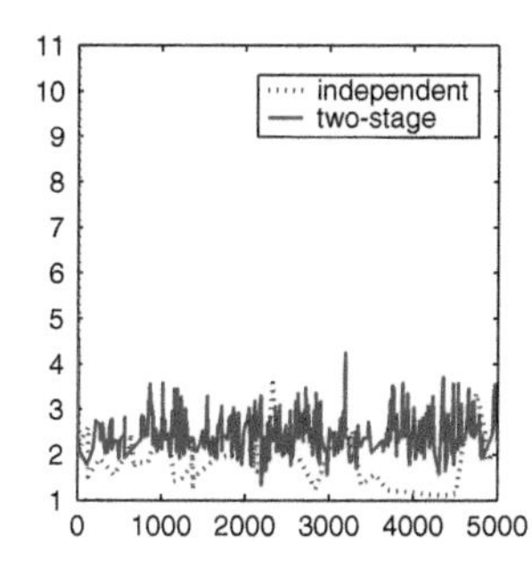

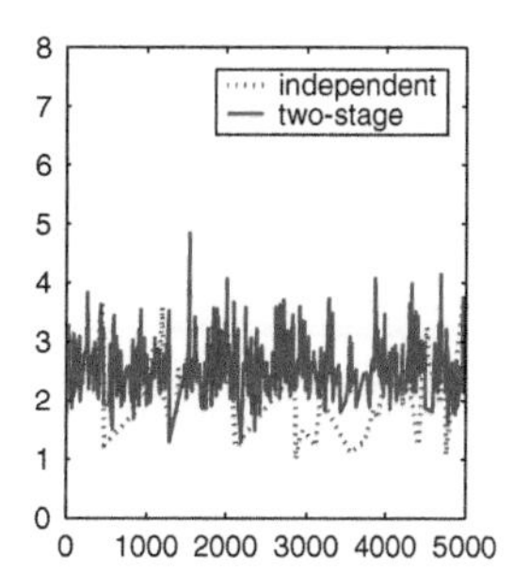

Figure 6.2 Water cut errors vs. iterations for $\sigma_f = 0.1$ (left), $\sigma_f = 0.15$ (middle) and $\sigma_f = 0.2$ (right).

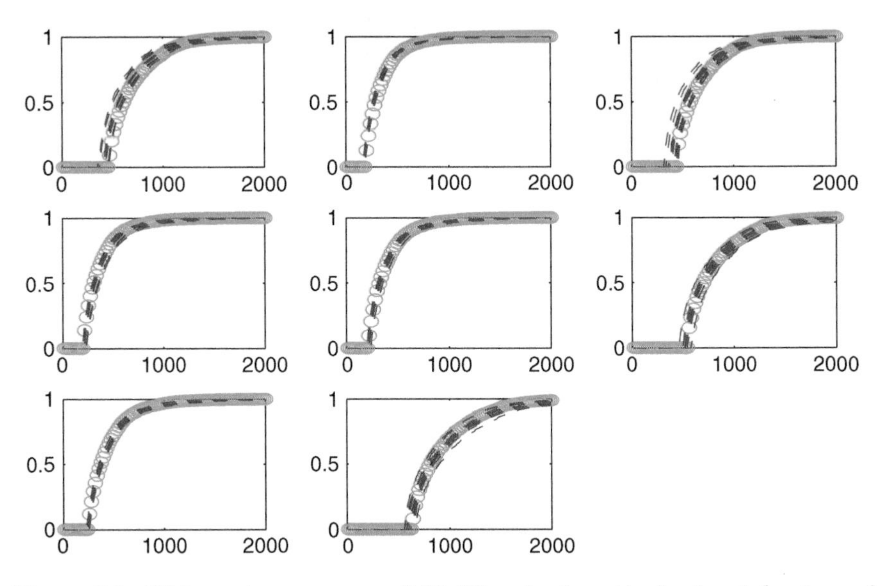

Figure 6.3 Water cut curves $\sigma_f = 0.15$. Time is along the horizontal axis, and water cut (the fraction of water produced at the wells) is along the vertical axis. Each figure represents the water-cut at different wells.

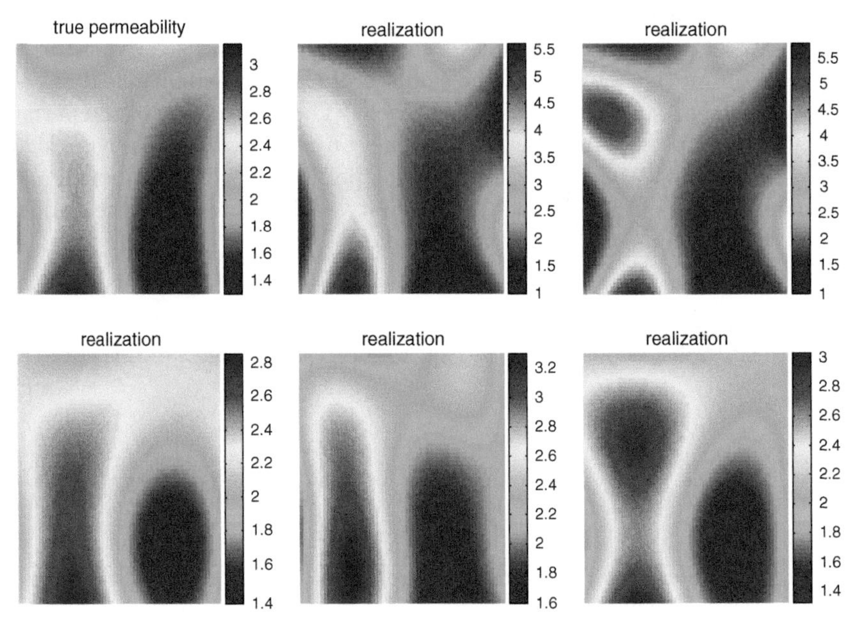

Figure 6.4 Realizations of permeability field.

For high dimensional example (with parameter space dimension 9) we have observed similar trend in numerical results. In general with less observation points, the approximate surface is not accurate, and the gain in two-stage MCMC method is lower.

6.5 Conclusions

In this chapter, we use Bayesian Partition Model (BPM) for dynamic data integration that arises in subsurface characterization. The main idea of the BPM is to split the parameter space into the regions and use the same probability distribution within each region. We use Voronoi tessellation to define distinct regions. The BPM procedure used in the paper allows the flexibility in the number of Voronoi tessellation by assuming the number of Voronoi tessellation is unknown. Because of possible approximation errors due to small sample size used in the BPM, we employ two-stage Markov chain Monte Carlo (MCMC), where at the first stage we use BPM approximation. If the proposal passes the first stage, we use direct simulations to compute the acceptance probability based on Metropolis-Hasting rule. Numerical examples are presented which demonstrate that the combined procedure can increase the acceptance rate by several times. Although the results presented in this chapter are encouraging, there is scope for further exploration. In particular, the rejected samples of two-stage MCMC can be used further to obtain more accurate approximation within BPM framework. These data points can allow us to get better approximations within a region, or possibly used to modify Voronoi tessellation locally. These issues will be studied in our future work.

Acknowledgment

This work is partly supported by NSF CMG DMS 0724704 and DOE NNSA under the predictive science academic alliances program by grant DE-FC52-08NA28616.

References

Behrens, R. and Tran, T. 1998 Incorporating seismic data of intermediate vertical resolution into 3d reservoir models. *SPE 49143 paper* presented at the 1998 SPE Annual Technical Conference and Exhibition, New Orleans, LA, October 4–7.

Christen, A. and Fox, C. 1992 MCMC using an approximation. *Technical report, Department of Mathematics, The University of Auckland, New Zealand.*

De, Finetti (1974) *Theory of Probability*, 2 vols. Translated by A. Machi, and A.F.M. Smith, Wiley, London.

Denison, D.G.T., Holmes, C.C., Mallick, B.K. and Smith, A.F.M. 2002 *Bayesian Methods for Nonlinear Classification and Regression*, Joh Wiley & Sons, Ltd.

Deutsch, C., Srinivasan, S. and Mo, Y. 1996 Geostatistical reservoir modeling accounting for precision and scale of seismic data. *SPE 36497 paper*, presented at the 1996 SPE Annual Technical Conference and Exhibition, Denver, Colorado, October 6–9.

Doyen, P., Psaila, D., den Boer, L. and Jans, D. 1997 Reconciling data at seismic and well log scales in 3-d earth modeling. *SPE 38698 paper*, presented at the 1997 SPE Annual Technical Conference and Exhibition, San Antonio, TX, October 5–8.

Efendiev, Y., Datta-Gupta, A., Ginting, V., Ma, X. and Mallick, B. 2005a An efficient two-stage Markov chain Monte Carlo method for dynamic data integration. *Water Resources Research*, **41**, W12423, doi: = 10.1029/2004WR003764

Efendiev, Y., Datta-Gupta, A., Osaka, I. and Mallick, B.K. 2005b Multiscale data integration using coarse-scale models. *Advances in Water Resources*, **28**, 303–314, 2005.

Frykman, P. and Deutsch, C. 1999 Geostatistical scaling laws applied to core and log data. *SPE 56822 paper*, presented at the 1999 SPE Annual Technical Conference and Exhibition, Houston, TX, October 3–6.

Glimm, J. and Sharp, D.H. 1999 Prediction and the quantification of uncertainty. *Phys. D*, **133**, pp. 152–170.

Green, P. 1995 Reversible jump Markov chain Monte Carlo computation and Bayesian model determination. *Biometrika*, **82**, 711–32, 1995.

Holmes, C.C., Denison, D.G.T. and Mallick, B.K. 2001 Bayesian Prediction via Partitioning. Technical report. Department of Mathematics, Imperial College, London. 2001.

Holmes, C.C., Denison, D.G.T., Ray, S. and Mallick, B.K. 2005 Bayesian partitioning for classification and regression. *Journal of computation and graphical statistics*, **14**, 2005, pp. 811–830.

Liu, J. 2001 *Monte Carlo Strategies in Scientific Computing*. Springer, New-York.

Loève, M. 1977 *Probability Theory*, 4th ed., Springer, Berlin.

Subbey, S., Christie, M. and Sambridge, M. 2004 Prediction under uncertainty in reservoir modeling. *Journal of Petroleum Science and Engineering*, **44**, 2004, pp. 143–153.

Wong, E. 1971 *Stochastic Processes in Information and Dynamical Systems*, McGraw-Hill, 1971.

Xu, X., Tran, T., Srivastava, R. and Deutsch, C. 1992 Integrating seismic data in reservoir modeling: The collocated cokriging approach. *SPE 24742 paper presented at the 1992 SPE Annual Technical Conference and Exhibition*, Washington, DC, October 4–7.

Xue, G. and Datta-Gupta, A. 1996 A new approach to seismic data integration during reservoir characterization using optimal nonparametric transformations. *SPE 36500 paper presented at the 1996 SPE Annual Technical Conference and Exhibition*, Denver, CO, October 6–9.

Chapter 7

Surrogate and Reduced-Order Modeling: A Comparison of Approaches for Large-Scale Statistical Inverse Problems

M. Frangos,[1] Y. Marzouk,[1] K. Willcox[1] and B. van Bloemen Waanders[2]

[1]*Massachusetts Institute of Technology, USA*

[2]*Sandia National Laboratories, USA*

7.1 Introduction

Solution of statistical inverse problems via the frequentist or Bayesian approaches described in earlier chapters can be a computationally intensive endeavor, particularly when faced with large-scale forward models characteristic of many engineering and science applications. High computational cost

Large-Scale Inverse Problems and Quantification of Uncertainty Edited by L. Biegler, G. Biros, O. Ghattas, M. Heinkenschloss, D. Keyes, B. Mallick, Y. Marzouk, L. Tenorio, B. van Bloemen Waanders and K. Willcox

arises in several ways. First, thousands or millions of forward simulations may be required to evaluate estimators of interest or to characterize a posterior distribution. In the large-scale setting, performing so many forward simulations is often computationally intractable. Second, sampling may be complicated by the large dimensionality of the input space – as when the inputs are fields represented with spatial discretizations of high dimension – and by nonlinear forward dynamics that lead to multimodal, skewed, and/or strongly correlated posteriors.

In this chapter, we present an overview of surrogate and reduced-order modeling methods that address these computational challenges. For illustration, we consider a Bayesian formulation of the inverse problem. Though some of the methods we review exploit prior information, they largely focus on simplifying or accelerating evaluations of a stochastic model for the data, and thus are also applicable in a frequentist context.

Methods to reduce computational cost of solving of a statistical inverse problem can be classed broadly in three groups. First, there are many ways to reduce the cost of a posterior evaluation (or more specifically, the cost of a forward simulation) through surrogate models, reduced-order models, multigrid and multiscale approaches, and stochastic spectral approaches. Second, the dimension of the input space can be reduced, through truncated Karhunen-Loève expansions, coarse grids, and parameter-space reductions. The third set of approaches targets a reduction in the number of forward simulations required to compute estimators of interest, i.e. more efficient sampling. In the Bayesian setting, these methods include a wide range of adaptive and multi-stage Markov chain Monte Carlo (MCMC) schemes.

This chapter summarizes the state of the art in methods to reduce the computational complexity of statistical inverse problems, provides some introductory material to support the technical chapters that follow, and offers insight into the relative advantages of different methods through a simple illustrative example. Section 7.2 presents an overview of state-of-the-art approaches. Section 7.3 then describes our specific problem setup, with a focus on the structure of the large-scale forward model that underlies the inverse problem. We then present a detailed comparison of reduced-order modeling and stochastic spectral approximations in Sections 7.4 and 7.5, respectively. Section 7.6 presents an illustrative example and Section 7.7 provides some concluding remarks.

7.2 Reducing the Computational Cost of Solving Statistical Inverse Problems

As discussed in Section 7.1, we consider three general classes of approaches to reducing the computational cost of solving statistical inverse problems: reducing the cost of a forward simulation, reducing the dimension of the input parameter space, and reducing the number of samples (forward simulations) required.

7.2.1 Reducing the Cost of Forward Simulations

Many attempts at accelerating inference in computationally intensive inverse problems have relied on surrogates for the forward model, typically constructed through repeated forward simulations that are performed in an offline phase. Eldred *et al.* (2004) categorize surrogates into three different classes: data-fit models, reduced-order models, and hierarchical models, all of which have been employed in the statistical inverse problem setting.

Data-fit models are generated using interpolation or regression of simulation data from the input/output relationships in the high-fidelity model. In the statistical literature, Gaussian processes have been used extensively as surrogates for complex computational models (Kennedy and O'Hagan 2001). These approaches treat the forward model as a black box, and thus require careful attention to experimental design and to modeling choices that specify the mean and covariance of the surrogate Gaussian process. Bliznyuk *et al.* (2008) use local experimental design combined with radial basis function approximations to approximate the posterior and demonstrate the effectiveness of their approach for a pollutant diffusion problem.

Reduced-order models are commonly derived using a projection framework; that is, the governing equations of the forward model are projected onto a subspace of reduced dimension. This reduced subspace is defined via a set of basis vectors, which, for general nonlinear problems, can be calculated via the proper orthogonal decomposition (POD) (Holmes *et al.* 1996; Sirovich 1987) or with reduced basis methods (Noor and Peters 1980). For both approaches, the empirical basis is pre-constructed using full forward problem simulations – or 'snapshots.' Wang and Zabaras (2005) use the POD to accelerate forward model calculations in a radiative source inversion problem. Galbally *et al.* (2010) combine POD with an empirical interpolation method (Barrault *et al.* 2004; Grepl *et al.* 2007) for Bayesian inference in a highly nonlinear combustion problem governed by an advection-diffusion-reaction partial differential equation. In both of these applications, the choice of inputs to the simulations – in particular, how closely the inputs must resemble the inverse solution – can be important (Wang and Zabaras 2005).

Hierarchical surrogate models span a range of physics-based models of lower accuracy and reduced computational cost. Hierarchical surrogates are derived from higher-fidelity models using approaches such as simplifying physics assumptions, coarser grids, alternative basis expansions, and looser residual tolerances. Arridge *et al.* (2006) use mesh coarsening for solving a linear inverse problem, employing spatial discretizations that are coarser than those necessary for accurate solution of the forward problem. Their

Bayesian formulation also includes the statistics of the error associated with the discretization, but unfortunately the level of mesh coarsening required for decreasing the computational cost of large-scale problems to acceptable levels may result in large errors that are difficult to quantify or that even yield unstable numerical schemes. Balakrishnan *et al.* (2003) introduce a polynomial chaos (PC) representation of the forward model in a groundwater transport parameter identification problem, and obtain the PC coefficients by linear regression; again, this process depends on a series of representative snapshots obtained from repeated forward simulations.

7.2.2 Reducing the Dimension of the Input Space

When the object of inference is a spatially distributed parameter, dimensionality of the input space may naively be tied to dimensionality of the spatial discretization. Often, however, there is knowledge of smoothness or structure in the parameter field that can lead to a more efficient basis. In particular, one can introduce a Karhunen-Loève (K-L) expansion based on the prior covariance, transforming the inverse problem to inference on a truncated sequence of weights of the K-L modes (Marzouk and Najm 2009). Li and Cirpka (2006) emphasize the role of K-L expansions in enabling geostatistical inversion on unstructured grids. Efendiev *et al.* (2006) use K-L expansions to parameterize a log-permeability field, introducing constraints among the weights in order to match known values of the permeability at selected spatial locations.

Recent work has also proposed a reduced-basis approach to reducing the dimension of the input space. In the same way that a reduced-order model is formed by projecting the state onto a reduced subspace, Lieberman (2010) considers projection of a high-dimensional parameter field onto an empirical parameter basis. The effectiveness of this approach is demonstrated for a groundwater inverse problem where MCMC sampling is carried out in the reduced parameter space.

7.2.3 Reducing the Number of Samples

A different set of approaches retain the full forward model but use simplified or coarsened models to guide and improve the efficiency of MCMC sampling. Christen and Fox (2005) use a local linear approximation of the forward model to improve the acceptance probability of proposed moves, reducing the number of times the likelihood must be evaluated with the full forward model. Higdon *et al.* (2003) focus on the estimation of spatially distributed inputs to a complex forward model. They introduce coarsened representations of the inputs and apply a Metropolis-coupled MCMC scheme (Geyer 1991) in which 'swap proposals' allow information from the coarse-scale formulation to influence the fine-scale chain. Efendiev *et al.* (2006) also develop a two-stage MCMC algorithm, using a coarse-scale model based on multiscale finite volume methods to improve the acceptance rate of MCMC proposals.

7.3 General Formulation

Here we describe a general inverse problem formulation and then focus on the structure of the forward problem. We consider the task of inferring inputs $\boldsymbol{\theta}$ from limited and imperfect observations $\mathbf{d}$. For simplicity, we let both $\mathbf{d}$ and $\boldsymbol{\theta}$ be finite-dimensional. The relationship between observables $\mathbf{d}$ and inputs $\boldsymbol{\theta}$ is *indirect*, and is denoted by $\mathbf{d} = \mathcal{G}(\boldsymbol{\theta}, \boldsymbol{\eta})$; the additional input $\boldsymbol{\eta}$ is a random variable encompassing measurement noise and/or modeling errors. The likelihood function $\pi(\mathbf{d}|\boldsymbol{\theta})$ describes the probability of measurements $\mathbf{d}$ given a particular value of $\boldsymbol{\theta}$. The likelihood function thus incorporates the forward model and the measurement noise, and results from the particular form of $\mathcal{G}$. In a Bayesian formulation of the inverse problem, $\boldsymbol{\theta}$ is treated as a random variable, endowed with a prior probability density $\pi_{\text{pr}}(\boldsymbol{\theta})$ that encodes any available prior knowledge about the inputs. Bayes' rule then yields the posterior probability density of the inputs, π_{post}:

$$\pi_{\text{post}}(\boldsymbol{\theta}) \equiv \pi(\boldsymbol{\theta}|\mathbf{d}) \propto \pi(\mathbf{d}|\boldsymbol{\theta})\, \pi_{\text{pr}}(\boldsymbol{\theta}). \tag{7.1}$$

The expression (7.1) casts the solution of the inverse problem as a probability density for the model inputs $\boldsymbol{\theta}$.

This chapter focuses on methods to approximate the input–output map $\mathcal{G}(\cdot)$. In physical problems, $\mathcal{G}$ usually contains a deterministic map from $\boldsymbol{\theta}$ to some idealized observations $\mathbf{d}_0$; interaction of $\mathbf{d}_0$ and $\boldsymbol{\theta}$ with the random variable $\boldsymbol{\eta}$ then yields the actual data. We set aside this interaction and concentrate on the deterministic map, which can be broken into two elements: the state equations that describe the evolution of the state $\mathbf{u}$ in response to the input $\boldsymbol{\theta}$, and the output equations that define the relationship between outputs of interest $\mathbf{y}$ – which in turn define the observations $\mathbf{d}$ – and state $\mathbf{u}$. We present a general nonlinear model in the semi-discrete form that might result from spatial discretization of a partial differential equation (using, for example, a finite element or finite volume method), or that might represent a set of differential-algebraic equations (such as arises in circuit modeling). The semi-discrete equations describing such a system are

$$\dot{\mathbf{u}} = \mathbf{f}(\mathbf{u}, \boldsymbol{\theta}, t), \tag{7.2}$$

$$\mathbf{y} = \mathbf{h}(\mathbf{u}, \boldsymbol{\theta}, t), \tag{7.3}$$

with initial condition

$$\mathbf{u}(\boldsymbol{\theta}, 0) = \mathbf{u}^0(\boldsymbol{\theta}). \tag{7.4}$$

In this semi-discrete model, $\mathbf{u}(\boldsymbol{\theta}, t) \in \mathbb{R}^N$ is the discretized state vector of N unknowns, and $\boldsymbol{\theta} \in \Theta \subseteq \mathbb{R}^p$ is the vector of p parametric inputs on some domain Θ. The vector $\mathbf{u}^0(\boldsymbol{\theta}) \in \mathbb{R}^N$ is the specified initial state, t is time, and the dot indicates a derivative with respect to time. The nonlinear discretized residual vector $\mathbf{f}: \mathbb{R}^N \times \mathbb{R}^p \times [0, \infty) \to \mathbb{R}^N$ is for generality written as a function of the state, parameter inputs, and time. Our observable outputs are represented in spatially discretized form with q components in the vector $\mathbf{y}(\boldsymbol{\theta}, t) \in \mathbb{R}^q$, and defined by the general nonlinear function $\mathbf{h} : \mathbb{R}^N \times \mathbb{R}^p \times [0, \infty) \to \mathbb{R}^q$.

If the governing equations are linear in the state (or if a linearized model of (7.2) and (7.3) is derived by linearizing about a steady state), then the system is written

$$\dot{\mathbf{u}} = \mathbf{A}(\boldsymbol{\theta})\mathbf{u} + \mathbf{g}(\boldsymbol{\theta}, t), \tag{7.5}$$

$$\mathbf{y} = \mathbf{H}(\boldsymbol{\theta})\mathbf{u}, \tag{7.6}$$

where $\mathbf{A} \in \mathbb{R}^{N \times N}$ is a matrix that possibly depends on the parameters but not on the state, and the general nonlinear function $\mathbf{g} \in \mathbb{R}^N$ represents the direct contributions of the parameters to the governing equations as well as forcing due to boundary conditions and source terms. For the linearized output equation, $\mathbf{H} \in \mathbb{R}^{q \times N}$ is a matrix that maps states to outputs, and we have assumed no direct dependence of outputs on parameters (without loss of generality – such a term could easily be incorporated).

Note that in both (7.3) and (7.6), the output vector $\mathbf{y}$ is continuous in time. In most practical situations and in the Bayesian formulation described above, the observational data are finite-dimensional on some domain $\mathcal{D}$; thus $\mathbf{d} \in \mathcal{D} \subseteq \mathbb{R}^M$ consists of $\mathbf{y}$ evaluated at a finite set of times, $\mathbf{d} = (\mathbf{y}(t_1), \ldots, \mathbf{y}(t_{n_t}))$ with $M = n_t q$. We denote the deterministic forward model mapping inputs $\boldsymbol{\theta}$ to finite-time observations $\mathbf{d}$ by $\mathbf{G}(\boldsymbol{\theta}) : \Theta \to \mathcal{D}$.

With nonlinear forward models, the inverse solution is typically represented by samples simulated from the posterior distribution (7.1). From these samples, posterior moments, marginal distributions, and other summaries can be evaluated. Solution of the statistical inverse problem thus requires many evaluations of the forward model (typically many thousands or even millions) – a computationally prohibitive proposition for large-scale systems. In the following two sections, we describe approaches to accelerate solution of the forward model – model reduction via state projection onto a reduced basis, and a stochastic spectral approach based on generalized PC.

7.4 Model Reduction

Model reduction seeks to derive a low-complexity model of the system that is fast to solve but preserves accurately the relationship between input parameters $\boldsymbol{\theta}$ and outputs $\mathbf{y}$, represented by the input–output map $\mathcal{G}$. If the system forward solves required to evaluate samples drawn from the posterior distribution are performed using a reduced-order model as a surrogate for the large-scale system, then it becomes tractable to carry out the many thousands of iterations required to solve the statistical inverse problem. Most large-scale model reduction frameworks are based on a projection approach, which is described in general terms in this section. We briefly describe computation of the basis via POD and reduced basis methods, and then discuss various options that have been proposed for sampling the parameter space.

7.4.1 General Projection Framework

The first step in creating a projection-based reduced-order model is to approximate the N-dimensional state $\mathbf{u}(\boldsymbol{\theta}, t)$ by a linear combination of n basis vectors,

$$\mathbf{u} \approx \mathbf{\Phi}\mathbf{u}_r, \tag{7.7}$$

where $n \ll N$. The projection matrix $\mathbf{\Phi} \in \mathbb{R}^{N\times n}$ contains as columns the basis vectors $\boldsymbol{\varphi}_i$, i.e., $\mathbf{\Phi} = [\boldsymbol{\varphi}_1\boldsymbol{\varphi}_2 \cdots \boldsymbol{\varphi}_n]$, and the vector $\mathbf{u}_r(\boldsymbol{\theta}, t) \in \mathbb{R}^n$ contains the corresponding modal amplitudes.

This approximation can be employed in the general nonlinear system (7.2) or the general linear system (7.5), in either case resulting in a residual (since in general the N equations cannot all be satisfied with $n \ll N$ degrees of freedom). We define a left basis $\mathbf{\Psi} \in \mathbb{R}^{N\times n}$ so that $\mathbf{\Psi}^T\mathbf{\Phi} = \mathbf{I}$. Using a Petrov-Galerkin projection, we require the residual to be orthogonal to the space spanned by the columns of $\mathbf{\Psi}$. In the nonlinear case, this yields the reduced-order model of (7.2)–(7.4) as

$$\dot{\mathbf{u}}_r = \mathbf{\Psi}^T\mathbf{f}(\mathbf{\Phi}\mathbf{u}_r, \boldsymbol{\theta}, t), \tag{7.8}$$

$$\mathbf{y}_r = \mathbf{h}(\mathbf{\Phi}\mathbf{u}_r, \boldsymbol{\theta}, t), \tag{7.9}$$

$$\mathbf{u}_r(\boldsymbol{\theta}, 0) = \mathbf{\Psi}^T\mathbf{u}^0(\boldsymbol{\theta}), \tag{7.10}$$

where $\mathbf{y}_r(\boldsymbol{\theta}, t) \in \mathbb{R}^q$ is the reduced model approximation of the output $\mathbf{y}(\boldsymbol{\theta}, t)$. For the linear system (7.5), (7.6) with initial condition (7.4), the reduced-order model is

$$\dot{\mathbf{u}}_r = \mathbf{A}_r(\boldsymbol{\theta})\mathbf{u}_r + \mathbf{\Psi}^T\mathbf{g}(\boldsymbol{\theta}, t), \tag{7.11}$$

$$\mathbf{y}_r = \mathbf{H}_r(\boldsymbol{\theta})\mathbf{u}_r, \tag{7.12}$$

$$\mathbf{u}_r(\boldsymbol{\theta}, 0) = \mathbf{\Psi}^T\mathbf{u}^0(\boldsymbol{\theta}), \tag{7.13}$$

where $\mathbf{A}_r(\boldsymbol{\theta}) = \mathbf{\Psi}^T\mathbf{A}(\boldsymbol{\theta})\mathbf{\Phi}$ and $\mathbf{H}_r(\boldsymbol{\theta}) = \mathbf{H}(\boldsymbol{\theta})\mathbf{\Phi}$.

As in the full problem described in Section 7.3, the reduced model output $\mathbf{y}_r$ is continuous in time. Following our notation for the full problem, we define the reduced model data vector $\mathbf{d}_r \in \mathcal{D} \subseteq \mathbb{R}^M$ to consist of $\mathbf{y}_r$ evaluated at a finite set of times, $\mathbf{d}_r = (\mathbf{y}_r(t_1), \ldots, \mathbf{y}_r(t_{n_t}))$ with $M = n_t q$. We denote the reduced forward model of state dimension n that maps $\boldsymbol{\theta}$ to $\mathbf{d}_r$ by $\mathbf{G}^n_{\text{ROM}} : \Theta \to \mathcal{D}$.

Our approach to reducing the computational cost of solving the inverse problem is to use the reduced model $\mathbf{G}^n_{\text{ROM}}$ in place of the full model $\mathbf{G}$

in the likelihood function $\pi(\mathbf{d}|\boldsymbol{\theta})$. This can lead to a dramatic reduction in the cost of an evaluation of the posterior distribution (7.1). However, care must be taken to ensure efficient construction and solution of the reduced-order models (7.8)–(7.10) or (7.11)–(7.13). In the case of general nonlinear parametric dependence, these models have low dimension but are not necessarily fast to solve, since for each new parameter $\boldsymbol{\theta}$, solution of the ROM requires evaluating the large-scale system matrices or residual, projecting those matrices/residual onto the reduced subspace, and then solving the resulting reduced-order model. Since many elements of these computations depend on N, the dimension of the large-scale system, in general this process will not be computationally efficient. One option is to employ linearization of the parametric dependence (Daniel *et al.* 2004; Grepl and Patera 2005; Veroy *et al.* 2003).

A more general approach is to employ the missing point estimation approach (Astrid *et al.* 2008), which approximates nonlinear terms in the reduced-order model with selective spatial sampling, or the coefficient-function approximation (Barrault *et al.* 2004; Grepl *et al.* 2007), which replaces nonlinear parametric dependencies with a reduced-basis expansion and then uses interpolation to efficiently compute the coefficients of that expansion for new parameter values.

7.4.2 Computing the Basis

The basis vectors can be calculated with several techniques. Methods to compute the basis in the large-scale setting include approximate balanced truncation (Gugercin and Antoulas 2004; Li and White 2002; Moore 1981; Penzl 2006; Sorensen and Antoulas 2002), Krylov-based methods (Feldmann and Freund 1995; Gallivan *et al.* 1994; Grimme 1997), proper orthogonal decomposition (POD) (Deane *et al.* 1991; Holmes *et al.* 1996; Sirovich 1987), and reduced basis methods (Fox and Miura 1971; Noor and Peters 1980). Balanced truncation and Krylov-based methods are largely restricted to linear systems, thus here we focus on a brief description of reduced basis methods and POD. While the development of these methods has been in the context of reduction of the forward problem for simulation and (to some extent) control, recent work has shown applicability to the inverse problem setting, by introducing strategies that use characteristics of the inverse problem (including Hessian and prior information) to inform computation of the basis (Galbally *et al.* 2010; Lieberman 2010; Nguyen 2005). We discuss some of these strategies in the next subsection.

In the reduced basis and POD methods, the basis is formed as the span of a set of state solutions, commonly referred to as snapshots. These snapshots are computed by solving the system (7.2) or (7.5) for selected values of the parameters $\boldsymbol{\theta}$. In the POD method of snapshots (Sirovich 1987), the resulting state solutions at selected times and/or parameter values are collected in the columns of the matrix $\mathbf{U} \in \mathbb{R}^{n \times n_s}$,

$$\mathbf{U} = \left[\mathbf{u}^1 \ \mathbf{u}^2 \ \ldots \ \mathbf{u}^{n_s}\right], \tag{7.14}$$

where $\mathbf{u}^i$ is the ith snapshot and n_s is the total number of snapshots, which depends on both the number of parameter values considered and the number of timesteps sampled for each parameter value.

The POD basis is given by the left singular vectors of the matrix $\mathbf{U}$ that correspond to the largest singular values. A basis of dimension n is thus

$$\mathbf{\Phi} = \left[\boldsymbol{\varphi}^1 \ \boldsymbol{\varphi}^2 \ \ldots \ \boldsymbol{\varphi}^n\right], \tag{7.15}$$

where $\boldsymbol{\varphi}^i$ is the ith left singular vector of $\mathbf{U}$, which has corresponding singular value σ_i. It can be shown that the POD basis is optimal in the sense that, for a basis of dimension n, it minimizes the least squares error of the representation of the snapshots in the reduced basis. This error is given by the sum of the squares of the singular values corresponding to those modes not included in the basis:

$$\sum_{i=1}^{n_s} \|\mathbf{u}^i - \mathbf{\Phi}\mathbf{\Phi}^T\mathbf{u}^i\|_2^2 = \sum_{j=n+1}^{n_s} \sigma_j^2. \tag{7.16}$$

It is however important to note that this optimality of representation of the snapshot set in the general case provides no corresponding rigorous error bound on the resulting POD-based reduced-order model.

Since the POD basis is orthogonal, a common choice for the left basis is $\mathbf{\Psi} = \mathbf{\Phi}$. Other choices for $\mathbf{\Psi}$ are also possible, such as one that minimizes a least-squares weighted-residual (Bui-Thanh *et al.* 2008; Maday *et al.* 2002; Rovas 2003; Rozza and Veroy 2006), or one that includes output information through use of adjoint solutions (Lall *et al.* 2002; Willcox and Peraire 2002).

7.4.3 Computing a Basis for Inverse Problem Applications: Sampling the Parameter Space

A critical issue in computing the reduced basis is sampling of the parameter space: the quality of the resulting reduced-order model is highly dependent on the choice of parameters for which snapshots are computed. This is particularly important in the statistical inverse problem setting, since solution of the inverse problem will likely require broad exploration of the parameter space. Furthermore, problems of interest may have high-dimensional parameter spaces and the range of parameters explored in solving the inverse problem may not be known a priori.

Sampling methods to build reduced-order models must thus address two challenges. First, a systematic strategy is needed to choose where and how many samples to generate. Second, the strategy must be scalable so that parameter spaces of high dimension can be effectively sampled with a small number of large-scale system solves. Standard sampling schemes such as uniform sampling (uniform gridding of the parameter space) or random sampling are one option for creating snapshots. However, if the dimension of the parameter space is large, uniform sampling will quickly become too computationally expensive due to the combinatorial explosion of samples needed to cover the parameter space. Random sampling, on the other hand, might fail to recognize important regions in the parameter space. One can use knowledge of the application at hand to determine representative parametric inputs, as has been done to sample the parameter space for the quasi-convex optimization relaxation method (Sou *et al.* 2005), and to generate a POD or Krylov basis for problems in which the number of input parameters is small. Some examples of applications of parameterized model reduction with a small number of parameters include structural dynamics (Allen *et al.* 2004), aeroelasticity (Amsallem *et al.* 2007), Rayleigh-Bénard convection (Ly and Tran 2001), design of interconnect circuits (Bond and Daniel 2005; Daniel *et al.* 2004), and parameters describing inhomogeneous boundary conditions for parabolic PDEs (Gunzburger *et al.* 2007). For optimal control applications, online adaptive sampling has been employed as a systematic way to generate snapshot information (Afanasiev and Hinze 2001; Fahl and Sachs 2003; Hinze and Volkwein 2005; Kunisch and Volkwein 1999). However, these methods have not been scaled to problems that contain more than a handful of parameters.

To address the challenge of sampling a high-dimensional parameter space to build a reduced basis, the greedy sampling method was introduced in Grepl (2005); Grepl and Patera (2005); Veroy and Patera (2005); Veroy *et al.* (2003) to adaptively choose samples by finding the location at which the estimate of the error in the reduced model is maximum. The greedy sampling method was applied to find reduced models for the parameterized steady incompressible Navier-Stokes equations (Veroy and Patera 2005). It was also combined with a posteriori error estimators for parameterized parabolic PDEs, and applied to several optimal control and inverse problems (Grepl 2005; Grepl and Patera 2005). In Bui-Thanh *et al.* (2008), the greedy sampling approach was formulated as a sequence of adaptive model-constrained optimization problems that were solved to determine appropriate sample locations. Unlike other sampling methods, this model-constrained optimization sampling approach incorporates the underlying physics and scales well to systems with a large number of parameters. In Lieberman (2010), the optimization-based greedy sampling approach was extended to the inverse problem setting by formulating the sequence of optimization problems to also include the prior probability density. Lieberman (2010) also addressed directly the challenge of high-dimensional parameter spaces in MCMC sampling by performing reduction in both state and parameter, and demonstrated the approach on a subsurface model with a distributed parameter representing the hydraulic conductivity over the domain.

7.5 Stochastic Spectral Methods

Based on PC representations of random variables and processes (Cameron and Martin 1947; Debusschere *et al.* 2004; Ghanem and Spanos 1991; Wan and Karniadakis 2005; Wiener 1938; Xiu and Karniadakis 2002), stochastic spectral methods have been used extensively for *forward* uncertainty propagation – characterizing the probability distribution of the output of a model given a known distribution on the input. These methods exploit regularity in the dependence of an output or solution field on uncertain parameters. They constitute attractive alternatives to Monte Carlo simulation in numerous applications: transport in porous media (Ghanem 1998), structural mechanics, thermo-fluid systems (Le Maître *et al.* 2001, 2002), electrochemistry (Debusschere *et al.* 2003), and reacting flows (Najm *et al.* 2009; Reagan *et al.* 2004).

Stochastic spectral methods have more recently been applied in the *inverse* context (Marzouk *et al.* 2007), for both point and spatially distributed parameters (Marzouk and Najm 2009). Here, the essential idea is to construct a stochastic forward problem whose solution approximates the deterministic forward model over the support of the prior. This procedure – effectively propagating prior uncertainty through the forward model – yields a polynomial approximation of the forward model's dependence on uncertain parameters θ. The polynomial approximation then enters the likelihood function, resulting in a 'surrogate' posterior density that is inexpensive to evaluate, often orders of magnitude less expensive than the original posterior.

7.5.1 Surrogate Posterior Distribution

Here we describe the construction of a PC based surrogate posterior in greater detail. Let us begin with (i) a finite-dimensional representation of the unknown quantity that is the object of inference, and (ii) a prior distribution on the parameters $\boldsymbol{\theta}$ of this representation. For instance, if the unknown quantity is a field endowed with a Gaussian process prior, the finite representation may be a truncated K-L expansion with mode strengths $\boldsymbol{\theta}$ and priors $\theta_i \sim N(0, 1)$. Let $\Theta \subseteq \mathbb{R}^p$ denote the support of the prior. The Bayesian formulation in Section 7.3 describes the inverse solution in terms of the posterior density of $\boldsymbol{\theta}$, which entails evaluations of the forward model $\mathbf{G}(\boldsymbol{\theta}) : \Theta \to \mathcal{D}$, with $\mathcal{D} \subseteq \mathbb{R}^M$.

Now define a random vector $\check{\boldsymbol{\theta}} = \mathbf{c}(\check{\boldsymbol{\xi}})$, each component of which is given by a PC expansion

$$\check{\theta}_i = c_i(\check{\boldsymbol{\xi}}) = \sum_{|\mathbf{k}| \leq P} c_{i\mathbf{k}} \Psi_{\mathbf{k}}(\check{\boldsymbol{\xi}}), \qquad i = 1, \ldots, p. \tag{7.17}$$

Here $\check{\boldsymbol{\xi}}$ is a vector of p independent and identically distributed (i.i.d.) random variables, $\mathbf{k} = (k_1, \ldots, k_p)$ is a multi-index with magnitude $|\mathbf{k}| \equiv k_1 + \ldots + k_p$, and $\Psi_{\mathbf{k}}$ are multivariate polynomials (of degree k_i in coordinate $\check{\xi}_i$) orthogonal with respect to the measure on $\check{\boldsymbol{\xi}}$ (Ghanem and Spanos 1991). The total polynomial order of the PC basis is truncated at P. The vector $\check{\boldsymbol{\theta}}$ will serve as an input to $\mathbf{G}$, thus specifying a *stochastic forward problem.*

Note that the distribution of $\check{\boldsymbol{\xi}}$ (e.g. standard normal, Beta, etc.) and the corresponding polynomial form of Ψ (e.g. Hermite, Jacobi, etc.) are intrinsic properties of the PC basis (Xiu and Karniadakis 2002). In the simplest construction, PC coefficients in (7.17) are then chosen such that $\check{\boldsymbol{\theta}}$ is distributed according to the prior on $\boldsymbol{\theta}$. This is not a strict requirement, however. A necessary condition on $\mathbf{c}$ is that $\Xi_\theta = \mathbf{c}^{-1}[\Theta]$, the inverse image of the support of the prior, be contained within the range of $\check{\boldsymbol{\xi}}$. This condition ensures that there is a realization of $\check{\boldsymbol{\xi}}$ corresponding to every feasible value of $\boldsymbol{\theta}$.

Having defined the stochastic forward problem, we can solve it with a Galerkin or collocation procedure (see Section 7.5.2 below), thus obtaining a PC representation for each component of the model output. Here G_i is the i-th component of $\mathbf{G}$ and G_i^P is its Pth-order PC approximation:

$$G_i^P(\check{\boldsymbol{\xi}}) = \sum_{|\mathbf{k}| \leq P} g_{i\mathbf{k}} \Psi_{\mathbf{k}}(\check{\boldsymbol{\xi}}), \qquad i = 1, \ldots, M. \tag{7.18}$$

The forward solution $\mathbf{G}^P$ obtained in this fashion is a polynomial function of $\check{\boldsymbol{\xi}}$. Evaluating $\mathbf{G}^P$ with a deterministic argument,[1] it can be viewed simply as a polynomial approximation of $\mathbf{G} \circ \mathbf{c}$, where $\circ$ denotes composition. We will use this approximation to replace $\mathbf{G}$ in the likelihood function $L(\boldsymbol{\theta})$.

Consider the simple case of additive noise, $\mathbf{d} = \mathcal{G}(\boldsymbol{\theta}, \boldsymbol{\eta}) = \mathbf{G}(\boldsymbol{\theta}) + \boldsymbol{\eta}$, such that $L(\boldsymbol{\theta}) = \pi_\eta(\mathbf{d} - \mathbf{G}(\boldsymbol{\theta}))$, with π_η being the probability density of $\boldsymbol{\eta}$. The likelihood function can be rewritten as a function of $\boldsymbol{\xi}$:

$$L(\mathbf{c}(\boldsymbol{\xi})) = \pi_\eta(\mathbf{d} - \mathbf{G}(\mathbf{c}(\boldsymbol{\xi}))) \approx \pi_\eta\left(\mathbf{d} - \mathbf{G}^P(\boldsymbol{\xi})\right). \tag{7.19}$$

This change of variables from $\boldsymbol{\theta}$ to $\boldsymbol{\xi}$ lets us define a posterior density for $\boldsymbol{\xi}$:

$$\pi_\xi(\boldsymbol{\xi}) \propto L(\mathbf{c}(\boldsymbol{\xi}))\, \pi_{\mathrm{pr}}(\mathbf{c}(\boldsymbol{\xi})) \det D\mathbf{c}(\boldsymbol{\xi}). \tag{7.20}$$

[1] In this exposition we have used ˇ to identify the random variables $\check{\boldsymbol{\theta}}$ and $\check{\boldsymbol{\xi}}$ in order to avoid confusion with deterministic arguments to probability density functions, e.g., $\boldsymbol{\theta}$ and $\boldsymbol{\xi}$. Elsewhere, we revert to the usual notational convention and let context make clear the distinction between the two.

In this expression, $D\mathbf{c}$ is the Jacobian of $\mathbf{c}$, det denotes the determinant, and π_{pr} is the prior density of $\boldsymbol{\theta}$. The last two factors on the right side, $\pi_{\text{pr}}(\mathbf{c}(\boldsymbol{\xi})) \det D\mathbf{c}(\boldsymbol{\xi})$, are the probability density on $\boldsymbol{\xi}$ that corresponds to the prior on θ. Replacing the forward model in the likelihood function via (7.19) then yields the *surrogate* posterior density π_ξ^P:

$$\pi_\xi(\boldsymbol{\xi}) \approx \pi_\xi^P(\boldsymbol{\xi}) \propto \pi_\eta\left(\mathbf{d} - \mathbf{G}^P(\boldsymbol{\xi})\right) \pi_{\text{pr}}(\mathbf{c}(\boldsymbol{\xi})) \left|\det D\mathbf{c}(\boldsymbol{\xi})\right|. \tag{7.21}$$

Despite the change of variables, it is straightforward to recover the posterior expectation of an arbitrary function f:

$$\mathbb{E}_{\pi_{\text{post}}} f = \mathbb{E}_{\pi_\xi}(f \circ \mathbf{c}) \tag{7.22}$$

where $\pi_{\text{post}} \equiv \pi(\boldsymbol{\theta}|\mathbf{d})$ is the posterior density on Θ, and π_ξ is the corresponding posterior density of $\boldsymbol{\xi}$.

The surrogate posterior distribution may be explored with any suitable sampling strategy, in particular MCMC. Evaluating the density for purposes of sampling may have negligible cost; nearly all the computational time may be spent in solving the stochastic forward problem, i.e. obtaining the PC expansions in (7.18). Depending on model nonlinearities, the necessary size of the PC basis, and the number of posterior samples required, this computational effort may be orders of magnitude less costly than exploring the posterior via direct sampling. Moreover, as it requires only the forward model and the prior, the stochastic forward solution may be obtained 'offline,' independently of the data. Accuracy of the surrogate posterior depends on the order and family of the PC basis, as well as on the choice of transformation $\mathbf{c}$ – for instance, whether the distribution of $\check{\boldsymbol{\theta}}$ assigns sufficient probability to regions of Θ favored by the posterior. A detailed discussion of these issues can be found in Marzouk *et al.* (2007). Some convergence results are summarized below.

7.5.2 Forward Solution Methodologies and Convergence Results

Solution of the stochastic forward problem (7.18) is an essential step in the inversion procedure outlined above. While a survey of PC methods for solving ODEs and PDEs with random inputs is beyond the scope of this chapter (see for instance Najm (2009); Xiu (2009)), we highlight two broad classes of approaches. Stochastic Galerkin methods (Ghanem and Spanos 1991; Le Maître *et al.* 2001; Matthies and Keese 2005) involve a reformulation of the governing equations, essentially creating a larger system of equations for the PC coefficients $g_{i\mathbf{k}}$; these equations are generally coupled, though one may take advantage of problem-specific structure in devising efficient solution schemes (Xiu and Shen 2009). Stochastic collocation methods (Babuška *et al.* 2007; Xiu and Hesthaven 2005), on the other hand, are 'non-intrusive';

these require only a finite number of uncoupled deterministic simulations, with no reformulation of the governing equations of the forward model. Collocation methods using sparse grids (Bieri *et al.* 2009; Ganapathysubramanian and Zabaras 2007; Ma and Zabaras 2009; Nobile *et al.* 2008; Smolyak 1963; Xiu and Hesthaven 2005), offer great efficiency and ease of implementation for higher-dimensional problems.

For systems with more complex dynamics – discontinuities or bifurcations with respect to uncertain parameters, or even limit cycles (Beran *et al.* 2006) – global bases may be unsuitable. Instead, piecewise polynomial (Wan and Karniadakis 2005, 2009) or multi-wavelet (Le Maître *et al.* 2004) generalizations of PC enable efficient propagation of uncertainty; such bases can also be used to construct surrogate posteriors. Indeed, the overall Bayesian inference scheme is quite flexible with regard to *how* one chooses to solve a stochastic forward problem. Error analysis of the Bayesian stochastic spectral framework (Marzouk and Xiu 2009) has reinforced this flexibility. The relevant convergence results can be summarized as follows. Consider the mean-square error, with respect to π_{pr}, in the forward solution: $e(P) \equiv \|\mathbf{G}^P(\boldsymbol{\xi}) - \mathbf{G}(\boldsymbol{\xi})\|_{L^2_{\pi_{\text{pr}}}}$. Suppose that observational errors η_i are additive and i.i.d. Gaussian. If $e(P)$ converges at a particular rate, $e(P) \leq CP^{-\alpha}$, then at sufficiently large P, the Kullback-Leibler (KL) divergence of the true posterior from the surrogate posterior maintains at least the same rate of convergence, $D\left(\pi_\xi^P \| \pi_\xi\right) \lesssim P^{-\alpha}$. In particular, exponential convergence of the forward solution implies exponential convergence of the surrogate posterior to the true posterior. (Recall that the Kullback-Leibler divergence quantifies the difference between probability distributions in information theoretic terms (Gibbs and Su 2002).) These results provide a guideline relevant to *any* approximation of the forward model.

7.6 Illustrative Example

We explore the relative advantages and disadvantages of model reduction and stochastic spectral approaches in the context of a simple transient source inversion problem. Consider a dimensionless diffusion equation on a square domain $\Omega = [0, 1] \times [0, 1]$ with adiabatic boundaries:

$$\frac{\partial u}{\partial t} = \nabla^2 u + \frac{s}{2\pi\gamma^2} \exp\left(-\frac{|\boldsymbol{\theta} - \mathbf{x}|^2}{2\gamma^2}\right) [1 - H(t - T)], \tag{7.23}$$

$$\nabla u \cdot \mathbf{n} = 0 \quad \text{on } \partial\Omega, \tag{7.24}$$

$$u(\mathbf{x}, 0) = 0 \quad \text{in } \Omega. \tag{7.25}$$

The solution field $u(\mathbf{x}, t)$ can represent temperature or the concentration of some contaminant species, with $\mathbf{x} \equiv (x_1, x_2) \in \Omega$ and time $t \geq 0$. $H(t)$ denotes the unit step function. Thus, the source term in (7.23) comprises a single localized source, active on the interval $t \in \mathcal{T} = [0, T]$ and centered at location $\boldsymbol{\theta} \in \Theta = \Omega$ with strength s and characteristic width γ.

The governing equations (7.23)–(7.25) are discretized on a uniform spatial grid using a second-order-accurate finite difference scheme. This spatial discretization and the application of the boundary conditions lead to a semi-discrete system of the form (7.5)–(7.6), linear in the state but nonlinear in the parameters $\boldsymbol{\theta}$. The state vector $\mathbf{u}(t)$ contains $u(\mathbf{x}, t)$ evaluated at the N grid points; the sparse matrix $A \in \mathbb{R}^{N \times N}$ reflects the spatial discretization and application of the Neumann boundary conditions; and $\mathbf{g}(\boldsymbol{\theta})$ is a nonlinear function representing the source term in Eq. (7.23). Note that $\mathbf{g}$ is here just a function of the parameters (the source location) and not the state.

In the inverse problem, we are given noisy observations of the solution $u(\mathbf{x}, t)$ at a few locations in space and a few instants in time. From these data, we wish to infer the source location $\boldsymbol{\theta} = (\theta_1, \theta_2)$. For simplicity, we assume that the shutoff time T, strength s, and source width γ are known. We assume that observations of u are available for $n_t = 3$ time instants, $t \in \{0.1, 0.2, 0.3\}$, at $q = 9$ locations on a uniform 3×3 grid covering the domain Ω. The forward model $\mathbf{G}(\boldsymbol{\theta})$ is thus a map from the source location $\boldsymbol{\theta}$ to noise-free observations $\mathbf{d}_0 \in \mathbb{R}^{qn_t}$. These observations are perturbed with additive Gaussian noise $\boldsymbol{\eta}$ to yield the data vector $\mathbf{d} = \mathbf{d}_0 + \boldsymbol{\eta}$. Components of $\boldsymbol{\eta}$ are i.i.d., $\boldsymbol{\eta} \sim N\left(0, \sigma^2 I\right)$. The likelihood function is therefore given by $L(\boldsymbol{\theta}) = \pi_\eta\left(\mathbf{d} - \mathbf{G}(\boldsymbol{\theta})\right)$. The prior on $\boldsymbol{\theta}$ reflects a uniform probability assignment over the entire domain of possible source locations, $\theta_i \sim U(0,1)$. The posterior density is then

$$\begin{aligned} \pi(\boldsymbol{\theta}|\mathbf{d}) &\propto \pi_\eta\left(\mathbf{d} - \mathbf{G}(\boldsymbol{\theta})\right) \mathbf{1}_\Omega(\boldsymbol{\theta}) \\ &= \begin{cases} \exp\left(-\frac{1}{2\sigma^2}\left(\mathbf{d} - \mathbf{G}(\boldsymbol{\theta})\right)^T\left(\mathbf{d} - \mathbf{G}(\boldsymbol{\theta})\right)\right) & \text{if } \boldsymbol{\theta} \in \Omega, \\ 0 & \text{otherwise.} \end{cases} \end{aligned} \tag{7.26}$$

Figure 7.1 shows an example forward solution, obtained on a 69×69 uniform grid. The plots show the solution field $u(\mathbf{x}, t)$ before and after the source shutoff time of $T = 0.2$. The source is located at $\boldsymbol{\theta} = (0.6, 0.9)$, with strength $s = 2$ and width $\gamma = 0.05$. The solution field at the earlier time is peaked around the source location and contains useful information for the inverse problem. After the shutoff time, however, the field tends to flatten out due to diffusion. Eventually observations of the u-field at times well after the shutoff will provide no useful information for inference of the source location.

We now consider solution of the inverse problem using two approximation approaches – a POD-based reduced order model and a PC surrogate obtained with pseudospectral stochastic collocation. Constructing the approximate posterior distribution in *either* case requires evaluating the forward model at a number of parameter values $\{\boldsymbol{\theta}_1, \ldots, \boldsymbol{\theta}_Q\}$. With POD, these forward model evaluations are used to construct the snapshot matrix, while for

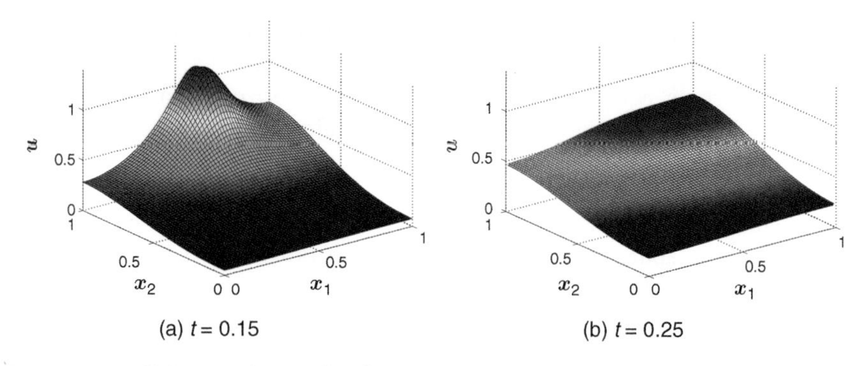

Figure 7.1 Solution field $u(\mathbf{x}, t)$ of the full forward model at two different times for source parameters $\boldsymbol{\theta} = (0.6, 0.9)$, $T = 0.20$, $s = 2$, and $\gamma = 0.05$.

stochastic collocation they are used to evaluate integrals for the PC coefficients:

$$g_{i\mathbf{k}} \langle \Psi_{\mathbf{k}}^2 \rangle = \int_{\Theta} \mathbf{G}_i (\boldsymbol{\theta}) \, \Psi_{\mathbf{k}} (\boldsymbol{\theta}) \, \pi_{\mathrm{pr}} (\boldsymbol{\theta}) \, d\boldsymbol{\theta} \tag{7.27}$$

$$\approx \sum_{j=1}^{Q} \mathbf{G}_i (\boldsymbol{\theta}_j) \, \Psi_{\mathbf{k}} (\boldsymbol{\theta}_j) \, w_j. \tag{7.28}$$

Since the prior π_{pr} is uniform on Θ, the polynomials $\Psi_{\mathbf{k}}$ are taken to be bivariate Legendre polynomials, while the nodes and weights $\{\boldsymbol{\theta}_j, w_j\}_{j=1}^{Q}$ are chosen according to a Gaussian quadrature rule. In particular, we use a tensor product of l-point Gauss-Legendre rules on [0,1], such that $Q = l^2$. In the comparison below, we use the same nodal set of Q parameter values to construct the snapshot matrix (7.14). This is certainly not the only choice (or even the best choice) of parameter values to employ for POD. Our selection is motivated mostly by simplicity, so that identical forward simulations support both approximation techniques. We revisit this choice in later remarks.

We first evaluate the accuracy of the different forward models as a function of the 'order' of approximation. The L^2 error of an approximate forward model $\widetilde{\mathbf{G}}$ is defined as

$$e = \int_{\Theta} \|\mathbf{G}(\boldsymbol{\theta}) - \widetilde{\mathbf{G}}(\boldsymbol{\theta})\|_2 \pi_{\mathrm{pr}}(\boldsymbol{\theta}) d\boldsymbol{\theta}. \tag{7.29}$$

In other words, this is the prior-weighted error in model predictions integrated over the parameter space. The precise meaning of 'order' depends on context, of course. With stochastic collocation, we take order to be the maximal polynomial degree P. In the projection approach, order is the dimension of the reduced model, i.e. the number of POD basis vectors n retained. Figure 7.2

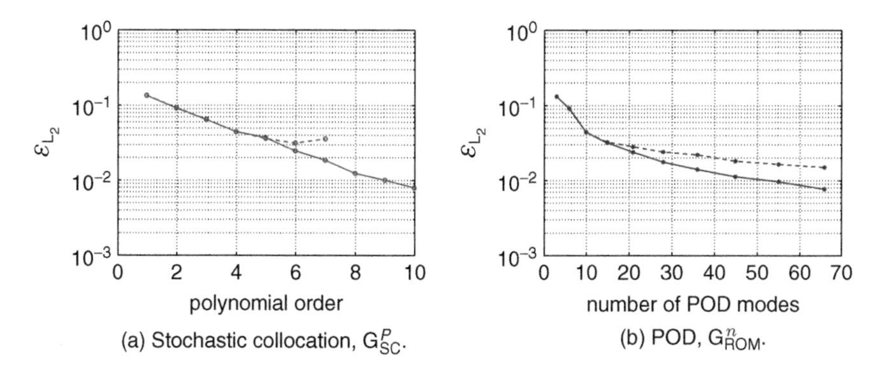

Figure 7.2 L^2 error in the approximate forward models, versus polynomial degree P for stochastic collocation and versus the number of modes n for POD. Dotted line represents $Q = 36$.

shows the L^2 error for both methods of approximation. In Figure 7.2(a), the error decreases more or less exponentially with polynomial degree, provided a sufficiently high order quadrature rule ($Q = 100$) is applied. Results with a 6-point quadrature rule in each direction ($Q = 36$) diverge for $P > 5$; this is understandable, as the degree of the integrand in (7.27) increases with $|\mathbf{k}|$ and aliasing errors corrupt the higher-degree polynomial coefficients. In Figure 7.2(b), error decreases with the number of POD modes n, but the accuracy of the reduced model at larger n depends on Q. For smaller Q – i.e., less dense coverage of the parameter space – the error begins to plateau at a larger value than for $Q = 100$. These results suggest that sufficiently large values of both Q and n are needed for an accurate reduced-order model.

Turning from the forward model approximation to the posterior distribution, Figure 7.3 shows the posterior density of the source location, $\pi(\boldsymbol{\theta}|\mathbf{d})$ (7.26), for various forward model approximations. The data $\mathbf{d}$ reflect the same source parameters used in Figure 7.1, i.e. $\boldsymbol{\theta} = (0.6, 0.9)$. Observations of the exact solution field are perturbed with Gaussian noise $\boldsymbol{\eta} \sim N\left(0, \sigma^2 I\right)$, with $\sigma = 0.2$. The noise magnitude is therefore roughly 20–40 % of the nominal values of $\mathbf{d}_0$. Figure 7.3(a) shows the baseline case: contours of the exact posterior density, obtained via evaluations of the full forward model $\mathbf{G}$. Plots (b) and (c) show contours of the approximate posterior density π^P_{SC} obtained by evaluation of the stochastic collocation model $\mathbf{G}^P_{\mathrm{SC}}$, at polynomial orders $P = 2$ and $P = 10$, respectively. Plots (d) and (e) show the contours of the approximate posterior density π^n_{ROM} obtained with POD models $\mathbf{G}^n_{\mathrm{ROM}}$ of dimension $n = 6$ and $n = 66$, respectively. Both the stochastic collocation and POD models were constructed with $Q = 100$ forward evaluations, at parameter values $\boldsymbol{\theta}$ chosen with a 10-point quadrature rule in each direction. At

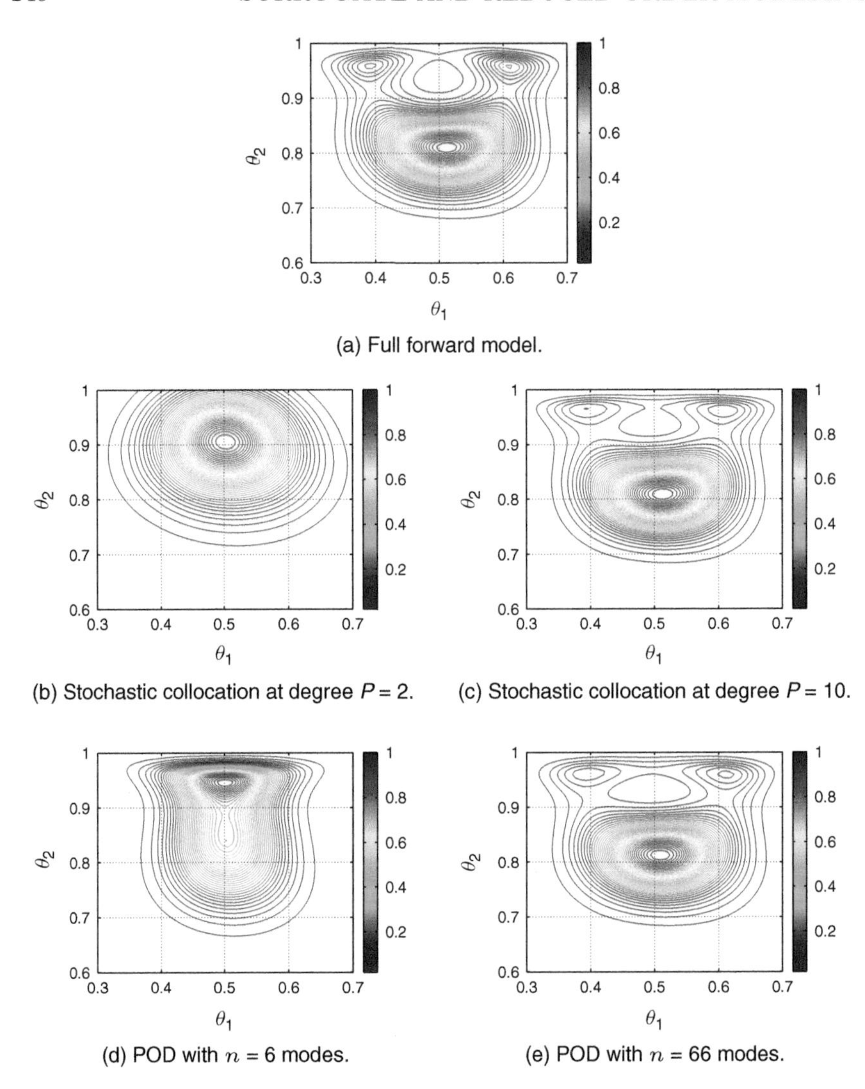

(a) Full forward model.

(b) Stochastic collocation at degree $P = 2$.

(c) Stochastic collocation at degree $P = 10$.

(d) POD with $n = 6$ modes.

(e) POD with $n = 66$ modes.

Figure 7.3 Contours of the posterior density $\pi(\boldsymbol{\theta}|\mathbf{d})$ using the full forward model and various approximations. Contour lines are plotted at 40 equally spaced contour levels in the range [0.02,1].

sufficiently high P or n, both types of models yield close agreement with the true posterior density. Note that the true posterior is multi-modal; all three modes are well captured by the surrogate posterior densities in Figure 7.3(d)–(e).

A more quantitative measurement of posterior error is the Kullback-Leibler (KL) divergence from the true posterior to the approximate posterior.

Letting $\widetilde{\pi}$ denote the approximate posterior density, the KL divergence of $\widetilde{\pi}(\boldsymbol{\theta})$ from $\pi(\boldsymbol{\theta})$ is:

$$D(\pi\|\widetilde{\pi}) = \int_{\Theta} \pi(\boldsymbol{\theta}) \log \frac{\pi(\boldsymbol{\theta})}{\widetilde{\pi}(\boldsymbol{\theta})} \mathrm{d}\boldsymbol{\theta}. \tag{7.30}$$

Figure 7.4 shows $D(\pi\|\widetilde{\pi})$ versus the order of approximation for the POD and stochastic collocation approaches. The true value of $\boldsymbol{\theta}$ and all of the other source parameters are identical to Figure 7.3; the same data vector $\mathbf{d}$ is used throughout. We contrast POD and collocation models constructed using either $Q = 36$ and $Q = 100$ nodes in the parameter space Θ. The integral in (7.30) was evaluated with the trapezoidal rule, using a uniform grid of dimension 69×69 on the set $\Theta' = [0.3, 0.7] \times [0.6, 1] \subset \Theta$. (Note that

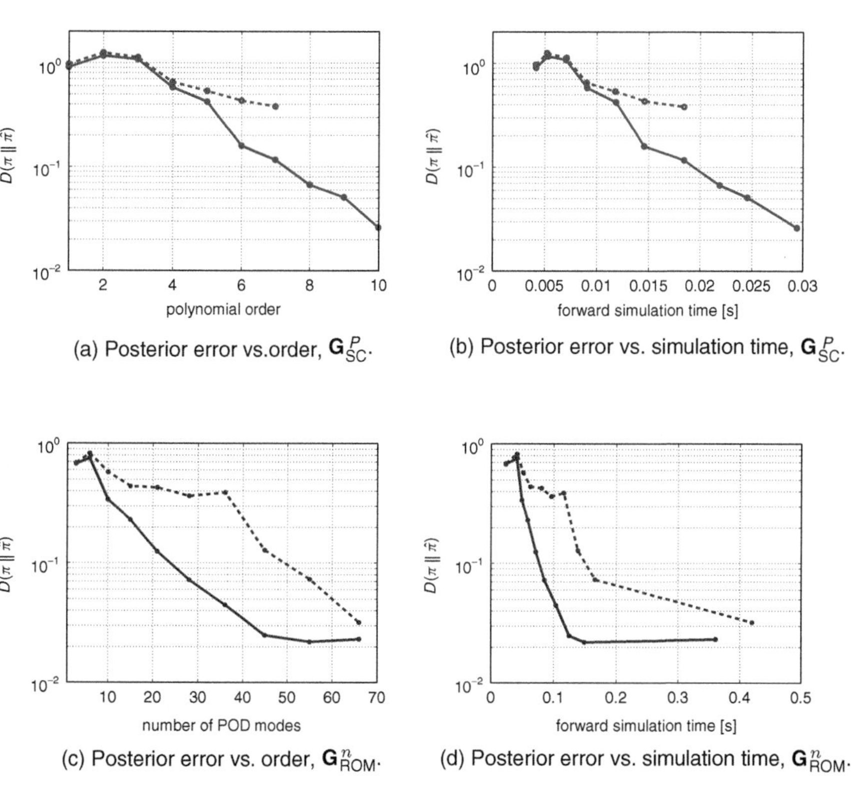

(a) Posterior error vs.order, $\mathbf{G}_{SC}^{P}$.

(b) Posterior error vs. simulation time, $\mathbf{G}_{SC}^{P}$.

(c) Posterior error vs. order, $\mathbf{G}_{\mathrm{ROM}}^{n}$.

(d) Posterior error vs. simulation time, $\mathbf{G}_{\mathrm{ROM}}^{n}$.

Figure 7.4 Kullback-Leibler divergence from the exact posterior to the approximate posterior. Plots (a) and (b) show the results for stochastic collocation models of different polynomial degree. Plots (c) and (d) show the results for POD models of different order. In both cases, models are constructed with forward simulations at either $Q = 36$ (dotted) or $Q = 100$ (solid) nodes in parameter space.

ignoring the remaining area of the parameter domain contributes negligible error, since the posterior density in these regions is nearly zero, as can be seen in Figure 7.3.)

Figures 7.4(a) and 7.4(c) show that, provided the value of Q is sufficiently high, both approaches achieve similar posterior accuracy. As in Figure 7.2, the accuracy of the PC-based surrogate posterior degrades due to aliasing errors when Q is too small; Q effectively limits the maximum polynomial degree that is meaningful to employ. When Q is larger, however, rapid convergence of $D(\pi\|\widetilde{\pi})$ with respect to polynomial degree is observed. Error in the POD-based surrogate posterior also decays rapidly with increasing model order in the $Q = 100$ case. When fewer $\boldsymbol{\theta}$ values are used to train the POD model ($Q = 36$), errors tend to be larger. Interestingly, though, these errors start to decay anew for $n > 40$. With $n = 66$ modes, the errors associated with the $Q = 36$ and $Q = 100$ surrogate posteriors π^n_{ROM} differ by less than a factor of two.

Computational speedups over the full model, which takes approximately six seconds per forward simulation, are significant in both cases. But at similar levels of accuracy, the POD models are several times more costly in forward simulation time than the stochastic collocation models. Indeed, in the present case it is faster to evaluate a polynomial expansion (7.18) than to integrate a reduced-order model (7.11)–(7.13). On the other hand, if Q is small, then the stochastic collocation approach cannot achieve high accuracy, but error in the POD-based posterior continues to decline as more modes are added. For the current problem, therefore, if only a limited number of full model evaluations can be performed (offline), the POD approach is a better choice; while if Q can be chosen sufficiently high, the stochastic collocation approach yields equivalent accuracy with much smaller online computational cost.

The computational times reported here and shown in Figure 7.4 were obtained on a desktop PC with an Intel Core 2 Duo processor at 3.16 GHz and 4 GB of RAM.

7.7 Conclusions

The simple numerical example above suggests that, while model reduction may require fewer offline forward simulations to achieve a certain accuracy, PC-based surrogate posteriors may be significantly cheaper to evaluate in the online phase. These conclusions are necessarily quite problem-specific, however. Our illustrative example was strongly nonlinear in the parameters, but these parameters were limited to two dimensions. The governing equations were also linear in the state $\mathbf{u}$. It is important to consider how the relative computational cost of these methods scales with dimensionality of the parameter space and with the dynamical complexity of the forward model, among other factors.

These questions underscore many challenges and open problems in surrogate and reduced-order modeling for statistical inverse problems. Techniques are needed to rigorously inform construction of the surrogate model with components of the inverse formulation, specifically incorporating both prior information and data. While some success has been demonstrated in this regard for simple problems, challenges remain in incorporating prior models for more complex fields (e.g. with discontinuities or other geometric structure) and in conditioning on data collected at multiple scales. High dimensionality also raises several open issues; successful surrogate modeling in this context should exploit the spectrum of the forward operator and any smoothing or structure provided by the prior in order to reduce the number of input parameters. Rigorous error bounds on the posterior computed using a surrogate or reduced-order model remain another outstanding challenge. Without some way of estimating the effects of using a surrogate in place of the full model, we cannot quantitatively answer questions such as how many samples are required to compute the basis.

Answers to these questions again may depend on the details of the algorithmic approaches. In the previous example, we focused on simplicity and did not apply more sophisticated approaches for either model reduction or PC approximation. With model reduction, for example, we did not employ greedy sampling in the parameter space; with stochastic collocation, we did not employ sparse grids (particularly anisotropic and adaptive sparse grids for high-dimensional problems), nor did we explore partitioning of the prior support and/or alternate polynomial bases. In tackling more complicated problems, one should certainly draw from the literature in all of these areas in developing forward model approximation schemes suited to inverse problems.

Beyond reducing the complexity of forward simulations, what is needed even more are *combinations* of the approaches discussed in Section 7.2 – simultaneously approximating the forward model, reducing the size of the input space, and reducing the requisite number of samples. Many of these methods will be described in the ensuing chapters.

Finally, we close by noting that this volume focuses on solution of the inverse problem – success is measured by our ability to accurately estimate some parameter of interest. However, in many scientific and engineering applications, the inverse problem is merely one step on the path to solving an ultimate design or control problem; that is, the inference task is often followed by a decision problem downstream. Potentially significant opportunities exist for an integrated consideration of inference and decision problems. For example, the downstream decision problem can inform solution of the inverse problem by defining the level of accuracy to which specific features of the parameter should be resolved. The decision problem might also guide the construction of more parsimonious surrogate or reduced-order models.

Acknowledgments

This material is based upon work supported in part by the Department of Energy under Award Number DE-FG02-08ER25858 and Award Number DE-SC00025217. The first and third authors also acknowledge support of the Singapore-MIT Alliance Computational Engineering Programme. The second and fourth authors also acknowledge support from the Laboratory Directed Research and Development (LDRD) program at Sandia National Laboratories. Sandia is a multiprogram laboratory operated by Sandia Corporation, a Lockheed-Martin Company, for the United States Department of Energy under Contract DE-AC04-94AL85000.

This report was prepared as an account of work sponsored in part by an agency of the United States Government. Neither the United States Government nor any agency thereof, nor any of their employees, makes any warranty, express or implied, or assumes any legal liability or responsibility for the accuracy, completeness, or usefulness of any information, apparatus, product, or process disclosed, or represents that its use would not infringe privately owned rights. Reference herein to any specific commercial product, process, or service by trade name, trademark, manufacturer, or otherwise does not necessarily constitute or imply its endorsement, recommendation, or favoring by the United States Government or any agency thereof. The views and opinions of authors expressed herein do no necessarily state or reflect those of the United States Government or any agency thereof.

References

Afanasiev, K. and Hinze, M. 2001 Adaptive control of a wake flow using proper orthogonal decomposition. *Lecture Notes in Pure and Applied Mathematics*. vol. 216 Marcel Dekker. pp. 317–332.

Allen, M., Weickum, G. and Maute, K. 2004 Application of reduced order models for the stochastic design optimization of dynamic systems. *10th AIAA/ISSMO Multidisciplinary Analysis and Optimization Conference*, pp. 1–19, Albany, NY, USA. AIAA 2004–4614.

Amsallem, D., Farhat, C. and Lieu, T. 2007 Aeroelastic analysis of F-16 and F-18/A configurations using adapted CFD-based reduced-order models. *48th AIAA/ASME/ASCE/AHS/ASC Structures, Structural Dynamics, and Materials Conference*, vol. AIAA 2007-2364, pp. 1–20, Honolulu, Hawaii.

Arridge, S., Kaipio, J., Kolehmainen, V., *et al.* 2006 Approximation errors and model reduction with an application in optical diffusion tomography.. *Inverse Problems*. **22**, 175–195.

Astrid, P., Weiland, S., Willcox, K. and Backx, T. 2008 Missing point estimation in models described by proper orthogonal decomposition. *IEEE Transactions on Automatic Control, to appear*.

Babuška, I., Nobile, F. and Tempone, R. 2007 A stochastic collocation method for elliptic partial differential equations with random input data. *SIAM Journal on Numerical Analysis*. **45**(3), 1005–1034.

Balakrishnan, S., Roy, A., Ierapetritou, M.G., Flach, G.P. and Georgopoulos, P.G. 2003 Uncertainty reduction and characterization for complex environmental fate and transport models: an empirical Bayesian framework incorporating the stochastic response surface method. *Water Resources Research*. **39**(12), 1350.

Barrault, M., Maday, Y., Nguyen, N. and Patera, A. 2004 An 'empirical interpolation' method: Application to efficient reduced-basis discretization of partial differential equations. *CR Acad Sci Paris Series*. **I**, 339–667.

Beran, P.S., Pettit, C.L. and Millman, D.R. 2006 Uncertainty quantification of limit-cycle oscillations. *J. Comput. Phys*. **217**(1), 217–47.

Bieri, M., Andreev, R. and Schwab, C. 2009 Sparse tensor discretizations of elliptic sPDEs. Technical report, ETH. Research Report No. 2009–07.

Bliznyuk, N., Ruppert, D., Shoemaker, C., Regis, R., Wild, S. and Mugunthan, P. 2008 Bayesian calibration of computationally expensive models using optimization and radial basis function approximation. *Journal of Computational and Graphical Statistics*. **17**(2), 1–25.

Bond, B. and Daniel, L. 2005 Parameterized model order reduction of nonlinear dynamical systems. *Proceedings of the IEEE Conference on Computer-Aided Design*. IEEE, San Jose, CA.

Bui-Thanh, T., Willcox, K. and Ghattas, O. 2008 Model reduction for large-scale systems with high-dimensional parametric input space. *SIAM Journal on Scientific Computing*. **30**(6), 3270–3288.

Cameron, R. and Martin, W. 1947 The orthogonal development of nonlinear functionals in series of Fourier-Hermite functionals. *Annals of Mathematics*. **48**, 385–392.

Christen, J.A. and Fox, C. 2005 MCMC using an approximation. *Journal of Computational and Graphical Statistics*. **14**(4), 795–810.

Daniel, L., Siong, O., Chay, L., Lee, K. and White, J. 2004 Multiparameter moment matching model reduction approach for generating geometrically parameterized interconnect performance models. *Transactions on Computer Aided Design of Integrated Circuits*. **23**(5), 678–693.

Deane, A., Kevrekidis, I., Karniadakis, G. and Orszag, S. 1991 Low-dimensional models for complex geometry flows: Application to grooved channels and circular cylinders. *Phys. Fluids*. **3**(10), 2337–2354.

Debusschere, B., Najm, H., Matta, A., Knio, O., Ghanem, R. and Le Maître O. 2003 Protein labeling reactions in electrochemical microchannel flow: Numerical simulation and uncertainty propagation. *Physics of Fluids*. **15**(8), 2238–2250.

Debusschere, B., Najm, H., Pébay, P., Knio, O., Ghanem, R. and Le Maître O. 2004 Numerical challenges in the use of polynomial chaos representations for stochastic processes. *SIAM J. Sci. Comp*. **26**(2), 698–719.

Efendiev, Y., Hou, T.Y. and Luo, W. 2006 Preconditioning Markov chain Monte Carlo simulations using coarse-scale models. *SIAM Journal on Scientific Computing*. **28**, 776–803.

Eldred, M., Giunta, S. and Collis, S. 2004 Second-order corrections for surrogate-based optimization with model hierarchies AIAA Paper 2004-4457, in *Proceedings of the 10th AIAA/ISSMO Multidisciplinary Analysis and Optimization Conference*.

Fahl, M. and Sachs, E. 2003 Reduced order modelling approaches to PDE – constrained optimization based on proper orthogonal decompostion. In *Large-Scale PDE-Constrained Optimization* (ed. Biegler, L., Ghattas, O., Heinkenschloss, M. and van Bloemen Waanders B.) Lecture Notes in Computational Science and Engineering, vol. 30. Springer-Verlag, Heidelberg.

Feldmann, P. and Freund, R. 1995 Efficient Linear Circuit Analysis by Padé Approximation via the Lanczos Process. *IEEE Transactions on Computer-Aided Design of Integrated Circuits and Systems*. **14**, 639–649.

Fox, R.L. and Miura, H. 1971 An approximate analysis technique for design calculations. *AIAA Journal*. **9**(1), 177–179.

Galbally, D., Fidkowski, K., Willcox, K. and Ghattas, O. 2010 Nonlinear model reduction for uncertainty quantification in large-scale inverse problems. *International Journal for Numerical Methods in Engineering*. **81**(12), 1581–1608.

Gallivan, K., Grimme, E. and Van Dooren, P. 1994 Padé approximation of large-scale dynamic systems with Lanczos methods. *Proceedings of the 33rd IEEE Conference on Decision and Control*.

Ganapathysubramanian, B. and Zabaras, N. 2007 Sparse grid collocation schemes for stochastic natural convection problems. *J. Comput. Phys*. **225**(1), 652–685.

Geyer, C.J. 1991 Markov chain Monte Carlo maximum likelihood. In *Computing Science and Statistics: Proceedings of the 23rd Symposium on the Interface* (ed. Keramidas, E.M.), vol. 23, pp. 156–163. Interface Foundation of North America.

Ghanem, R. 1998 Probabilistic characterization of transport in heterogeneous media. *Comput. Methods Appl. Mech. Engrg*. **158**, 199–220.

Ghanem, R. and Spanos, P. 1991 *Stochastic Finite Elements: A Spectral Approach*. Springer Verlag, New York.

Gibbs, A.L. and Su, F.E. 2002 On choosing and bounding probability metrics. *International Statistical Review*. **70**(3), 419–435.

Grepl, M. 2005 *Reduced-Basis Approximation and A Posteriori Error Estimation for Parabolic Partial Differential Equations*. PhD thesis MIT Cambridge, MA.

Grepl, M. and Patera, A. 2005 *A posteriori* error bounds for reduced-basis approximations of parametrized parabolic partial differential equations. *ESAIM-Mathematical Modelling and Numerical Analysis (M2AN)*. **39**(1), 157–181.

Grepl, M., Maday, Y., Nguyen, N. and Patera, A. 2007 Efficient reduced-basis treatment of nonaffine and nonlinear partial differential equations. *Mathematical Modelling and Numerical Analysis (M2AN)*. **41**(3), 575–605.

Grimme, E. 1997 *Krylov Projection Methods for Model Reduction*. PhD thesis. Coordinated-Science Laboratory, University of Illinois at Urbana-Champaign.

Gugercin, S. and Antoulas, A. 2004 A survey of model reduction by balanced truncation and some new results. *International Journal of Control*. **77**, 748–766.

Gunzburger, M., Peterson, J. and Shadid, J. 2007 Reduced-order modeling of time-dependent PDEs with multiple parameters in the boundary data. *Computer Methods in Applied Mechanics and Engineering*. **196**, 1030–1047.

Higdon, D., Lee, H. and Holloman, C. 2003 Markov chain Monte Carlo-based approaches for inference in computationally intensive inverse problems. *Bayesian Statistics*. **7**, 181–197.

Hinze, M. and Volkwein, S. 2005 Proper orthogonal decomposition surrogate models for nonlinear dynamical systems: Error estimates and suboptimal control. In *Dimension Reduction of Large-Scale Systems* (ed. Benner, P., Mehrmann, V. and Sorensen, D.), pp. 261–306 Lecture Notes in Computational and Applied Mathematics.

Holmes, P., Lumley, J. and Berkooz, G. 1996 *Turbulence, Coherent Structures, Dynamical Systems and Symmetry*. Cambridge University Press, Cambridge, UK.

Kennedy, M.C. and O'Hagan A. 2001 Bayesian calibration of computer models. *Journal of the Royal Statistical Society: Series B*. **63**(3), 425–464.

Kunisch, K. and Volkwein, S. 1999 Control of Burgers' equation by reduced order approach using proper orthogonal decomposition. *Journal of Optimization Theory and Applications*. **102**, 345–371.

Lall, S., Marsden, J. and Glavaski, S. 2002 A subspace approach to balanced truncation for model reduction of nonlinear control systems. *International Journal on Robust and Nonlinear Control*. **12**(5), 519–535.

Le Maître O., Knio, O., Najm, H. and Ghanem, R. 2001 A stochastic projection method for fluid flow I. Basic formulation. *J. Comput. Phys*. **173**, 481–511.

Le Maître O., Najm, H., Ghanem, R. and Knio, O. 2004 Multi-resolution analysis of Wiener-type uncertainty propagation schemes. *J. Comput. Phys*. **197**, 502–531.

Le Maître O., Reagan, M., Najm, H., Ghanem, R. and Knio, O. 2002 A stochastic projection method for fluid flow II. Random process. *J. Comput. Phys*. **181**, 9–44.

Li, J. and White, J. 2002 Low rank solution of Lyapunov equations. *SIAM Journal on Matrix Analysis and Applications*. **24**(1), 260–280.

Li, W. and Cirpka, O.A. 2006 Efficient geostatistical inverse methods for structured and unstructured grids. *Water Resources Research*. **42**, W06402.

Lieberman, C., Willcox, K. and Ghattas, O. 2010 Parameter and state model reduction for large-scale statistical inverse problems. *SIAM Journal on Scientific Computing*, forthcoming.

Ly, H. and Tran, H. 2001 Modeling and control of physical processes using proper orthogonal decomposition. *Mathematical and Computer Modeling*. **33**(1–3), 223–236.

Ma, X. and Zabaras, N. 2009 An adaptive hierarchical sparse grid collocation algorithm for the solution of stochastic differential equations. *J. Comp. Phys*. **228**, 3084–3113.

Maday, Y., Patera, A. and Rovas, D. 2002 A blackbox reduced-basis output bound method for noncoercive linear problems. In *Studies in Mathematics and its Applications* (ed. Cioranescu, D. and Lions, J.), vol. 31.

Marzouk, Y.M. and Najm, H.N. 2009 Dimensionality reduction and polynomial chaos acceleration of Bayesian inference in inverse problems. *J. Comp. Phys.* **228**(6), 1862–1902.

Marzouk, Y.M. and Xiu, D. 2009 A stochastic collocation approach to Bayesian inference in inverse problems. *Comm. Comput. Phys.* **6**(4), 826–847.

Marzouk, Y.M., Najm, H.N. and Rahn, L.A. 2007 Stochastic spectral methods for efficient Bayesian solution of inverse problems. *J. Comp. Phys.* **224**, 560–586.

Matthies, H.G. and Keese, A. 2005 Galerkin methods for linear and nonlinear elliptic stochastic partial differential equations. *Computer Methods in Applied Mechanics and Engineering.* **194**, 1295–1331.

Moore, B. 1981 Principal component analysis in linear systems: controllability, observability, and model reduction. *IEEE Transactions on Automatic Control.* **AC-26**(1), 17–31.

Najm, H.N. 2009 Uncertainty quantification and polynomial chaos techniques in computational fluid dynamics. *Ann. Rev. Fluid Mech.* **41**, 35–52.

Najm, H.N., Debusschere, B.J., Marzouk, Y.M., Widmer, S. and Maître O.L. 2009 Uncertainty quantification in chemical systems. *Int. J. Numer. Meth. Eng.* In press.

Nguyen, C. 2005 *Reduced basis approximation and a posteriori error bounds for non-affine and nonlinear partial differential equations: Application to inverse analysis.* PhD thesis Singapore-MIT Alliance.

Nobile, F., Tempone, R. and Webster, C. 2008 An anisotropic sparse grid stochastic collocation method for partial differential equations with random input data. *SIAM J. Num. Anal.* **46**(5), 2411–2442.

Noor, A. and Peters, J. 1980 Reduced basis technique for nonlinear analysis of structures. *AIAA Journal.* **18**(4), 455–462.

Penzl, T. 2006 Algorithms for model reduction of large dynamical systems. *Linear Algebra and its Applications.* **415**(2–3), 322–343.

Reagan, M., Najm, H., Debusschere, B., Maître O.L., Knio, O. and Ghanem, R. 2004 Spectral stochastic uncertainty quantification in chemical systems. *Comb. Theo. Mod.* **8**, 607–632.

Rovas, D. 2003 *Reduced-basis Output Bound Methods for Parametrized Partial Differential Equations.* PhD thesis Massachusetts Institute of Technology.

Rozza, G. and Veroy, K. 2006 On the stability of the reduced basis method for Stokes equations in parametrized domains. *Submitted to Elsevier Science.*

Sirovich, L. 1987 Turbulence and the dynamics of coherent structures. Part 1: Coherent structures. *Quarterly of Applied Mathematics.* **45**(3), 561–571.

Smolyak, S. 1963 Quadrature and interpolation formulas for tensor products of certain classes of functions. *Soviet Math. Dokl.* **4**, 240–243.

Sorensen, D. and Antoulas, A. 2002 The Sylvester equation and approximate balanced reduction. *Linear Algebra and its Applications*. **351–352**, 671–700.

Sou, K., Megretski, A. and Daniel, L. 2005 A quasi-convex optimization approach to parameterized model-order reduction. *IEEE/ACM Design Automation Conference*, Anaheim, CA.

Veroy, K. and Patera, A. 2005 Certified real-time solution of the parametrized steady incompressible Navier-Stokes equations: Rigorous reduced-basis *a posteriori* error bounds. *International Journal for Numerical Methods in Fluids*. **47**, 773–788.

Veroy, K., Prud'homme C., Rovas, D. and Patera, A. 2003 A posteriori error bounds for reduced-basis approximation of parametrized noncoercive and nonlinear elliptic partial differential equations. *Proceedings of the 16th AIAA Computational Fluid Dynamics Conference*, Orlando, FL. AIAA Paper 2003–3847.

Wan, X. and Karniadakis, G.E. 2005 An adaptive multi-element generalized polynomial chaos method for stochastic differential equations. *J. Comput. Phys*. **209**, 617–642.

Wan, X. and Karniadakis, G.E. 2009 Error control in multi-element generalized polynomial chaos method for elliptic problems with random coefficients. *Comm. Comput. Phys*. **5**(2–4), 793–820.

Wang, J. and Zabaras, N. 2005 Using Bayesian statistics in the estimation of heat source in radiation. *International Journal of Heat and Mass Transfer*. **48**, 15–29.

Wiener, N. 1938 The homogeneous chaos. *Am. J. Math*. **60**, 897–936.

Willcox, K. and Peraire, J. 2002 Balanced model reduction via the proper orthogonal decomposition. *AIAA Journal*. **40**(11), 2223–2230.

Xiu, D. 2009 Fast numerical methods for stochastic computations: a review. *Comm. Comput. Phys*. **5**, 242–272.

Xiu, D. and Hesthaven, J.S. 2005 High-order collocation methods for differential equations with random inputs. *SIAM J. Sci. Comp*. **27**(3), 1118–1139.

Xiu, D. and Karniadakis, G. 2002 The Wiener-Askey polynomial chaos for stochastic differential equations. *SIAM J. Sci. Comp*. **24**(2), 619–644.

Xiu, D. and Shen, J. 2009 Efficient stochastic Galerkin methods for random diffusion equations. *J. Comp. Phys*. **228**(2), 266–281.

Chapter 8

Reduced Basis Approximation and A Posteriori Error Estimation for Parametrized Parabolic PDEs: Application to Real-Time Bayesian Parameter Estimation

N.C. Nguyen, G. Rozza, D.B.P. Huynh and A.T. Patera

Massachusetts Institute of Technology, USA

Large-Scale Inverse Problems and Quantification of Uncertainty Edited by L. Biegler, G. Biros, O. Ghattas, M. Heinkenschloss, D. Keyes, B. Mallick, Y. Marzouk, L. Tenorio, B. van Bloemen Waanders and K. Willcox

8.1 Introduction

In this chapter we consider reduced basis (RB) approximation and a posteriori error estimation for linear functional outputs of affinely parametrized linear parabolic partial differential equations. The essential ingredients are Galerkin projection onto a low-dimensional space associated with a smooth parametrically induced manifold – dimension reduction; efficient and effective POD-Greedy$_{\text{RB}}$ sampling methods for identification of optimal and numerically stable approximations spaces – rapid convergence; rigorous and sharp a posteriori error bounds for the linear-functional outputs of interest – certainty; and Offline-Online computational decomposition strategies – minimum *marginal cost*. The RB approach is effective in the *real–time* context in which the expensive Offline stage is deemed unimportant – for example parameter estimation and control; the RB approach is also effective in the *many–query* context in which the expensive Offline stage is asymptotically negligible – for example design optimization, uncertainty quantification (Boyaval *et al.* 2008), and multi-scale analysis (Boyaval 2008; Nguyen 2008).

There are two parts to this chapter. In the first part we present a rather general RB formulation for linear parabolic equations: our development combines earlier work in the parabolic case (Grepl and Patera 2005; Haasdonk and Ohlberger 2008) and elliptic context (Prud'homme *et al.* 2002; Rozza *et al.* 2008) with new advances in sampling procedures and in particular the POD–Greedy$_{\text{RB}}$ approach. In the second part we develop an RB Bayesian framework for parameter estimation which exploits the rapid response and reliability of the certified reduced basis method; as an example we consider detection and characterization of a delamination crack by transient thermal analysis (Grepl 2005; Starnes 2002). In summary, the first part (Section 8.2) emphasizes the essential RB ingredients in a general linear context – the state of the art; the second part (Section 8.3) emphasizes the integration and impact of RB technology in applications – real–time and many–query applications; each section contains both background material and new contributions. Brief concluding remarks (Section 8.4) discuss the possibilities for the future.

8.2 Linear Parabolic Equations

8.2.1 Reduced Basis Approximation

We first introduce several notations required for the remainder of the chapter. Our parameter domain, a closed subset of $\mathbb{R}^P$, shall be denoted $\mathcal{D}$; a typical parameter value in $\mathcal{D}$ shall be denoted μ. Our time domain shall be denoted by $I = [0, t_f]$ with t_f the final time. Our physical domain in d space dimensions shall be denoted Ω with boundary $\partial\Omega$; a typical point in Ω shall be denoted $x = (x_1, \ldots, x_d)$. We can then define the function space $X = X(\Omega)$ such that $(H_0^1(\Omega))^{\mathcal{V}} \subset X \subset (H^1(\Omega))^{\mathcal{V}}$; here $H^1(\Omega) = \{v | v \in L^2(\Omega), \nabla v \in (L^2(\Omega))^d\}$, $H_0^1(\Omega) = \{v \in H^1(\Omega) | v_{\partial\Omega} = 0\}$, $L^2(\Omega)$ is the space of square integrable functions over Ω, and $\mathcal{V} = 1$ (respectively, d) for scalar (respectively, vector) problems. We denote by $(\cdot,\cdot)_X$ the inner product associated with the Hilbert space X; this inner product in turn induces a norm $\|\cdot\|_X = \sqrt{(\cdot,\cdot)_X}$ equivalent to the usual $(H^1(\Omega))^{\mathcal{V}}$ norm. Similarly, we denote by $(\cdot,\cdot)$ and $\|\cdot\|$ the $L^2(\Omega)$ inner product and induced norm, respectively.

We consider a generalized convection-diffusion equation (expressed in weak form): Given $\mu \in \mathcal{D}$, we find $u(t;\mu)$ such that

$$m(u_t(t;\mu), v; \mu) + a(u(t;\mu), v; \mu) = f(v; t; \mu), \quad \forall v \in X, \quad \forall t \in I, \tag{8.1}$$

subject to initial condition $u(t = 0; \mu) = u_0 \in L^2(\Omega)$. We then evaluate our output as

$$s(t;\mu) = \ell(u(t;\mu); t; \mu), \quad \forall t \in I. \tag{8.2}$$

We shall assume that a – which represents convection and diffusion – is time–invariant, continuous over X, and coercive over X with coercivity constant

$$\alpha(\mu) = \inf_{v \in X} \frac{a(w, w; \mu)}{\|w\|_X^2}, \quad \forall \mu \in \mathcal{D};$$

we assume that m – which represents 'mass' or inertia – is time–invariant, symmetric, and continuous and coercive over $L^2(\Omega)$ with coercivity constant

$$\sigma(\mu) = \inf_{v \in X} \frac{m(w, w; \mu)}{\|w\|^2}, \quad \forall \mu \in \mathcal{D};$$

we assume that f and ℓ are linear continuous functionals over X and $L^2(\Omega)$, respectively.

Finally, to effect our Offline–Online decomposition we shall require that our bilinear and linear forms are 'affine in parameter' (more precisely, affine in functions of the parameter): for some finite Q_a and Q_m, a and m may be expressed as

$$a(w, v; \mu) = \sum_{q=1}^{Q_a} \Theta_a^q(\mu) a^q(w, v), \qquad m(w, v; \mu) = \sum_{q=1}^{Q_m} \Theta_m^q(\mu) m^q(w, v) \tag{8.3}$$

for given parameter-dependent functions $\Theta_a^q, 1 \le q \le Q_a$, $\Theta_m^q, 1 \le q \le Q_m$, and continuous parameter-independent bilinear forms $a^q, 1 \le q \le Q_a$, $m^q, 1 \le q \le Q_m$; furthermore, for some finite Q_f and Q_ℓ, f and ℓ may be expressed as

$$f(v; t; \mu) = \sum_{q=1}^{Q_f} \Theta_f^q(\mu) f^q(v; t), \qquad \ell(v; t; \mu) = \sum_{q=1}^{Q_\ell} \Theta_\ell^q(\mu) \ell^q(v; t), \tag{8.4}$$

for given time/parameter-dependent functions $\Theta_f^q, 1 \le q \le Q_f, \Theta_\ell^q, 1 \le q \le Q_\ell$, and continuous parameter–independent linear forms $f^q, 1 \le q \le Q_f, \ell^q, 1 \le q \le Q_\ell$.

We now describe a general class – through not *the* most general class – of problems which honors these hypotheses; for simplicity we consider a scalar field ($\mathcal{V} = 1$) in two space dimension ($d = 2$). We shall first define an 'original' problem (subscript o) for a field variable $u_o(t;\mu) \in X_o(\mu)$ over a *parameter-dependent* domain $\Omega_o(\mu) \subset \mathbb{R}^2$,

$$m_o(u_{o\,t}(t;\mu), v; \mu) + a_o(u_o(t;\mu), v; \mu) = f_o(v; t; \mu), \quad \forall v \in X_o(\mu), \tag{8.5}$$

$$s_o(t;\mu) = \ell_o(u_o(t;\mu); t; \mu); \tag{8.6}$$

we will then map $\Omega_o(\mu)$ to a *parameter-independent* reference domain $\Omega = \Omega(\mu_{\text{ref}}), \mu_{\text{ref}} \in \mathcal{D}$, to arrive at the 'transformed' problem (8.1), (8.2) – which is the point of departure of our reduced basis approach. It remains to place restrictions on both geometry ($\Omega_o(\mu)$) and operators (a_o, m_o, f_o, ℓ_o) such that (upon mapping) this transformed problem satisfies our hypotheses – in particular, the affine assumption (8.3), (8.4). Towards this end, a domain decomposition shall prove indispensable.

We first consider the class of admissible geometries. We may consider $\Omega_o(\mu)$ ($\Rightarrow \Omega = \Omega(\mu_{\text{ref}})$) of the form

$$\overline{\Omega}_o(\mu) = \bigcup_{j=1}^{J_{\text{dom}}} \overline{\Omega}_o^j(\mu) \quad \left(\Rightarrow \overline{\Omega} = \bigcup_{j=1}^{J_{\text{dom}}} \overline{\Omega}^j \right)$$

for which $\Omega_o^j(\mu) = \mathcal{T}^j(\Omega^j;\mu), 1 \le j \le J_{\text{dom}}, \forall \mu \in \mathcal{D}$; here the Ω_o^j (respectively, Ω^j), $1 \le j \le J_{\text{dom}}$, constitute a conforming triangulation of $\Omega_o(\mu)$ (respectively, Ω), and the $\mathcal{T}^j, 1 \le j \le J_{\text{dom}}$, are *affine* mappings. We next consider the class of admissible operators. We may consider

$$a_o(w, v; \mu) = \sum_{j=1}^{J_{\text{dom}}} \int_{\Omega_o^j(\mu)} \begin{bmatrix} w_{,1} & w_{,2} & w \end{bmatrix} \mathcal{K}_o^j(\mu) \begin{bmatrix} v_{,1} \\ v_{,2} \\ v \end{bmatrix},$$

$$m_o(w, v; \mu) = \sum_{j=1}^{J_{\text{dom}}} \int_{\Omega_o^j(\mu)} w \mathcal{M}_o^j(\mu) v, \tag{8.7}$$

$$f_o(v; t; \mu) = \sum_{j=1}^{J_{\text{dom}}} \int_{\Omega_o^j(\mu)} \mathcal{F}_o^j(t;\mu) v, \quad \ell_o(v; t; \mu) = \sum_{j=1}^{J_{\text{dom}}} \int_{\Omega_o^j(\mu)} \mathcal{L}_o^j(\mu) v; \tag{8.8}$$

here $w_{,i}$ refers to differentiation with respect to the ith spatial coordinate, and the $\mathcal{K}_o^j : \mathcal{D} \to \mathbb{R}^{3\times 3}$, $\mathcal{M}_o^j : \mathcal{D} \to \mathbb{R}$, $\mathcal{F}_o^j : \mathcal{D} \to \mathbb{R}, \mathcal{L}_o^j : \mathcal{D} \to \mathbb{R}$, $1 \le j \le J_{\text{dom}}$, are prescribed coefficients. (There are additional standard restriction on $\mathcal{K}_o^j, \mathcal{M}_o^j, 1 \le j \le J_{\text{dom}}$, related to coercivity.)

The process by which we map this original problem to the transformed problem – in which the $\Theta_a^q, \Theta_m^q, \Theta_f^q, \Theta_\ell^q$ reflect the geometric and coefficient parametric dependence – can be largely automated (Huynh *et al.* 2007-2009; Rozza *et al.* 2008). There are many ways in which we can relax our assumptions and thus treat an even broader class of problems. For example, we may consider 'elliptical' or 'curvy' triangular subdomains (Rozza *et al.* 2008); we may consider a which satisfy only a weak coercivity (Garding) inequality (Knezevic and Patera 2009); we may consider non–time–invariant bilinear forms a and m; we may consider coefficient functions $\mathcal{K}, \mathcal{M}$ which are polynomial in the spatial coordinate (or more generally approximated by the Empirical Interpolation Method (Barrault *et al.* 2004; Grepl *et al.* 2007a)). These generalizations can be pursued, with no loss in rigor, by modification of the methods – the error estimators, the sampling procedures, and the Offline–Online decompositions – presented in this chapter. However, it is important to recognize that, in general, increased complexity in geometry and operator will result in more terms in our affine expansions – larger Q_a, Q_m, Q_f, Q_ℓ – with corresponding detriment to the reduced basis (Online) computational performance; we return to this point in the context of the operation counts provided in Section 8.3 and then again for our particular example of Section 8.3.

We next introduce the finite difference in time and finite element (FE) in space discretization of this parabolic problem (Quarteroni and Valli 1997). We first divide the time interval I into K subintervals of equal length $\Delta t = t_f/K$ and define $t^k \equiv k\Delta t$, $0 \le k \le K$. We next define the finite element approximation space $X^{\mathcal{N}} \subset X$ of dimension $\mathcal{N}$. Then, given $\mu \in \mathcal{D}$, we look for $u^{\mathcal{N}\,k}(\mu) \in X^{\mathcal{N}}, 0 \le k \le K$, such that

$$\frac{1}{\Delta t} m(u^{\mathcal{N}\,k}(\mu) - u^{\mathcal{N}\,k-1}(\mu), v; \mu) + a(u^{\mathcal{N}\,k}(\mu), v; \mu) = f(v; t^k; \mu),$$
$$\forall v \in X^{\mathcal{N}}, 1 \le k \le K, \tag{8.9}$$

subject to initial condition $(u^{\mathcal{N}\,0}, v) = (u_0, v), \forall v \in X^{\mathcal{N}}$. We then evaluate the output: for $0 \le k \le K$,

$$s^{\mathcal{N}\,k}(\mu) = \ell(u^{\mathcal{N}\,k}(\mu); t^k; \mu). \tag{8.10}$$

We shall sometimes denote $u^{\mathcal{N}\,k}(\mu)$ as $u^{\mathcal{N}}(t^k;\mu)$ and $s^{\mathcal{N}\,k}(\mu)$ as $s^{\mathcal{N}}(t^k;\mu)$ to more clearly identify the discrete time levels. Equation (8.9) – Euler-Backward Galerkin discretization of (8.1) – shall be our point of departure: we shall presume that Δt is sufficiently small and $\mathcal{N}$ is sufficiently large such that $u^{\mathcal{N}}(t^k;\mu)$ and $s^{\mathcal{N}}(t^k;\mu)$ are effectively indistinguishable from $u(t^k;\mu)$ and $s(t^k;\mu)$, respectively. (The development readily extends to Crank-Nicolson discretization; for purposes of exposition, we consider the simple Euler Backward approach.) Our goal is to accelerate the 'truth' discretization (8.9), (8.10), in the real-time and many-query contexts: we shall build our RB approximation on this truth discretization; and we shall measure the RB computational performance and RB accuracy relative to this truth discretization.

We now introduce the reduced basis (RB) approximation (Almroth *et al.* 1978; Fink and Rheinboldt 1983; Noor and Peters 1980; Porsching 1985). Given a set of mutually $(\cdot,\cdot)_X$–orthogonal basis functions $\xi_n \in X^{\mathcal{N}}, 1 \leq n \leq N_{\max}$, the $N_{\max}$ hierarchical RB spaces are given by

$$X_N \equiv \operatorname{span}\{\xi_n, 1 \leq n \leq N\}, \quad 1 \leq N \leq N_{\max}. \tag{8.11}$$

In actual practice (see Section 8.2.3), the spaces $X_N \in X^{\mathcal{N}}$ will be generated by a POD–Greedy$_{\text{RB}}$ sampling procedure which combines spatial snapshots in time and parameter – $u^{\mathcal{N}\,k}(\mu)$ – in an optimal fashion. Given $\mu \in \mathcal{D}$, we now look for $u_N^k(\mu) \in X_N, 0 \leq k \leq K$, such that

$$\begin{aligned} &\frac{1}{\Delta t} m(u_N^k(\mu) - u_N^{k-1}(\mu), v; \mu) + a(u_N^k(\mu), v; \mu) = f(v; t^k; \mu), \\ &\quad \forall v \in X_N, 1 \leq k \leq K, \end{aligned} \tag{8.12}$$

subject to $(u_N^0(\mu), v) = (u^{\mathcal{N}\,0}, v), \forall v \in X_N$. We then evaluate the associated output: for $0 \leq k \leq K$,

$$s_N^k(\mu) = \ell(u_N^k(\mu); t^k, \mu). \tag{8.13}$$

We shall sometimes denote $u_N^k(\mu)$ as $u_N(t^k;\mu)$ and $s_N^k(\mu)$ as $s_N(t^k;\mu)$ to more clearly identify the discrete time levels; note that the RB approximation inherits the timestep of the truth discretization – there is no reduction in the 'temporal' dimension. The RB quantities should in fact bear a $\mathcal{N}$–$X_N^{\mathcal{N}}, u_N^{\mathcal{N}\,k}(\mu), s_N^{\mathcal{N}\,k}(\mu)$ – since the RB approximation is defined in terms of a particular truth discretization: for clarity of exposition, we shall typically suppress the 'truth' superscript; we nevertheless insist on stability/uniformity as $\mathcal{N} \to \infty$ (and $\Delta t \to 0$).

In general we can choose our domain decomposition to ensure that the functions $\Theta^q_{a,m,f,\ell}$ are very smooth; it can then be demonstrated that the field variable $u(\mu)$ is very smooth with respect to the parameter. It is then plausible (and in certain cases may be proven theoretically (Maday *et al.* 2002; Rozza *et al.* 2008)) that a properly chosen RB approximation – a Galerkin – optimal linear combination of 'good' snapshots on a smooth manifold – will converge very rapidly (even exponentially) with increasing N. Numerical results for a large number of coercive elliptic and parabolic problems (Grepl and Patera 2005; Rozza *et al.* 2008) – including the example of this chapter – support this conjecture: typically $N \approx O(10 - 100)$ and hence $N \ll \mathcal{N}$; the 'sparse samples' identified by the POD–Greedy$_{\text{RB}}$ of Section 2.3 play an important role. Of course performance will degrade as the number of parameters and the ranges of the parameters increase; we return to this point in our computational summary below.

8.2.2 A Posteriori Error Estimation

Rigorous, sharp, and inexpensive a posteriori error bounds are crucial for informed application of the reduced basis method: a posteriori error bounds confirm *in practice* the dimension N required for the desired accuracy. To construct the a posteriori error bounds for the RB approximation, we need two ingredients. The first ingredient is the dual norm of the residual

$$\varepsilon_N(t^k;\mu) = \sup_{v\in X^{\mathcal{N}}} \frac{r_N(v;t^k;\mu)}{\|v\|_X}, \quad 1\le k\le K, \tag{8.14}$$

where $r_N(v;t^k;\mu)$ is the residual associated with the RB approximation (8.12)

$$\begin{aligned} r_N(v;t^k;\mu) = f(v;t^k;\mu) - \frac{1}{\Delta t} m\big(u_N^k(\mu) - u_N^{k-1}(\mu), v;\mu\big) \\ -a\big(u_N^k(\mu), v;\mu\big), \quad \forall v\in X^{\mathcal{N}}, \quad 1\le k\le K. \end{aligned} \tag{8.15}$$

The second ingredient is a lower bound $0 < \alpha_{\mathrm{LB}}^{\mathcal{N}}(\mu) \le \alpha^{\mathcal{N}}(\mu), \forall \mu\in\mathcal{D}$, for the coercivity constant $\alpha^{\mathcal{N}}(\mu)$ defined as

$$\alpha^{\mathcal{N}}(\mu) = \inf_{v\in X^{\mathcal{N}}} \frac{a(v,v;\mu)}{\|v\|_X^2}, \quad \forall\mu\in\mathcal{D}, \tag{8.16}$$

and a lower bound $0 < \sigma_{\mathrm{LB}}^{\mathcal{N}}(\mu) \le \sigma^{\mathcal{N}}(\mu), \forall\mu\in\mathcal{D}$, for the coercivity constant $\sigma^{\mathcal{N}}(\mu)$ defined as

$$\sigma^{\mathcal{N}}(\mu) = \inf_{v\in X^{\mathcal{N}}} \frac{m(v,v;\mu)}{\|v\|^2}, \quad \forall\mu\in\mathcal{D}. \tag{8.17}$$

Note that since $X^{\mathcal{N}}\subset X$, $\alpha^{\mathcal{N}}(\mu)\ge\alpha(\mu)>0$ and $\sigma^{\mathcal{N}}(\mu)\ge\sigma(\mu)>0$, $\forall\mu\in\mathcal{D}$; we shall ensure by construction that $\alpha_{\mathrm{LB}}^{\mathcal{N}}(\mu)>0$ and $\sigma_{\mathrm{LB}}^{\mathcal{N}}(\mu)>0$, $\forall\mu\in\mathcal{D}$.

We can now define our error bounds in terms of the dual norm of the residual and the lower bounds for the coercivity constants. In particular, it can readily be proven (Grepl and Patera 2005; Haasdonk and Ohlberger 2008; Nguyen *et al.* 2009) that for all $\mu\in\mathcal{D}$ and all N,

$$\begin{aligned} \|u^{\mathcal{N}k}(\mu) - u_N^k(\mu)\| \le \Delta_N^k(\mu), \\ |s^{\mathcal{N}k}(\mu) - s_N^k(\mu)| \le \Delta_N^{s\,k}(\mu), \quad 1\le k\le K, \end{aligned} \tag{8.18}$$

where $\Delta_N^k(\mu)\equiv\Delta_N(t^k;\mu)$ (the L^2 error bound) and $\Delta_N^{s\,k}(\mu)\equiv\Delta_N^s(t^k;\mu)$ (the 'output error bound') are given by

$$\begin{aligned} \Delta_N^k(\mu) \equiv \sqrt{\frac{\Delta t}{\alpha_{\mathrm{LB}}^{\mathcal{N}}(\mu)\,\sigma_{\mathrm{LB}}^{\mathcal{N}}(\mu)} \sum_{k'=1}^{k} \varepsilon_N^2(t^{k'};\mu)}, \\ \Delta_N^{s\,k}(\mu) \equiv \left(\sup_{v\in X^{\mathcal{N}}} \frac{\ell(v)}{\|v\|}\right) \Delta_N^k(\mu). \end{aligned} \tag{8.19}$$

(We assume for simplicity that $u^{\mathcal{N}0} \in X_N$; otherwise there will be an additional contribution to $\Delta_N^k(\mu)$.) Note again that the RB error is measured relative to the 'truth' discretization.

It should be clear that our error bound for the output is rather crude. We may pursue primal-dual RB approximations (Grepl and Patera 2005; Pierce and Giles 2000; Rozza *et al.* 2008) that provide both more rapid convergence of the output and also more robust (sharper) estimation of the output error. However, in cases in which *many* outputs are of interest, for example inverse problems, the primal-only approach described above can be more efficient and also more adaptive – efficiently expanded to include additional outputs.

8.2.3 Offline–Online Computational Approach

The arguments above indicate that a reduced basis space of greatly reduced dimension $N \ll \mathcal{N}$ may suffice for accurate approximation of the field and output. However, each member of this space – and in particular each basis function $\xi_n, 1 \leq n \leq N$ – will be represented as a vector in $\mathbb{R}^{\mathcal{N}}$ corresponding to (say) the finite element nodal values. It is thus not clear how we can efficiently compute our reduced basis solution and output without appeal to high – dimensional objects – more generally, how we can translate the reduced dimension into reduced computational effort. The error estimator is even more problematic: the dual norm of the residual requires the truth solution of a (Poisson–like) problem which will be almost as expensive as the original PDE. In both cases, the affine assumption is the crucial enabler that permits an efficient implementation in the real-time or many–query contexts: we can pre-compute components of the mass and stiffness matrices and residual dual norm. We now provide the details.

Construction-Evaluation Decomposition

The affine representation (8.3) permits a 'Construction-Evaluation' decomposition (Balmes 1996; Prud'homme *et al.* 2002) of computational effort that greatly reduces the marginal cost – relevant in the real-time and many-query contexts – of both the RB output evaluation, (8.13), *and* the associated error bounds, (8.19). The expensive Construction stage, performed once, provides the foundation for the subsequent very inexpensive Evaluation stage, performed many times for each new desired $\mu \in \mathcal{D}$. We first consider the Construction-Evaluation decomposition for the output and then address the error bounds. For simplicity, in this section we assume that $f(v; t^k; \mu) = g(t^k) f(v)$ for some control $g(t)$, and that $\ell(v; t^k; \mu) = \ell(v)$.

We represent $u_N^k(\mu)$ as $u_N^k(\mu) = \sum_{n=1}^{N} \omega_{N\,n}^k(\mu)\xi_n$, where we recall that the $\xi_n, 1 \leq n \leq N$, are the basis functions for our RB space X_N. We may then evaluate the RB output as

$$s_N^k(\mu) = L_N^T \omega_N^k(\mu), \quad 1 \leq k \leq K, \tag{8.20}$$

where $L_{N\,n} = \ell(\xi_n)$, $1 \le n \le N$. To find the $\omega^k_{N\,j}(\mu), 1 \le j \le N, 1 \le k \le K$, we insert $u^k_N(\mu) = \sum_{n=1}^N \omega^k_{N\,n}(\mu)\xi_n$, $u^{k-1}_N(\mu) = \sum_{n=1}^N \omega^{k-1}_{N\,n}(\mu)\xi_n$, and $v = \xi_m$ in (8.12) to obtain the discrete system

$$(M_N(\mu) + \Delta t A_N(\mu))\omega^k_N(\mu) = \Delta t g(t^k) F_N + M_N(\mu)\omega^{k-1}_N(\mu), \quad 1 \le k \le K \tag{8.21}$$

where $A_N(\mu) \in \mathbb{R}^{N\times N}, M_N(\mu) \in \mathbb{R}^{N\times N}$, and $F_N(\mu) \in \mathbb{R}^N$ are given by $A_{N\,m,n}(\mu) = a(\xi_n, \xi_m; \mu)$, $M_{N\,m,n}(\mu) = m(\xi_n, \xi_m; \mu)$, $1 \le m, n \le N$, and $F_{N\,n} = f(\xi_n)$, $1 \le n \le N$, respectively. We next note that $A_N(\mu)$ and $M_N(\mu)$ can be expressed, thanks to (8.3), as

$$A_N(\mu) = \sum_{q=1}^{Q_a} \Theta^q_a(\mu) A^q_N, \quad M_N(\mu) = \sum_{q=1}^{Q_m} \Theta^q_m(\mu) M^q_N \tag{8.22}$$

where the $A^q_{N\,m,n} \equiv a^q(\xi_n, \xi_m), 1 \le m, n \le N$, $1 \le q \le Q_a$, $M^q_{N\,m,n} \equiv m^q(\xi_n, \xi_m)$, $1 \le m, n \le N$, $1 \le q \le Q_m$, are parameter-*independent*. We can now readily identify the Construction–Evaluation decomposition.

In the *Construction* stage we first form and store the time–independent and μ–independent matrices/vectors $\mathrm{A}^q_{N_{\max}\,ij}$, $M^{q'}_{N_{\max}\,ij}$, $F_{N_{\max}\,i}$, and $L_{N_{\max}\,i}$, $1 \le i, j \le N_{\max}$, $1 \le q \le Q_a$, $1 \le q' \le Q_m$. The operation count in the Construction stage of course depends on $\mathcal{N}$ – even once the $\xi_i, 1 \le i \le N_{\max}$, are *known* (obtained by the sampling procedure of the next section), it remains to compute $O(N^2_{\max})$ finite element quadratures over the $O(\mathcal{N})$ triangulation. Note that, thanks to the hierarchical nature of the RB spaces, the stiffness matrices/vectors $A^q_{N\,ij}, M^{q'}_{N\,ij}, F_{N\,i}$, and $L_{N\,i}, 1 \le i, j \le N, 1 \le q \le Q_a, 1 \le q' \le Q_m$, for any $N \le N_{\max}$ can be extracted as principal subarrays of the corresponding $N_{\max}$ quantities. (For non-hierarchical RB spaces the storage requirements are much higher.)

In the *Evaluation* stage, we first form the left–hand side of (8.21) in $O((Q_a + Q_m)N^2)$ operations; we then invert the resulting $N \times N$ matrix in $O(N^3)$ operations (in general, we must anticipate that the RB matrices will be dense); finally, we compute $\omega^k_{N\,j}, 1 \le j \le N, 1 \le k \le K$, in $O(KN^2)$ operations – $O(KN^3)$ operations for non-LTI systems – by matrix-vector multiplication. Note that $g(t^k)$ need only be specified in the Online stage; we return to this point in our sampling strategy below. Once the $\omega^k_{N\,j}, 1 \le j \le N, 1 \le k \le K$, are obtained – $O((Q_a + Q_m + N + K)N^2)$ operations in total – we evaluate our output from (8.20) in $O(NK)$ operations. The storage and operation count in the Evaluation phase is clearly independent of $\mathcal{N}$, and we can thus anticipate – presuming $N \ll \mathcal{N}$ – very rapid RB response in the real-time and many-query contexts.

The Construction-Evaluation procedure for the output error bound is a bit more involved. There are three components to this bound: the dual norm of ℓ (readily computed, once, in the Construction phase); the lower bound for the coercivity constants, $\alpha_{\mathrm{LB}}^{\mathcal{N}}(\mu)$ and $\sigma_{\mathrm{LB}}^{\mathcal{N}}(\mu)$, computed Offline–Online by the Successive Constraint Method (SCM) as described in detail in Huynh *et al.* (2007); Rozza *et al.* (2008), and not discussed further here; and the dual norm of the residual $\varepsilon_N(t^k;\mu)$. We consider here the Construction-Evaluation decomposition for the dual norm of the residual (Grepl and Patera 2005). We first note from duality arguments that $\varepsilon_N(t^k;\mu)$ can be expressed as

$$\varepsilon_N^2(t^k;\mu) = \|\hat{e}_N(t^k;\mu)\|_X^2, \quad 1 \le k \le K, \tag{8.23}$$

where $\hat{e}_N(t^k;\mu)$ is the Riesz representation of the residual,

$$(\hat{e}_N(t^k;\mu), v)_X = r_N(v;t^k;\mu), \quad \forall v \in X^{\mathcal{N}}. \tag{8.24}$$

Here $r_N(v;t^k;\mu)$ is the residual defined in (8.15) (with $f(v;t^k;\mu) = g(t^k)f(v)$), which we may further write – exploiting the reduced basis representation $u_N^k(\mu) = \sum_{n=1}^N \omega_{N\,n}^k(\mu)\xi_n$ and affine assumption (8.3) – as

$$\begin{aligned} r_N(v;t^k;\mu) = g(t^k)f(v) &- \frac{1}{\Delta t}\sum_{q=1}^{Q_m}\sum_{j=1}^{N}\Theta_m^q(\mu)(\omega_{N\,j}^k(\mu) - \omega_{N\,j}^{k-1}(\mu))m^q(\xi_j, v) \\ &- \sum_{q=1}^{Q_a}\sum_{j=1}^{N}\Theta_a^q(\mu)\omega_{N\,j}^k(\mu)a^q(\xi_j, v), \end{aligned} \tag{8.25}$$

for $1 \le k \le K$.

It now follows directly from (8.24) and (8.25) that

$$\begin{aligned} \hat{e}_N(t^k;\mu) = g(t^k)\Gamma_N &+ \frac{1}{\Delta t}\sum_{q=1}^{Q_m}\sum_{j=1}^{N}\Theta_m^q(\mu)(\omega_{N\,j}^k(\mu) - \omega_{N\,j}^{k-1}(\mu))\Lambda_N^{qj} \\ &+ \sum_{q=1}^{Q_a}\sum_{j=1}^{N}\Theta_a^q(\mu)\omega_{N\,j}^k(\mu)\Upsilon_N^{qj}, \quad 1 \le k \le K, \end{aligned} \tag{8.26}$$

where

$$\begin{aligned} (\Gamma_N, v)_X &= f(v), && \forall v \in X^{\mathcal{N}}, \\ (\Lambda_N^{qj}, v)_X &= -m^q(\xi_j, v), && \forall v \in X^{\mathcal{N}}, \quad 1 \le q \le Q_m, 1 \le j \le N, \\ (\Upsilon_N^{qj}, v)_X &= -a^q(\xi_j, v), && \forall v \in X^{\mathcal{N}}, \quad 1 \le q \le Q_a, 1 \le j \le N. \end{aligned} \tag{8.27}$$

It then follows from (8.28) that

$$\begin{aligned} &\varepsilon_N^2(t^k;\mu) \\ &= g(t^k)g(t^k)C_{N\,1}^{ff}(\mu) + \sum_{j=1}^{N}\sum_{j'=1}^{N}\omega_{N\,j}^k(\mu)\omega_{N\,j'}^k(\mu)C_{N\,jj'}^{aa}(\mu) \end{aligned}$$

$$+\frac{1}{\Delta t^2}\sum_{j=1}^{N}\sum_{j'=1}^{N}(\omega_{N\,j}^{k}(\mu)-\omega_{N\,j}^{k-1}(\mu))(\omega_{N\,j'}^{k}(\mu)-\omega_{N\,j'}^{k-1}(\mu))C_{N\,jj'}^{mm}(\mu)$$

$$+2g(t^k)\sum_{j=1}^{N}\omega_{N\,j}^{k}(\mu)C_{N\,j}^{fa}(\mu)+\frac{2g(t^k)}{\Delta t}\sum_{j=1}^{N}(\omega_{N\,j}^{k}(\mu)-\omega_{N\,j}^{k-1}(\mu))C_{N\,j}^{fm}(\mu)$$

$$+\frac{2}{\Delta t}\sum_{j=1}^{N}\sum_{j'=1}^{N}(\omega_{N\,j}^{k}(\mu)-\omega_{N\,j}^{k-1}(\mu))\omega_{N\,j'}^{k}(\mu)C_{N\,jj'}^{am}(\mu), \quad 1\le k\le K, \tag{8.28}$$

where, for $1 \le j, j' \le N$,

$$\begin{aligned}
C_{N\,1}^{ff}(\mu) &= (\Gamma_N,\Gamma_N)_X,\\
C_{N\,jj'}^{aa}(\mu) &= \sum_{q=1}^{Q_a}\sum_{q'=1}^{Q_a}\Theta_a^q(\mu)\Theta_a^{q'}(\mu)(\Upsilon_N^{qj},\Upsilon_N^{q'j'})_X,\\
C_{N\,jj'}^{mm}(\mu) &= \sum_{q=1}^{Q_m}\sum_{q'=1}^{Q_m}\Theta_m^q(\mu)\Theta_m^{q'}(\mu)(\Lambda_N^{qj},\Lambda_N^{q'j'})_X,\\
C_{N\,j}^{fa}(\mu) &= \sum_{q=1}^{Q_a}\Theta_a^q(\mu)(\Upsilon_N^{qj},\Gamma_N)_X,\\
C_{N\,j}^{fm}(\mu) &= \sum_{q=1}^{Q_m}\Theta_m^q(\mu)(\Lambda_N^{qj},\Gamma_N)_X,\\
C_{N\,jj'}^{am}(\mu) &= \sum_{q=1}^{Q_m}\sum_{q'=1}^{Q_a}\Theta_m^q(\mu)\Theta_a^{q'}(\mu)(\Lambda_N^{qj},\Upsilon_N^{q'j'})_X.
\end{aligned} \tag{8.29}$$

The Construction–Evaluation decomposition is now clear. We emphasize that in infinite precision (8.28) and (8.23) are equivalent: (8.28) is a reformulation of (8.23) that admits an Offline-Online decomposition.[1]

In the *Construction* stage, we find the $\Gamma_{N_{\max}}, \Lambda_{N_{\max}}^{qj}, 1 \le q \le Q_m, 1 \le j \le N_{\max}, \Upsilon_{N_{\max}}^{qj}, 1 \le q \le Q_a, 1 \le j \le N_{\max}$, and form the inner products $(\Gamma_{N_{\max}},\ \Gamma_{N_{\max}})_X$, $(\Lambda_{N_{\max}}^{qj},\ \Lambda_{N_{\max}}^{q'j'})_X$, $1 \le q, q', \le Q_m$, $1 \le j, j' \le N_{\max}$, $(\Upsilon_{N_{\max}}^{qj}, \Upsilon_{N_{\max}}^{q'j'})_X$, $1 \le q, q', \le Q_a, 1 \le j, j' \le N_{\max}$, $(\Lambda_{N_{\max}}^{qj}, \Upsilon_{N_{\max}}^{q'j'})_X, 1 \le q \le Q_m, 1 \le q' \le Q_a, 1 \le j, j' \le N_{\max}$, $(\Lambda_{N_{\max}}^{qj}, \Gamma_{N_{\max}})_X, 1 \le q \le Q_m, 1 \le j \le N_{\max}$, $(\Upsilon_{N_{\max}}^{qj}, \Gamma_{N_{\max}})_X, 1 \le q \le Q_a, 1 \le j \le N_{\max}$. The operation count for the Construction stage clearly depends on $\mathcal{N} - 1 + (Q_a + Q_m)N$ finite element 'Poisson' problems (8.27) and $(1 + (Q_a + Q_m)N)^2$ finite element quadratures over the triangulation. (The temporary storage associated with

[1] In finite precision (8.28) and (8.23) are not equivalent: $\varepsilon_N(t^k;\mu)$ computed from (8.28) will only be accurate to the square root of machine precision; $\varepsilon_N(t^k;\mu)$ computed from (8.23) will be accurate to machine precision. The former is rarely a limitation for actual error tolerances of interest.

the latter can be excessive for higher–dimensional problems: it is simple to develop procedures that balance temporary storage and re–computation.) Note that, thanks to the hierarchical nature of the reduced basis spaces, these inner products for any $N \leq N_{\max}$ can be directly extracted from the corresponding $N_{\max}$ quantities. (As already noted, for non-hierarchical reduced basis spaces the storage requirements will be considerably higher.)

In the *Evaluation* stage, given the reduced basis coefficients $\omega_{N\,j}(t^k;\mu)$, $1 \leq j \leq N, 1 \leq k \leq K$, and coefficient functions $\Theta_a^q(\mu), 1 \leq q \leq Q_a$, $\Theta_m^q(\mu)$, $1 \leq q \leq Q_m$: we can readily compute the coefficient functions (8.29) from the stored inner products in $O((Q_a + Q_m)^2 N^2)$ operations; we then simply perform the sum (8.28) in $O(N^2)$ operations per time step and hence $O(KN^2)$ operations in total. The operation count for the Evaluation stage is thus (roughly) $(K + (Q_a + Q_m)^2)N^2$; note that the operation count for the Evaluation stage is $O(K(Q_a + Q_m)^2 N^2)$ operations for *non*-LTI systems since the coefficient functions (8.29) must be evaluated for each timestep. The crucial point, again, is that the cost and storage in the Evaluation phase – the *marginal* cost for each new value of μ – is independent of $\mathcal{N}$: thus we can not only evaluate our output prediction but also our rigorous output error bound very rapidly in the parametrically interesting contexts of real-time or many-query investigation. In short, we inherit the high fidelity and certainty of the FE approximation but at the low cost of a reduced-order model.

This concludes the discussion of the Construction – Evaluation decomposition. The Construction stage is performed Offline; the Evaluation stage is invoked Online – for each new μ of interest in the real–time or many–query contexts. However, there is another component to the Offline stage: we must construct a good (rapidly convergent) reduced basis space and associated basis functions $\xi_i, 1 \leq i \leq N_{\max}$, by a POD-Greedy$_{\rm RB}$ procedure; this sampling process in fact relies on the Construction–Evaluation decomposition to greatly reduce the requisite number of (expensive) 'candidate' finite element calculations over an (extensive) Greedy$_{\rm RB}$ training sample, $\Xi_{\rm train,RB}$, as we now describe. (In actual practice there is also an Offline-Component to the SCM construction of $\alpha_{\rm LB}^{\mathcal{N}}(\mu)$ and $\sigma_{\rm LB}^{\mathcal{N}}(\mu)$ as reported in Huynh *et al.* (2007); Rozza *et al.* (2008).)

POD-Greedy$_{\rm RB}$ Procedure

We address here the generation of our reduced basis space X_N. Our sampling procedure combines, as first proposed in Haasdonk and Ohlberger (2008), the POD (Proper Orthogonal Decomposition) in t^k – to capture the causality associated with our evolution equation – with a Greedy procedure (Grepl and Patera 2005; Rozza *et al.* 2008; Veroy *et al.* 2003b;) in μ – to treat efficiently the higher dimensions and more extensive ranges of parameter variation. (For an alternative 'interpolation' approach to reduced order time-parameter spaces see Amsallem and Farhat (2008); Amsallem *et al.* (2009).)

To begin, we summarize the well-known optimality property of the POD (Kunisch and Volkwein 2002). Given J elements of $X^{\mathcal{N}}$, $w_j \in X^{\mathcal{N}}, 1 \le j \le J$, and any positive integer $M \le J$, $\text{POD}(\{w_1, \ldots, w_J\}, M)$ returns M $(\cdot,\cdot)_X$-orthogonal functions $\{\chi_m, 1 \le m \le M\}$ such that the space $V_M = \text{span}\{\chi_m, 1 \le m \le M\}$ is optimal in the sense that

$$V_M = \arg \inf_{Y_M \subset \text{span}\{w_j, 1 \le j \le J\}} \left(\frac{1}{J} \sum_{j=1}^{J} \inf_{v \in Y_M} \|w_j - v\|_X^2 \right)^{1/2}$$

where Y_M denotes a M-dimensional linear space. We also recall that to find the χ_p we first form the correlation matrix C with entries $C_{ij} = (w_i, w_j)_X$, $1 \le i, j \le J$; we then find the largest M eigenvalues $\lambda^m, 1 \le m \le M$, and associated eigenvectors $v^m \in \mathbb{R}^J, 1 \le m \le M$, of the system $Cv^m = \lambda^m v^m$ with normalization $(v^m)^T v^m = 1$; finally we form $\chi_m = \sum_{j=1}^{J} v_j^m w_j, 1 \le m \le M$. Note that the χ_m thus satisfy the orthogonality condition $(\chi_m, \chi_n)_X = \lambda^m \delta_{mn}, 1 \le m, n \le M$.

To initiate the POD-Greedy$_{\text{RB}}$ sampling procedure we must specify a very large (exhaustive) 'training' sample of $n_{\text{train,RB}}$ points in $\mathcal{D}$, $\Xi_{\text{train,RB}}$, and an initial (say, random) RB parameter sample $S^* = \{\mu_0^*\}$. Typically we choose $\Xi_{\text{train,RB}}$ by Monte Carlo sampling over $\mathcal{D}$ with respect to a prescribed (usually uniform) density, however for P small (few parameters) often a uniform or log-uniform deterministic distribution is preferred. The algorithm is then given by

Set $\mathcal{Z} = \emptyset$;
Set $\mu^* = \mu_0^*$;
While $N \le N_{\max}$
  $\{\chi_m, 1 \le m \le M_1\} = \text{POD}(\{u^{\mathcal{N}}(t^k; \mu^*), 1 \le k \le K\}, M_1)$;
  $\mathcal{Z} \leftarrow \{\mathcal{Z}, \{\chi_m, 1 \le m \le M_1\}\}$;
  $N \leftarrow N + M_2$;
  $\{\xi_n, 1 \le n \le N\} = \text{POD}(\mathcal{Z}, N)$;
  $X_N = \text{span}\{\xi_n, 1 \le n \le N\}$;
  $\mu^* = \arg\max_{\mu \in \Xi_{\text{train,RB}}} \Delta_N(t^K = t_f; \mu)$;
  $S^* \leftarrow \{S^*, \mu^*\}$;
end.
Set $X_N = \text{span}\{\xi_n, 1 \le n \le N\}, 1 \le N \le N_{\max}$.

In actual practice, we typically exit the POD-Greedy sampling procedure at $N = N_{\max} \le N_{\max,0}$ for which a prescribed error tolerance is satisfied: to wit, we define

$$\varepsilon_{N,\max}^* = \max_{\mu \in \Xi_{\text{train,RB}}} \frac{\Delta_N(t^K; \mu)}{\|u_N(t^K; \mu)\|},$$

and terminate when $\varepsilon_{N,\max}^* \le \varepsilon_{\text{tol}}$. Note, by virtue of the final re-definition, the POD-Greedy generates *hierarchical* spaces X_N, $1 \le N \le N_{\max}$, which is computationally very advantageous.

There are two 'tuning' variables in the POD-Greedy$_{\rm RB}$ procedure, M_1 and M_2. We choose M_1 to satisfy an internal POD error criterion based on the usual sum of eigenvalues; we choose $M_2 \leq M_1$ to minimize duplication in the reduced basis space – though typically we prefer $M_2 > 1$ in order to reduce the number of Greedy$_{\rm RB}$ iterations and hence Offline cost. We make two observations. First, the POD – Greedy$_{\rm RB}$ method readily accommodates a repeat μ^* in successive Greedy$_{\rm RB}$ cycles – new information will always be available and old information rejected; in contrast, a pure Greedy$_{\rm RB}$ approach in both t and μ (Grepl and Patera 2005), though often generating good spaces, can 'stall.' Second, thanks to the POD normalization $(\chi_m, \chi_n)_X = \lambda^m \delta_{mn}, 1 \leq m, n \leq M_1$, the modes generated in the first POD at any parameter value μ^* are automatically scaled by their respective importance in representing $u(t^k; \mu^*), 1 \leq k \leq K$; the inputs to the second POD (of $\mathcal{Z}$) are thus correctly weighted to accommodate modes from different parameter values. (An alternative single-stage POD-Greedy$_{\rm RB}$ procedure is proposed in Knezevic and Patera (2009).)

The procedure remains computationally feasible even for large parameter domains and very extensive training samples (and in particular in higher parameter dimensions $P > 1$): the POD is conducted in only one (time) dimension and the Greedy$_{\rm RB}$ addresses the remaining (parameter) dimensions. The crucial point to note is that the operation count for the POD-Greedy$_{\rm RB}$ algorithm is additive and not multiplicative in $n_{\rm train,RB}$ and $\mathcal{N}$: in searching for the next parameter value μ^*, we invoke the Construction–Evaluation decomposition to inexpensively calculate the a posteriori error bound at the $n_{\rm train,RB}$ candidate parameter values; $\Delta_N(t^k; \mu)$ over $\Xi_{\rm train,RB}$ is computed by first *constructing* the necessary parameter-independent inner products at cost which depends on $\mathcal{N}$ but not on $n_{\rm train,RB}$ and then *evaluating* the error over all training points at cost $n_{\rm train,RB}(K + (Q_a + Q_m)^2)N^2$ (independent of $\mathcal{N}$) – hence the additive and not multiplicative dependence on $\mathcal{N}$ and $n_{\rm train,RB}$. In contrast, in a pure POD approach, we would need to evaluate the finite element 'truth' solution at the $n_{\rm train,RB}$ candidate parameter values at cost $O(n_{\rm train,RB}\mathcal{N}^{\bullet})$. (Of course, much of the computational economies are due not to the Greedy$_{\rm RB}$ *per se*, but rather to the accommodation within the Greedy$_{\rm RB}$ of the inexpensive error bounds.) As a result, in the POD–Greedy$_{\rm RB}$ approach we can take $n_{\rm train,RB}$ relatively large: we can thus anticipate reduced basis spaces and approximations that provide rapid convergence *uniformly* over the entire parameter domain. (Note that more sophisticated and hence efficient search algorithms can be exploited in the Greedy$_{\rm RB}$ context, see for example Bui-Thanh *et al.* (2007).)

We pursue the POD-Greedy$_{\rm RB}$ sampling procedure – which involves both the Construction and Evaluation phases – in an Offline stage. Then, in the Online stage, we invoke only the very inexpensive Evaluation phase: $\mu \rightarrow s_N^k(\mu), \Delta_N^k(\mu), 1 \leq k \leq K$. Note that in the POD-Greedy$_{\rm RB}$ procedure we choose for $g(t)$ the impulse function Grepl and Patera (2005); the resulting RB space will thus have good approximation properties for any $g(t)$, and hence $g(t)$ can be specified in the Online stage. (The latter property is of course lost for non-LTI problems.)

Summary

We briefly summarize here the various steps in the full algorithm. First a prerequisite: as described in Section 8.2.1, the original problem (potentially posed over a parameter–dependent domain) must be mapped to a transformed problem (over a reference domain); we must then confirm our hypotheses, including the affine assumption (8.3). Not all (original) problems will yield transformed problems that honor our hypotheses; and not all transformed problems can be efficiently treated by our RB approach – in particular if P or Q is too large, as discussed further below.

We then conduct the Offline stage. First, the (Offline component) of the SCM procedure is executed in order to provide the small database invoked by the Online component of the SCM lower bound for the coercivity constants, $\alpha_{\mathrm{LB}}^{\mathcal{N}}(\mu)$ of (8.16) and $\alpha_{\mathrm{LB}}^{\mathcal{N}}(\mu)$ of (8.17); we have chosen not to emphasize this algorithmic ingredient given extensive details in Huynh *et al.* (2007); Rozza *et al.* (2008). Second, the POD-Greedy$_{\mathrm{RB}}$ procedure is executed to provide the small database invoked by the Online component of the RB output and output error bound prediction; the database comprises $A_{N_{\max}}^q \in \mathbb{R}^{N_{\max} \times N_{\max}}, 1 \leq q \leq Q_a$, $M_{N_{\max}}^q \in \mathbb{R}^{N_{\max} \times N_{\max}}$, $1 \leq q \leq Q_m, F_{N_{\max}}^q \in \mathbb{R}^{N_{\max}}$, $1 \leq q \leq Q_f, L_{N_{\max}}^q \in \mathbb{R}^{N_{\max}}$, $1 \leq q \leq Q_\ell$ as defined in (8.22) for the output; the L^2 norm of ℓ and $(\Gamma_{N_{\max}}, \Gamma_{N_{\max}})_X$, $(\Lambda_{N_{\max}}^{qj}, \Lambda_{N_{\max}}^{q'j'})_X$, $1 \leq q, q', \leq Q_m, 1 \leq j, j' \leq N_{\max}$, $(\Upsilon_{N_{\max}}^{qj}, \Upsilon_{N_{\max}}^{q'j'})_X$, $1 \leq q, q', \leq Q_a, 1 \leq j, j' \leq N_{\max}$, $(\Lambda_{N_{\max}}^{qj}, \Upsilon_{N_{\max}}^{q'j'})_X, 1 \leq q \leq Q_m, 1 \leq q' \leq Q_a, 1 \leq j, j' \leq N_{\max}$, $(\Lambda_{N_{\max}}^{qj}, \Gamma_{N_{\max}})_X, 1 \leq q \leq Q_m, 1 \leq j \leq N_{\max}$, $(\Upsilon_{N_{\max}}^{qj}, \Gamma_{N_{\max}})_X$, $1 \leq q \leq Q_a, 1 \leq j \leq N_{\max}$, for the output error bound. (Recall that the necessary quantities for $N \leq N_{\max}$ can be extracted as subarrays of the corresponding $N_{\max}$ quantities.)

We may then exercise the Online stage. Given a parameter value $\mu \in \mathcal{D}$ of interest, we first calculate the output prediction: solution of (8.21) for the RB field coefficients followed by evaluation of the sum (8.20). We next calculate the output error bound: evaluation of the sums first of (8.29) and subsequently of (8.28) for the dual norm of the residual; computation of the SCM lower bounds $\alpha_{\mathrm{LB}}^{\mathcal{N}}$, $\sigma_{\mathrm{LB}}^{\mathcal{N}}$; and finally assembly of the final result, (8.19).

We shall focus here on the operation count for the Online stage. It is clear from our earlier discussions that to leading order the output and output error bound can be evaluated in $O(N^3 + (K + (Q_a + Q_m)^2)N^2)$ operations (the additional contribution of the SCM is typically negligible in the Online stage) – independent of $\mathcal{N}$. (The latter in turn implies that, at least as regards the Online stage, $\mathcal{N}$ may be chosen conservatively large.) We do not have a strong a priori theory which permits us to forecast the requisite N as a function of desired accuracy. However, at least for parametrically smooth problems, the POD-Greedy$_{\rm RB}$ spaces and Galerkin projection should yield – and our posteriori error bounds will *confirm* – RB output predictions which are highly accurate for $N \ll \mathcal{N}$. Nevertheless, it is important to emphasize that the Online operation count will certainly increase with the difficulty of the problem: N will increase with the number of parameters P and the extent of variation in the parameters as reflected in $\mathcal{D}$; Q_a, Q_m will increase with the geometric and operator complexity of the problem. In actual practice, for coercive problems with as many as $O(10)$ parameters $N = O(100)$ suffices (see Sen *et al.* (2006) for an elliptic example); even more parameters may be considered if the parametric representation is effective – $\mathcal{D}$ of increasingly limited extent in the higher parameter dimensions (Boyaval *et al.* 2008). Our example of Section 8.3 shall serve as additional calibration of computational effort associated with both the Online and also Offline stages.

8.3 Bayesian Parameter Estimation

8.3.1 Bayesian Approach

In parameter estimation problems we would like to infer the unknown parameter $\mu_\star \in \mathcal{D} \subset \mathbb{R}^P$ from the measurements of outputs of interest, $s^{(m)}(t;\mu_\star), 1 \le m \le M_{\rm out}$, collected for $t = t^{k_j^{\rm exp}} = k_j^{\rm exp}\Delta t \in [0, t_f], 1 \le j \le J$; here $M_{\rm out}$ is the number of outputs and J is the number of measurements per output. (In actual practice, some of the P parameters – for example, measurement system design variables – may be specified (or optimized) rather than inferred.) In our case the outputs are expressed as functionals of the solution of the forward problem (8.1) – $s^{(m)}(t;\mu_\star) = \ell^{(m)}(u(t;\mu_\star))$ for $1 \le m \le M_{\rm out}$. In order to assess our approach to parameter estimation we create 'synthetic' data as

$$G_{mj}^{\rm exp}(\mu_\star;\varepsilon_{\rm exp}) = s^{(m)\mathcal{N}}(t^{k_j^{\rm exp}};\mu_\star) + \varepsilon_{mj}^{\rm exp}, \quad 1 \le m \le M_{\rm out}, 1 \le j \le J, \tag{8.30}$$

where the $s^{(m)\mathcal{N}}(t^{k_j^{\rm exp}};\mu_\star)$ are the finite element approximation to the exact output $s^{(m)}(t^{k_j^{\rm exp}};\mu_\star)$ and the $\varepsilon_{mj}^{\rm exp}$ represent the 'experimental' error. We assume the $\varepsilon_{mj}^{\rm exp}$ to be independent identically distributed (i.i.d.) Gaussian random variables (hence white in time) with zero mean and known variance $\sigma_{\rm exp}^2$; our formulation can in fact treat any desired probability distribution.

We apply the Bayesian approach to parameter estimation (Mosegaard and Tarantola 2002) to the truth discretization of the forward problem (8.1). The expected value[2] $\mathsf{E}^{\mathcal{N}}[\mu_\star|G^{\rm exp}]$ of the unknown parameter $\mu_\star$ conditional on

[2]For brevity we consider only the expectation; our methodology also applies to the variance and indeed the full empirical posterior distribution function.

the data G^{exp} is given by

$$\mathsf{E}^{\mathcal{N}}[\mu_\star|G^{\text{exp}}] = \frac{\int_{\mathcal{D}} \mu \Pi^{\mathcal{N}}(G^{\text{exp}}|\mu)\Pi_0(\mu)d\mu}{\int_{\mathcal{D}} \Pi^{\mathcal{N}}(G^{\text{exp}}|\mu')\Pi_0(\mu')d\mu'}. \tag{8.31}$$

Here the likelihood function $\Pi^{\mathcal{N}}(G^{\text{exp}}|\mu)$ is given by

$$\Pi^{\mathcal{N}}(G^{\text{exp}}|\mu) = \left(\frac{1}{2\pi\sigma_{\text{exp}}^2}\right)^{M_{\text{out}}J/2} \exp\left(-\frac{(G^{\text{exp}} - F^{\mathcal{N}}(\mu))^T(G^{\text{exp}} - F^{\mathcal{N}}(\mu))}{2\sigma_{\text{exp}}^2}\right), \tag{8.32}$$

where, for $1 \leq m \leq M_{\text{out}}$ and $1 \leq j \leq J$, $F_{mj}^{\mathcal{N}} : \mu \in \mathcal{D} \to s^{(m)\mathcal{N}}(t^{k_j^{\text{exp}}}; \mu)$ denotes the finite element evaluation of the mth output at time $t^{k_j^{\text{exp}}}$ at any given μ in our parameter domain $\mathcal{D}$. The prior distribution on the parameter μ, $\Pi_0(\mu)$, is also assumed Gaussian[3]

$$\Pi_0(\mu) = \left(\frac{1}{2\pi\sigma_0^2}\right)^{P/2} \exp\left(-\frac{(\mu - \mu_0)^T(\mu - \mu_0)}{2\sigma_0^2}\right), \tag{8.33}$$

where $\mu_0 \in \mathcal{D}$ is the prior mean and σ_0^2 is the associated variance (more generally a covariance); our approach is not limited to any particular prior. Note that $\mathsf{E}^{\mathcal{N}}[\mu_\star|G^{\text{exp}}]$ in (8.31) is an expectation with respect to the 'random' parameter μ: for any given measurement, G^{exp}, $\mathsf{E}^{\mathcal{N}}[\mu_\star|G^{\text{exp}}]$ is our estimator for $\mu_\star$; properly speaking, $\mathsf{E}^{\mathcal{N}}[\mu_\star|G^{\text{exp}}]$ is a realization of a random variable – a function of G^{exp}. (To avoid cumbersome notation, G^{exp} refers both to the measurement random variable and to associated realizations.)

The expected value in (8.31) necessitates the computation of multidimensional integrals, which in turn require numerous evaluations of the truth outputs; in particular, in the remainder of this section we shall interpret

$$\int_{\mathcal{D}} \Phi(\mu) \equiv \sum_{i=1}^{n_{\text{quad}}} w_i^{\text{quad}} \Phi(\mu_i^{\text{quad}}), \tag{8.34}$$

for $w_i^{\text{quad}} \in \mathbb{R}_+$, $\mu_i^{\text{quad}} \in \mathcal{D}$, $1 \leq i \leq n_{\text{quad}}$.[4] As a consequence, the parameter estimation procedure can be very expensive. To reduce the computational cost of Bayesian inverse analysis Wang and Zabaras (2005) introduce POD-based model reduction. Our emphasis here is a posteriori error

[3]In theory, we must multiply (8.33) by a pre-factor reflecting the bounded $\mathcal{D}$. In practice, we shall consider small σ_0 and large μ_0 such that μ outside $\mathcal{D}$ are highly improbable – and hence $\mathcal{D}$ is effectively $\mathbb{R}^P$.

[4]In this chapter we consider an adaptive piecewise Gauss–Legendre technique: we first create a domain decomposition selectively refined near an approximate $\mu_\star$; we then apply standard tensor–product Gauss–Legendre quadrature within each subdomain. We denote by n_{quad} the total number of integrand evaluations required. For problems with more parameters, Monte Carlo techniques would be necessary.

estimation (absent in earlier Bayesian model reduction approaches): our error bounds shall ensure that our Bayesian inferences are (*i*) certifiably accurate (relative to the truth), and (*ii*) as efficient as possible – through optimal choice of N for a given error tolerance. In the subsequent subsection, we incorporate our a posteriori error bounds into the Bayesian approach to permit rapid and reliable parameter estimation. (See also Grepl (2005); Grepl *et al.* (2007b) for an alternative approach to RB inverse analysis which more explicitly characterizes parameter uncertainty.)

8.3.2 A Posteriori Bounds for the Expected Value

We develop here *inexpensive, rigorous* lower and upper bounds for the expected value (8.31) (with quadratures evaluated as (8.34)) based on the RB outputs and associated output error bounds. Toward this end, we first introduce $F_{N\,mj}(\mu) = s_N^{(m)}(t^{k_j^{\exp}};\mu)$ and $\Delta F_{N\,mj}(\mu) = \Delta_N^{s\,(m)}(t^{k_j^{\exp}};\mu)$ for $1 \leq m \leq M_{\text{out}}$ and $1 \leq j \leq J$, and then $F_N^{\pm}(\mu) = F_N(\mu) \pm \Delta F_N(\mu)$; here $s_N^{(m)}(t^k;\mu)$ and $\Delta_N^{s\,(m)}(t^k;\mu)$ are the RB prediction and associated error bound for the mth output. We then define, for $1 \leq m \leq M_{\text{out}}$ and $1 \leq j \leq J$,

$$B_{N\,mj}(\mu) = \max\{|G_{mj}^{\exp} - F_{N\,mj}^{-}(\mu)|, |G_{mj}^{\exp} - F_{N\,mj}^{+}(\mu)|\},$$

and

$$D_{N\,mj}(\mu) = \begin{cases} 0, & \text{if } G_{mj}^{\exp} \in [F_{N\,mj}^{-}(\mu), F_{N\,mj}^{+}(\mu)], \\ \min\{|G_{mj}^{\exp} - F_{N\,mj}^{-}(\mu)|, & \\ \qquad |G_{mj}^{\exp} - F_{N\,mj}^{+}(\mu)|\}, & \text{otherwise}. \end{cases} \tag{8.35}$$

Note that $G^{\exp} \in \mathbb{R}^{M_{\text{out}}J}$, $F_N^{\pm}(\mu) \in \mathbb{R}^{M_{\text{out}}J}$, $D_N(\mu) \in \mathbb{R}^{M_{\text{out}}J}$, and $B_N(\mu) \in \mathbb{R}^{M_{\text{out}}J}$.

We now introduce two new likelihood functions

$$\begin{aligned} \Pi_N^a(G^{\exp}|\mu) &= \left(\frac{1}{2\pi\sigma_{\exp}^2}\right)^{M_{\text{out}}J/2} \exp\left(-\frac{D_N^T(\mu)D_N(\mu)}{2\sigma_{\exp}^2}\right), \\ \Pi_N^b(G^{\exp}|\mu) &= \left(\frac{1}{2\pi\sigma_{\exp}^2}\right)^{M_{\text{out}}J/2} \exp\left(-\frac{B_N^T(\mu)B_N(\mu)}{2\sigma_{\exp}^2}\right), \end{aligned} \tag{8.36}$$

from which we may evaluate

$$\begin{aligned} \mathsf{E}_N^{\text{LB}}[\mu_\star|G^{\exp}] &= \frac{\int_{\mathcal{D}} \mu \Pi_N^b(G^{\exp}|\mu)\Pi_0(\mu)d\mu}{\int_{\mathcal{D}} \Pi_N^a(G^{\exp}|\mu')\Pi_0(\mu')d\mu'}, \\ \mathsf{E}_N^{\text{UB}}[\mu_\star|G^{\exp}] &= \frac{\int_{\mathcal{D}} \mu \Pi_N^a(G^{\exp}|\mu)\Pi_0(\mu)d\mu}{\int_{\mathcal{D}} \Pi_N^b(G^{\exp}|\mu')\Pi_0(\mu')d\mu'}. \end{aligned} \tag{8.37}$$

(If μ takes on negative values then (8.37) must be modified slightly.) We shall take

$$\mathsf{E}_N^{\mathrm{AV}}[\mu_\star|G^{\mathrm{exp}}] = \frac{1}{2}(\mathsf{E}_N^{\mathrm{LB}}[\mu_\star|G^{\mathrm{exp}}] + \mathsf{E}_N^{\mathrm{UB}}[\mu_\star|G^{\mathrm{exp}}])$$

as our RB approximation to $\mathsf{E}^{\mathcal{N}}[\mu_\star|G^{\mathrm{exp}}]$.

It can be shown that the expected values defined in (8.37) satisfy

$$\mathsf{E}_N^{\mathrm{LB}}[\mu_\star|G^{\mathrm{exp}}] \leq \mathsf{E}^{\mathcal{N}}[\mu_\star|G^{\mathrm{exp}}] \leq \mathsf{E}_N^{\mathrm{UB}}[\mu_\star|G^{\mathrm{exp}}], \tag{8.38}$$

and hence

$$\begin{aligned}|\mathsf{E}^{\mathcal{N}}[\mu_\star|G^{\mathrm{exp}}] - \mathsf{E}_N^{\mathrm{AV}}[\mu_\star|G^{\mathrm{exp}}]| &\leq \frac{1}{2}\Delta\mathsf{E}_N[\mu_\star|G^{\mathrm{exp}}] \\ &\equiv \frac{1}{2}(\mathsf{E}_N^{\mathrm{UB}}[\mu_\star|G^{\mathrm{exp}}] - \mathsf{E}_N^{\mathrm{LB}}[\mu_\star|G^{\mathrm{exp}}]).\end{aligned}$$

We sketch the proof: we first note that, since $|s^{(m)\mathcal{N}}(t^{k_j^{\mathrm{exp}}};\mu) - s_N^{(m)}(t^{k_j^{\mathrm{exp}}};\mu)| \leq \Delta_N^{s\,(m)}(t^{k_j^{\mathrm{exp}}};\mu)$,

$$F_N^-(\mu) \leq F^{\mathcal{N}}(\mu) \leq F_N^+(\mu), \quad \forall \mu \in \mathcal{D}; \tag{8.39}$$

it thus follows that

$$D_N(\mu)^T D_N(\mu) \leq (G^{\mathrm{exp}} - F^{\mathcal{N}}(\mu))^T (G^{\mathrm{exp}} - F^{\mathcal{N}}(\mu)) \leq B_N(\mu)^T B_N(\mu), \tag{8.40}$$

and hence

$$\Pi_N^b(G^{\mathrm{exp}}|\mu) \leq \Pi^{\mathcal{N}}(G^{\mathrm{exp}}|\mu) \leq \Pi_N^a(G^{\mathrm{exp}}|\mu). \tag{8.41}$$

The bound result (8.38) is a direct consequence of the definitions (8.37) and inequality (8.41), and the non-negativity of Π^a, Π^b, Π_0, (here) $\mu \in \mathcal{D}$, and finally the quadrature weights.

In words, Π_N^a is an upper bound for $\Pi^{\mathcal{N}}$ since we exploit the reduced basis error bounds to ensure that for each quadrature point the argument of the Π_N^a Gaussian is of smaller magnitude than the argument of the $\Pi^{\mathcal{N}}$ – we *underestimate* the difference between the experimental data and the model prediction; similar arguments demonstrate that Π_N^b constitutes a lower bound for $\Pi^{\mathcal{N}}$ – now we *overestimate* the difference between the experimental data and the model prediction. Thus (given our non-negativity hypotheses) we can selectively choose upper bounds or lower bounds for the numerator and denominator of (8.31) as provided in (8.37). Note that for probability distributions that do not decay monotonically away from the (assumed zero) mean the same procedure can be applied but D_N and B_N will now be slightly more complicated (though still inexpensive to evaluate). We also emphasize that our error estimator $\Delta\mathsf{E}_N[\mu_*|G^{\mathrm{exp}}]$ is a rigorous bound for the difference between the expectation as calculated from the truth and the expectation as calculated from the reduced basis approximation *for the same quadrature formula* (8.34). Finally, we can not yet propose a similar error bound for the case of a Markov Chain Monte Carlo approach to the Bayesian estimation problem; we defer this topic to future work.

In the Offline stage the RB is constructed: the POD-Greedy$_{\rm RB}$ sampling procedure is invoked and all necessary Online quantities are computed and stored. Then, in the Online stage (which involves only the Evaluation phase), for each new identification ($\mu_\star$) – and hence for each new $G^{\rm exp}$ provided – we evaluate in 'real–time' the expectation lower and upper bounds (8.37). (Note for given $G^{\rm exp}$ the RB outputs and associated error bounds are computed (only once) and stored on the quadrature grid; we can then evaluate the several requisite integrals without further appeal to the RB approximation.) It is clear that the RB approach will be much faster than direct FE evaluation (of the requisite integrals) even for a single identification, and even more efficient for multiple identifications: in the limit that $n_{\rm quad}$ and/or the number of identifications tends to infinity, the RB Offline effort is negligible – only the very fast ($\mathcal{N}$–independent) RB Online evaluations are relevant. Equivalently, if our emphasis is on real–time identification, again only the very fast RB Online evaluations are important.

8.3.3 Numerical Example

We consider the application of transient thermal analysis to detection of flaws/defects in a Fiber-Reinforced Polymer (FRP) composite bonded to a concrete (C) slab (Grepl 2005; Starnes 2002). Since debonds or delaminations at the composite-concrete interface often occur (even at installation), effective and real-time quality control – providing reliable information about the thickness and fiber content of the composite, and the location and size of defects – is vital to safety.

We show the FRP-concrete system in Figure 8.1. The FRP layer is of thickness $h_{\rm FRP}$ and (truncated) lateral extent $10h_{\rm FRP}$; the concrete layer is of (truncated) depth and lateral extent $5h_{\rm FRP}$ and $10h_{\rm FRP}$, respectively. We presume that a delamination crack of *unknown* length $w_{\rm del}$ centered at $x_1 = 0$ is present at the FRP–concrete interface. The FRP thermal conductivity, specific heat, and density are given by k, c, and ρ with subscripts FRP and C, respectively. We shall assume that the FRP and concrete share the same *known* values for both the density and specific heat. We assume that the FRP (respectively, concrete) conductivity is *unknown* (respectively, *known*); we denote the (*unknown*) conductivity ratio as $\kappa = k_{\rm FRP}/k_{\rm C}$. (In practice, the FRP conductivity depends on fiber orientation and content – and hence somewhat unpredictable.)

We nondimensionalize all lengths by $h_{\rm FRP}/2$ and all times by $h^2_{\rm FRP}\rho_C c_C/4k_C$. The nondimensional temperature u is given by $(T - T_0)/(T_{\rm FRP,max} - T_0)$, where T is the dimensional temperature, T_0 is the initial

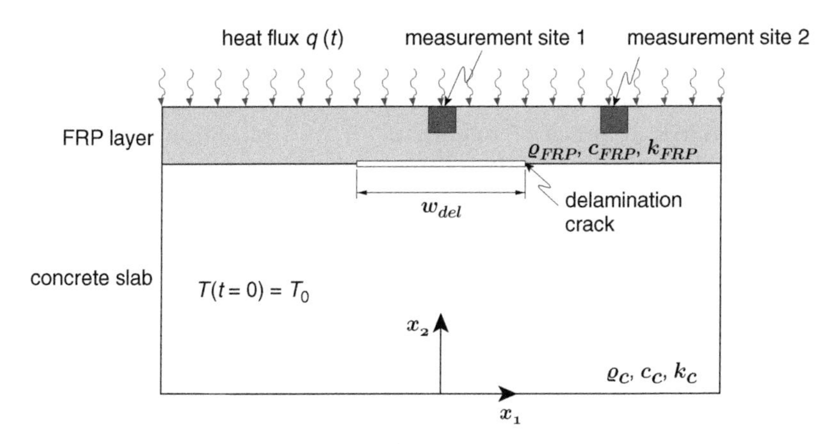

Figure 8.1 Delamination of a FRP layer bonded to a concrete slab.

temperature (uniform in both the FRP and concrete), and $T_{\mathrm{FRP,max}}$ is the maximum allowable FRP temperature. The nondimensional flux – imposed at the FRP exposed surface, as shown in Figure 8.1 – $g(t)$ is given by $q(t)h_{\mathrm{FRP}}/(2k_C(T_{\mathrm{FRP,max}} - T_0))$, where $q(t)$ is the dimensional flux. We presume that the nondimensional surface heat flux $g(t)$ – the stimulus – is unity for $0 \leq t \leq 5$ and zero for all $t > 5$. Henceforth, we refer only to non-dimensional quantities (and thus now w_{del} should be re-interpreted as the dimensional quantity normalized by $h_{\mathrm{FRP}}/2$.)

Upon application of our mapping procedures (to a reference domain with crack length $w_{\mathrm{del}} = 3$) as in Rozza *et al.* (2008) we arrive at the transformed problem statement (1) with affine expansions (2) for $Q_a = 15, Q_m = 2$. (In fact, due to symmetry, we consider only half the domain: $x_1 > 0$.) Our initial condition is $u = 0$; we integrate to a final time $t_f = 10.0$. Our $P = 2$ (both 'unknown') parameters are $\mu \equiv (\mu_1, \mu_2) \equiv (w_{\mathrm{del}}/2, \kappa)$ assumed to reside in the parameter domain $\mathcal{D} \equiv [1, 5] \times [0.5, 2]$. Finally, we introduce our truth discretization: we consider Euler backward discretization in time with $\Delta t = 0.05$ and hence $K = 200$ time levels $t^k = k\Delta t, 0 \leq k \leq K$; we consider a linear truth finite element approximation space $X^{\mathcal{N}}$ of dimension $\mathcal{N} = 3581$. (The triangulation provides high resolution in the vicinity of the surface and near the crack tip, the two regions which suffer sharp spatial gradients.) Finally, we consider $M_{\mathrm{out}} = 2$ outputs: as shown in Figure 8.1, each output functional corresponds to the average of the (temperature) field[5] over a 'small' square of side – length 1 (flush with the exposed FRP surface); the square for the first output is centered at (measurement site 1) $x_1 = 0$, while the square for the second output is centered at (measurement site 2) $x_1 = 6.5$.

[5]Note that we must consider a small area average (rather than pointwise measurement) to ensure that our output functionals remain bounded over $L^2(\Omega)$ (indeed, even over $H^1(\Omega)$); the $L^2(\Omega)$ norm of these 'area averaging' functionals increases as the inverse of the square root of the area.

We first briefly discuss the RB approximation and error bounds, and then turn to the inverse problem. This PDE is not too difficult: we need an RB space of dimension only $N = 50$ to ensure – based on $\Delta_N^{s\,(m)}(t^k, \mu), m = 1, 2$ – a 'certified' accuracy of roughly 0.5 % in both outputs. In fact, the effectivity – the ratio of the output error bound to the true output error – is rather large, $O(100)$, and hence the *actual* accuracy for $N = 50$ is less than 10^{-4}; however, since in the Online stage our inferences are based on the (inexpensive) error bound, we must construct an RB approximation for which the *error bound* is sufficiently accurate. For $N = 50$ the Online RB calculation $\mu \to s_N^{(m)}(t^k; \mu), \Delta_N^{s\,(m)}(t^k; \mu)$, $0 \le k \le K$, is effected in 0.21 seconds; in contrast, direct truth evaluation requires 22 seconds. All computations in this section are carried out on a 1.73 GHz Pentium IV processor with 1GB memory.

We now turn to parameter estimation. We focus on the sensitivity of the parameter estimation procedure to the RB dimension N as (inexpensively but rigorously) quantified by our expectation error bounds. In this experiment, we set $\mu_\star = (\mu_{1\star}, \mu_{2\star}) = (w_{\text{del}\star}, \kappa_\star) = (2.8, 0.9)$ and $\sigma^2_{\text{exp}} = 0.0025$; we choose for the prior mean and variance $\mu_0 = (3.3, 1.2)$ and $\sigma_0^2 = 0.04$, respectively. The synthetic experimental data (8.30) is generated by adding i.i.d. Gaussian random variables to our $M_{\text{out}} = 2$ outputs evaluated at $J = 20$ time levels $t^{k_j^{\text{exp}}}$, $k_j^{\text{exp}} = 10j, 1 \le j \le J$. We then apply our adaptive piecewise Gauss-Legendre quadrature algorithm with $n_{\text{quad}} = 10,000$ points.

We present in Table 8.1 the lower bound, $\mathsf{E}_N^{\text{LB}}[\mu_{p\star}]$, upper bound, $\mathsf{E}_N^{\text{UB}}[\mu_{p\star}]$, and bound gap $\Delta\mathsf{E}_N[\mu_{p\star}], p = 1, 2$, for the expected value of the unknown parameter $\mu_\star$; we consider a single realization G^{exp}. We observe that the bound gaps $\Delta\mathsf{E}_N[\mu_{p\star}] = \mathsf{E}_N^{\text{UB}}[\mu_{p\star}] - \mathsf{E}_N^{\text{LB}}[\mu_{p\star}], p = 1, 2$, decrease rapidly: as N increases, $\Delta_N^{s\,(m)}(t^k; \mu) \to 0$ and hence $D_N(\mu) \to B_N(\mu)$ rapidly. The parameter estimator is quite accurate: the expectation bounds for larger N are within the white noise (5.0 %) of the true parameter value $\mu_\star = (2.8, 0.9)$, biased toward μ_0 as expected. The RB Online computation (for $N = 50$) of the lower and upper bounds for the expected value is completed in approximately

Table 8.1 Lower bound, upper bound, and bound gap for the expected value of the delamination half-width μ_1 and conductivity ratio μ_2 as a function of N. The true parameter value is $\mu_{1\star} = 2.8$ and $\mu_{2\star} = 0.9$.

	Delamination half-width			Conductivity ratio		
N	$\mathsf{E}_N^{\text{LB}}[\mu_{1\star}]$	$\mathsf{E}_N^{\text{UB}}[\mu_{1\star}]$	$\Delta\mathsf{E}_N[\mu_{1\star}]$	$\mathsf{E}_N^{\text{LB}}[\mu_{2\star}]$	$\mathsf{E}_N^{\text{UB}}[\mu_{2\star}]$	$\Delta\mathsf{E}_N[\mu_{2\star}]$
10	1.0527	7.5175	6.4648	0.3427	2.4468	2.1041
20	2.3896	3.3120	0.9224	0.7764	1.0759	0.2996
30	2.7417	2.8836	0.1419	0.8917	0.9378	0.0461
40	2.8008	2.8236	0.0228	0.9111	0.9185	0.0074
50	2.8096	2.8192	0.0096	0.9136	0.9171	0.0035

35 minutes – arguably 'real-time' for this particular application – as opposed to 61 hours for direct FE evaluation. The RB Offline time is roughly 2.8 hours, and hence even for one identification the RB approach 'pays off'; for several identifications, the RB Offline effort will be negligible. (If real-time response 'in the field' is imperative, then even for one identification the RB Offline effort is not important.) In short, we are guaranteed the fidelity of the truth FE approximation but at the cost of a low order model.

8.4 Concluding Remarks

In this chapter we have developed a framework for reduced basis approximation and a posteriori error estimation for parametrized linear parabolic partial differential equations. We have argued, and computationally confirmed, that the reduced basis approach can provide highly accurate, very inexpensive, rigorously certified predictions in the real–time and many–query contexts. We have further demonstrated that the certified reduced basis method can be integrated into a Bayesian framework to provide very rapid yet reliable parameter estimation procedures; similar advances should be possible for optimization and control applications as well as multi-scale analyses.

Certainly the most important outstanding issue (at least within the context of parabolic partial differential equations) is generality: given an 'original' problem of interest (*i*) is there an effective parametrization such that the resulting 'transformed' problem is amenable to efficient and rigorous reduced basis treatment, and (*ii*) can this effective parametrization be automatically deduced and subsequently implemented within a general framework? At present we know of large classes of linear problems and much smaller classes of nonlinear problems (Nguyen *et al.* 2009) which can be and have been successfully addressed by the certified reduced basis approach; future work must focus both on theoretical advances to identify important impediments and computational advances to address these restrictions.

Acknowledgement

This work was supported by AFOSR Grant FA9550-07-1-0425 and the Singapore-MIT Alliance.

References

Almroth, B.O., Stern, P. and Brogan, F.A. 1978 Automatic choice of global shape functions in structural analysis. *AIAA Journal* **16**, 525–528.

Amsallem, D. and Farhat, C. 2008 Interpolation method for adapting reduced-order models and application to aeroelasticity. *AIAA Journal* **46**(7), 1803–1813.

Amsallem, D., Cortial, J. and Farhat, C. 2009 On-demand CFD-based aeroelastic predictions using a database of reduced-order bases and models. *47th AIAA Aerospace Sciences Meeting, Including the New Horizons Forum and Aerospace Exposition, 5–8 January 2009, Orlando, Florida*. Paper 2009-800.

Balmes, E. 1996 Parametric families of reduced finite element models: Theory and applications. *Mechanical Systems and Signal Processing* **10**(4), 381–394.

Barrault, M., Maday, Y., Nguyen, N.C. and Patera, A.T. 2004 An 'empirical interpolation' method: Application to efficient reduced-basis discretization of partial differential equations. *C. R. Acad. Sci. Paris, Série I*. **339**, 667–672.

Boyaval, S. 2008 Reduced-basis approach for homogenization beyond the periodic setting. *SIAM Multiscale Modeling & Simulation* **7**(1), 466–494.

Boyaval, S., Le Bris, C., Nguyen, N.C. and Patera, A.T. 2008 A reduced-basis approach for variational problems with stochastic parameters: Application to heat conduction with variable robin coefficient. *Comp. Meth. Appl. Mech. and Eng*.

Bui-Thanh, T., Willcox, K. and Ghattas, O. 2007 Model reduction for large-scale systems with high-dimensional parametric input space (AIAA Paper 2007-2049) *Proceedings of the 48th AIAA/ASME/ASCE/AHS/ASC Structures, Structural Dynamics and Material Conference*.

Cancès, E., Le Bris, C., Nguyen, N.C., Maday, Y., Patera, A.T. and Pau, G.S.H. 2007 Feasibility and competitiveness of a reduced basis approach for rapid electronic structure calculations in quantum chemistry. *Proceedings of the Workshop for High-dimensional Partial Differential Equations in Science and Engineering (Montreal)*.

Christensen, E.A., Brons, M. and Sorensen, J.N. 2000 Evaluation of pod-based decomposition techniques applied to parameter-dependent non-turbulent flows. *SIAM J. Sci. Comput*. **21**, 1419–1434.

Constantin, P. and Foias, C. 1988 *Navier-Stokes Equations*. Chicago Lectures in Mathematics. University of Chicago Press, Chicago, IL.

Deane, A., Kevrekidis, I., Karniadakis, G. and Orszag, S. 1991 Low-dimensional models for complex geometry flows: Application to grooved channels and circular cylinders. *Phys. Fluids* **10**, 2337–2354.

Deparis, S. 2008 Reduced basis error bound computation of parameter-dependent Navier–Stokes equations by the natural norm approach. *SIAM Journal of Numerical Analysis* **46**, 2039–2067.

Fink, J.P. and Rheinboldt, W.C. 1983 On the error behavior of the reduced basis technique for nonlinear finite element approximations. *Z. Angew. Math. Mech.* **63**(1), 21–28.

Grepl, M. 2005 *Reduced-Basis Approximations and A Posteriori Error Estimation for Parabolic Partial Differential Equations*. PhD thesis. Massachusetts Institute of Technology.

Grepl, M.A. and Patera, A.T. 2005 *A posteriori* error bounds for reduced-basis approximations of parametrized parabolic partial differential equations. *M2AN (Math. Model. Numer. Anal.)* **39**(1), 157–181.

Grepl, M.A., Maday, Y., Nguyen, N.C. and Patera, A.T. 2007a Efficient reduced-basis treatment of nonaffine and nonlinear partial differential equations. *M2AN (Math. Model. Numer. Anal.)* **41**, 575–605.

Grepl, M.A., Nguyen, N.C., Veroy, K., Patera, A.T. and Liu, G.R. 2007b Certified rapid solution of partial differential equations for real-time parameter estimation and optimization. In *Proceedings of the 2nd Sandia Workshop of PDE-Constrained Optimization: Real-Time PDE-Constrained Optimization* (ed. Biegler, L.T., Ghattas, O., Heinkenschloss, M., Keyes, D. and van B. Wandeers, B.), pp. 197–216 SIAM Computational Science and Engineering Book Series.

Gunzburger, M.D. 1989 *Finite Element Methods for Viscous Incompressible Flows.* Academic Press.

Gunzburger, M.D., Peterson, J. and Shadid, J.N. 2007 Reduced-order modeling of time-dependent PDEs with multiple parameters in the boundary data. *Comp. Meth. Applied Mech.* **196**, 1030–1047.

Haasdonk, B. and Ohlberger, M. 2008 Reduced basis method for finite volume approximations of parametrized evolution equations. *M2AN (Math. Model. Numer. Anal.)* **42**(2), 277–302.

Hinze, M. and Volkwein, S. 2005 Proper orthogonal decomposition surrogate models for nonlinear dynamical systems: error estimates and suboptimal control. *Lecture Notes in Computational Science and Engineering* vol. 45 Springer.

Huynh, D.B.P., Nguyen, N.C., Rozza, G. and Patera, A.T. 2007-2009 *rbMIT Software:* http://augustine.mit.edu/methodology/methodology_rbMIT_System.htm. Copyright MIT, Technology Licensing Office, case 12600, Cambridge, MA.

Huynh, D.B.P., Rozza, G., Sen, S. and Patera, A.T. 2007 A successive constraint linear optimization method for lower bounds of parametric coercivity and inf-sup stability constants. *C. R. Acad. Sci. Paris, Analyse Numérique* **345**(8), 473–478.

Ito, K. and Ravindran, S.S. 1998a A reduced basis method for control problems governed by PDEs. In *Control and Estimation of Distributed Parameter Systems* (ed. Desch, W., Kappel, F. and Kunisch, K.) Birkhäuser pp. 153–168.

Ito, K. and Ravindran, S.S. 1998b A reduced-order method for simulation and control of fluid flows. *Journal of Computational Physics* **143**(2), 403–425.

Ito, K. and Ravindran, S.S. 2001 Reduced basis method for optimal control of unsteady viscous flows. *International Journal of Computational Fluid Dynamics* **15**(2), 97–113.

Johansson, P.S., Andersson, H. and Rønquist, E. 2006 Reduced-basis modeling of turbulent plane channel flow. *Computers and Fluids* **35**(2), 189–207.

Johnson, C., Rannacher, R. and Boman, M. 1995 Numerical and hydrodynamic stability: Towards error control in computational fluid dynamics. *SIAM Journal of Numerical Analysis* **32**(4), 1058–1079.

Joseph, D. 1976 *Stability of fluid motions. I. & II.* vol. 27 & 28 of *Springer Tracts in Natural Philosophy*. Springer-Verlag, New York.

Karniadakis, G.E., Mikic, B.B. and Patera, A.T. 1988 Minimum-dissipation transport enhancement by flow destabilization – Reynolds' analogy revisited. *Journal of Fluid Mechanics* **192**, 365–391.

Knezevic, D. and Patera, A.T. 2009 A certified reduced basis method for fokker–planck equation of dilute polymeric fluids: Fene dumbbells in extensional flow. *SIAM J. Sci. Comput.* Submitted.

Kunisch, K. and Volkwein, S. 2002 Galerkin proper orthogonal decomposition methods for a general equation in fluid dynamics. *SIAM J. Num. Analysis* **40**(2), 492–515.

Kunish, K. and Volkwein, S. 1999 Control of Burgers' equation by a reduced order approach using proper orthogonal decomposition. *J. Optimization Theory and Applications* **102**, 345–371.

Maday, Y., Patera, A. and Turinici, G. 2002 *A Priori* convergence theory for reduced-basis approximations of single-parameter elliptic partial differential equations. *Journal of Scientific Computing* **17**(1–4), 437–446.

Mosegaard, K. and Tarantola, A. 2002 Probabilistic approach to inverse problems. *International Handbook of Earthquake and Engineering Seismology, Part A* pp. 237–265.

Nguyen, N.C. 2008 Efficient and robust parameter estimation using the reduced basis method: Application to the inverse scattering problem. *SIAM J. Sci. Comput.*

Nguyen, N.C. 2008 A multiscale reduced-basis method for parametrized elliptic partial differential equations with multiple scales. *Journal of Computational Physics* **227**(23), 9807–9822.

Nguyen, N.C., Rozza, G. and Patera, A.T. 2009 Reduced basis approximation and a posteriori error estimation for the time-dependent viscous burgers equation. *Calcolo* **46**, 157–185.

Nguyen, N.C., Veroy, K. and Patera, A.T. 2005 Certified real-time solution of parametrized partial differential equations. In *Handbook of Materials Modeling* (ed. Yip, S.) Springer pp. 1523–1558.

Noor, A.K. and Peters, J.M. 1980 Reduced basis technique for nonlinear analysis of structures. *AIAA Journal* **18**(4), 455–462.

Pierce, N. and Giles, M.B. 2000 Adjoint recovery of superconvergent functionals from PDE approximations. *SIAM Review* **42**(2), 247–264.

Porsching, T.A. 1985 Estimation of the error in the reduced basis method solution of nonlinear equations. *Mathematics of Computation* **45**(172), 487–496.

Prud'homme, C., Rovas, D., Veroy, K., Maday, Y., Patera, A. and Turinici, G. 2002 Reliable real-time solution of parametrized partial differential equations: Reduced-basis output bounds methods. *Journal of Fluids Engineering* **124**(1), 70–80.

Quarteroni, A. and Valli, A. 1997 *Numerical Approximation of Partial Differential Equations*, 2nd edn. Springer.

Rozza, G., Huynh, D.B.P. and Patera, A.T. 2008 Reduced basis approximation and a posteriori error estimation for affinely parametrized elliptic coercive partial differential equations: Application to transport and continuum mechanics. *Archives Computational Methods in Engineering* **15**(3), 229–275.

Sen, S., Veroy, K., Huynh, D.B.P., Deparis, S., Nguyen, N.C. and Patera, A.T. 2006 'Natural norm' *a posteriori* error estimators for reduced basis approximations. *Journal of Computational Physics* **217**, 37–62.

Starnes, M. 2002 *Development of Technical Bases for Using Infrared Thermography for Nondestructive Evaluation of Fiber Reinforced Polymer Composites Bonded to Concrete* PhD thesis Massachusetts Institute of Technology.

Veroy, K. and Patera, A.T. 2005 Certified real-time solution of the parametrized steady incompressible Navier-Stokes equations; Rigorous reduced-basis *a posteriori* error bounds. *International Journal for Numerical Methods in Fluids* **47**, 773–788.

Veroy, K., Prud'homme, C. and Patera, A.T. 2003a Reduced-basis approximation of the viscous Burgers equation: Rigorous *a posteriori* error bounds. *C. R. Acad. Sci. Paris, Série I* **337**(9), 619–624.

Veroy, K., Prud'homme, C., Rovas, D.V. and Patera, A.T. 2003b *A Posteriori* error bounds for reduced-basis approximation of parametrized noncoercive and nonlinear elliptic partial differential equations. *Proceedings of the 16th AIAA Computational Fluid Dynamics Conference*. Paper 2003-3847.

Wang, J. and Zabaras, N. 2005 Using Bayesian statistics in the estimation of heat source in radiation. *International Journal of Heat and Mass Transfer* **48**, 15–29.

Chapter 9

Calibration and Uncertainty Analysis for Computer Simulations with Multivariate Output

J. McFarland[1] and L. Swiler[2]
[1]*Southwest Research Institute, USA*
[2]*Sandia National Laboratories, USA*

9.1 Introduction

The significance of uncertainty in the modeling and simulation process is often overlooked. No model perfectly represents reality, so it is important to ask how imperfect a model is before it is applied for prediction. The scientific community relies heavily on modeling and simulation tools for forecasting, parameter studies, design, and decision making. These are all activities that can strongly benefit from meaningful representations of modeling uncertainty. For example, forecasts can contain error bars, designs can be made more robust, and decision makers can be better informed when modeling uncertainty is quantified to support these activities.

One opportunity to quantify contributors to the uncertainty in model predictions arises in the calibration process. Here calibration refers to the estimation of uncertain simulator inputs observations of the simulator output.

Large-Scale Inverse Problems and Quantification of Uncertainty Edited by L. Biegler, G. Biros, O. Ghattas, M. Heinkenschloss, D. Keyes, B. Mallick, Y. Marzouk, L. Tenorio, B. van Bloemen Waanders and K. Willcox

As such, model calibration can be viewed as an inverse problem. However, this type of calibration analysis poses several problems in practice:

1. The simulation is often expensive, rendering an exhaustive exploration of the parameter spacc infeasible.
2. Various ranges and/or combinations of input parameters may yield comparable fits to the observed data.
3. The observed data contain some degree of error or uncertainty.

In this chapter we illustrate how Bayesian inference can be used for model calibration analysis, addressing the above challenges. Because Bayesian calibration analysis is expensive, requiring thousands of evaluations of the simulation, we adopt the use of Gaussian process surrogates to approximate the relationship between the model inputs and outputs; the surrogate modeling approach we use is outlined in Section 9.2. Section 9.3 discusses the theory underlying the Bayesian calibration approach, and Section 9.4 presents a case study based on the thermal simulation of decomposing foam.

9.2 Gaussian Process Models

Gaussian process (GP) interpolation (which is in most cases equivalent to the family of methods that go by the name of 'kriging' predictors) is a powerful technique based on spatial statistics. Not only can Gaussian process models be used to fit a wide variety of functional forms, but they also provide a direct estimate of the uncertainty associated with their predictions.

The basic idea of the GP model is that the response values, Y, are modeled as a group of multivariate normal random variables (Rasmussen 1996; Santner *et al.* 2003). A parametric covariance function is then constructed as a function of the inputs, $\boldsymbol{x}$. The covariance function is based on the idea that when the inputs are close together, the correlation between the outputs will be high. As a result, the uncertainty associated with the model's predictions is small for input values that are close to the training points, and large for input values that are not close to the training points. In addition, the mean function of the GP may capture large-scale variations, such as a linear or quadratic regression of the inputs. However, for this work, we employ GP models having a constant mean function.

Thus, we denote by Y a Gaussian process with mean and covariance given by

$$\mathrm{E}[Y(\boldsymbol{x})] = \beta \tag{9.1}$$

and

$$\mathrm{Cov}[Y(\boldsymbol{x}), Y(\boldsymbol{x}^*)] = \lambda c(\boldsymbol{x}, \boldsymbol{x}^* \mid \boldsymbol{\xi}), \tag{9.2}$$

where $c(\boldsymbol{x}, \boldsymbol{x}^* \mid \boldsymbol{\xi})$ is the correlation between $\boldsymbol{x}$ and $\boldsymbol{x}^*$, $\boldsymbol{\xi}$ is the vector of parameters governing the correlation function, and λ is the process variance.

Note that the correlation function, $c(\cdot, \cdot)$, is a scalar function, and the process variance, λ, is a scalar as well.

Consider that we have observed the process at m locations (the training or design points) $\boldsymbol{x}_1, \ldots, \boldsymbol{x}_m$ of a p-dimensional input variable, so that we have the resulting observed random vector $\boldsymbol{Y} = (Y(\boldsymbol{x}_1), \ldots, Y(\boldsymbol{x}_m))^T$. By definition, the joint distribution of $\boldsymbol{Y}$ satisfies

$$\boldsymbol{Y} \sim \boldsymbol{N}_m\left(\beta\mathbf{1},\, \lambda\boldsymbol{R}\right), \tag{9.3}$$

where $\boldsymbol{R}$ is the $m \times m$ matrix of correlations among the training points, and $\mathbf{1}$ is a vector of m ones. Under the assumption that the parameters governing both the trend function and the covariance function are known, the expected value and variance (uncertainty) at any location $\boldsymbol{x}$ are calculated as

$$\mathrm{E}[Y(\boldsymbol{x}) \mid \boldsymbol{Y}] = \beta + \boldsymbol{r}^T(\boldsymbol{x})\boldsymbol{R}^{-1}(\boldsymbol{Y} - \beta\mathbf{1}) \tag{9.4}$$

and

$$\mathrm{Var}\left[Y(\boldsymbol{x}) \mid \boldsymbol{Y}\right] = \lambda\left(1 - \boldsymbol{r}^T\boldsymbol{R}^{-1}\boldsymbol{r}\right), \tag{9.5}$$

where $\boldsymbol{r}$ is the vector of correlations between $\boldsymbol{x}$ and each of the training points. Further, the full covariance matrix associated with a vector of predictions can be constructed using the following equation for the pairwise covariance elements:

$$\mathrm{Cov}\left[Y(\boldsymbol{x}), Y(\boldsymbol{x}^*) \mid \boldsymbol{Y}\right] = \lambda\left[c(\boldsymbol{x}, \boldsymbol{x}^*) - \boldsymbol{r}^T\boldsymbol{R}^{-1}\boldsymbol{r}_*\right], \tag{9.6}$$

where $\boldsymbol{r}$ is as above, and $\boldsymbol{r}_*$ is the vector of correlations between $\boldsymbol{x}^*$ and each of the training points.

There are a variety of possible parametrizations of the correlation function (Ripley 1981; Santner *et al.* 2003). Statisticians have traditionally recommended the Matérn family (Ripley 1981; Stein 1999), while engineers often use the squared-exponential formulation (Martin and Simpson 2005; Simpson *et al.* 2001) for its ease of interpretation and because it results in a smooth, infinitely differentiable function (Santner *et al.* 2003). This work uses the squared-exponential form, which is given by

$$c(\boldsymbol{x}, \boldsymbol{x}^*) = \exp\left[-\sum_{i=1}^{p} \xi_i (x_i - x_i^*)^2\right], \tag{9.7}$$

where p is the dimension of $\boldsymbol{x}$, and the p parameters ξ_i must be non-negative.

9.2.1 Estimation of Parameters Governing the GP

Before applying the Gaussian process model for prediction, values of the parameters $\boldsymbol{\xi} = (\xi_1, \ldots, \xi_p)$ and β must be chosen. Further, the value for λ also must be selected if Equation (9.5) is to be used for uncertainty estimation. One of the most commonly used methods for parameter estimation with

Gaussian process models is the method of maximum likelihood estimation (MLE; Martin and Simpson 2005, Rasmussen 1996).

Maximum likelihood estimation involves finding those parameters that maximize the likelihood function. The likelihood function describes the probability of observing the training data for a particular set of parameters. Under the Gaussian process framework, the likelihood function is based on the multivariate normal distribution. For computational reasons, the problem is typically formulated as a minimization of the negative log of the likelihood function:

$$-\log l(\boldsymbol{\xi}, \beta, \lambda) = m \log \lambda + \log |\boldsymbol{R}| + \lambda^{-1}(\boldsymbol{Y} - \beta \mathbf{1})^T \boldsymbol{R}^{-1}(\boldsymbol{Y} - \beta \mathbf{1}). \quad (9.8)$$

The numerical minimization of Eq. (9.8) can be an expensive task, since the $m \times m$ matrix $\boldsymbol{R}$ must be inverted for each evaluation. Fortunately, the gradients are available in analytic form (Martin and Simpson 2005; Rasmussen 1996). Further, the optimal values of the process mean and variance, conditional on the correlation parameters $\boldsymbol{\xi} = (\xi_1, \ldots, \xi_p)$, can be computed exactly. The optimal value of β is equivalent to the generalized least squares estimator:

$$\hat{\beta} = \left(\mathbf{1}^T \boldsymbol{R}^{-1} \mathbf{1}\right)^{-1} \mathbf{1}^T \boldsymbol{R}^{-1} \boldsymbol{Y}. \quad (9.9)$$

The conditional optimum for the process variance is given by

$$\hat{\lambda} = \frac{1}{m} (\boldsymbol{Y} - \beta \mathbf{1})^T \boldsymbol{R}^{-1} (\boldsymbol{Y} - \beta \mathbf{1}). \quad (9.10)$$

9.2.2 Modeling Time Series Output

In certain cases one may want a surrogate model that captures the relationship between some computer simulation inputs and a response, as a function of time. For example, one run of the computer simulation for a certain combination of inputs may produce a time-series output. From the perspective of surrogate modeling, it is natural to simply view time as an additional input, but doing so can be problematic when working with Gaussian process models.

When time is treated as an input, each individual run of the computer simulation can produce a large number of training points, each corresponding to a particular value of time. Naturally, most of these training points provide redundant information about the behavior of the response over time, as they are likely closely spaced over time. Unfortunately, Gaussian process models have trouble with redundant training points, because they often result in a highly ill-conditioned (near singular) correlation matrix, $\boldsymbol{R}$.

One option for handling this problem is to remove the redundant training points. McFarland *et al.* (2008) propose an iterative 'greedy point selection' process in which an initial GP surrogate model is built using a small fraction of the available points, and then points are iteratively added to the model at those locations where the error is the greatest. This process allows one to construct a more efficient surrogate model, while avoiding problems with an ill-conditioned correlation matrix, by selecting among the available training points in an optimal fashion.

9.3 Bayesian Model Calibration

Model calibration is a particular type of inverse problem in which one is interested in finding values for a set of computer model inputs that result in outputs that agree well with observed data. Bayesian analysis is particularly well-suited for model calibration inference because it allows the analyst to quantify the uncertainty in the estimated parameters in a meaningful and rigorous manner.

The fundamental concept of Bayesian analysis is that unknowns are treated as random variables. The power of this approach is that the established mathematical methods of probability theory can then be applied to develop informative representations of the state of knowledge regarding the unknowns. Uncertain variables are given 'prior' probability distribution functions, and these distribution functions are refined based on the available data, so that the resulting 'posterior' distributions represent the new state of knowledge, in light of the observed data. While the Bayesian approach can be computationally intensive in many situations, it is attractive because it provides a comprehensive treatment of uncertainty.

Bayesian analysis is founded on Bayes' theorem, which is a fundamental relationship among conditional probabilities. For continuous variables, Bayes' theorem is expressed as

$$f(\boldsymbol{\theta} \mid \boldsymbol{d}) = \frac{\pi(\boldsymbol{\theta}) f(\boldsymbol{d} \mid \boldsymbol{\theta})}{\int \pi(\boldsymbol{\theta}) f(\boldsymbol{d} \mid \boldsymbol{\theta})\, d\boldsymbol{\theta}}, \tag{9.11}$$

where $\boldsymbol{\theta}$ is the vector of unknowns, $\boldsymbol{d}$ contains the observations, $\pi(\boldsymbol{\theta})$ is the prior distribution, $f(\boldsymbol{d} \mid \boldsymbol{\theta})$ is the likelihood function, and $f(\boldsymbol{\theta} \mid \boldsymbol{d})$ is the posterior distribution. Note that the likelihood function is commonly written $L(\boldsymbol{\theta})$ because the data in $\boldsymbol{d}$ hold a fixed value once observed.

The primary computational difficulty in applying Bayesian analysis is the evaluation of the integral in the denominator of Equation (9.11), particularly when dealing with multidimensional unknowns. When closed form solutions are not available, computational sampling techniques such as Markov chain Monte Carlo (MCMC) sampling are often used. In particular, this work employs the component-wise scheme (Hastings 1970) of the Metropolis algorithm (Chib and Greenberg 1995; Metropolis *et al.* 1953); see McFarland *et al.* (2008) for more details.

Consider that we are interested in making inference about a set of computer model inputs $\boldsymbol{\theta}$. Now let the simulation be represented by the operator $G(\boldsymbol{\theta}, \boldsymbol{s})$, where the vector of inputs $\boldsymbol{s}$ represents a set of independent variables, which may typically represent boundary conditions, initial conditions, geometry, etc. Kennedy and O'Hagan (2001) term these inputs 'variable inputs', because they take on different values for different realizations of the system. Thus, $y = G(\boldsymbol{\theta}, \boldsymbol{s})$ is the response quantity of interest associated with the simulation.

Now consider a set of n experimental measurements,

$$\boldsymbol{d} = (d_1, \ldots, d_n)^T,$$

which are to be used to calibrate the simulation. Note that each experimental measurement corresponds to a particular value of the scenario-descriptor inputs, $\boldsymbol{s}$, and we assume that these values are known for each experiment. Thus, we are interested in finding those values of $\boldsymbol{\theta}$ for which the simulation outputs $(G(\boldsymbol{\theta}, \boldsymbol{s}_1), \ldots, G(\boldsymbol{\theta}, \boldsymbol{s}_n))$ agree well with the observed data in $\boldsymbol{d}$. But as mentioned above, we are interested in more than simply a point estimate for $\boldsymbol{\theta}$: we would like a comprehensive assessment of the uncertainty associated with this estimate.

First, we define a statistical relationship between the model output, $G(\boldsymbol{\theta}, \boldsymbol{s})$, and the observed data, $\boldsymbol{d}$:

$$d_i = G(\boldsymbol{\theta}, \boldsymbol{s}_i) + \varepsilon_i, \tag{9.12}$$

where ε_i is a random variable that can encompass both measurement errors on d_i and modeling errors associated with the simulation $G(\boldsymbol{\theta}, \boldsymbol{s})$. The most frequently used assumption for the ε_i is that they are i.i.d $N(0, \sigma^2)$, which means that the ε_i are independent, zero-mean Gaussian random variables, with variance σ^2. Of course, more complex models may be applied, for instance enforcing a parametric dependence structure among the errors.

The probabilistic model defined by Equation (9.12) results in a likelihood function for $\boldsymbol{\theta}$ that is the product of n normal probability density functions:

$$L(\boldsymbol{\theta}) = f(\boldsymbol{d} \mid \boldsymbol{\theta}) = \prod_{i=1}^{n} \frac{1}{\sigma\sqrt{2\pi}} \exp\left[-\frac{(d_i - G(\boldsymbol{\theta}, \boldsymbol{s}_i))^2}{2\sigma^2}\right]. \tag{9.13}$$

We can now apply Bayes' theorem (Equation (9.11)) using the likelihood function of Equation (9.13) along with a prior distribution for $\boldsymbol{\theta}$, $\pi(\boldsymbol{\theta})$, to compute the posterior distribution, $f(\boldsymbol{\theta} \mid \boldsymbol{d})$, which represents our belief about $\boldsymbol{\theta}$ in light of the data $\boldsymbol{d}$:

$$f(\boldsymbol{\theta} \mid \boldsymbol{d}) \propto \pi(\boldsymbol{\theta}) L(\boldsymbol{\theta}). \tag{9.14}$$

The posterior distribution for $\boldsymbol{\theta}$ represents the complete state of knowledge, and may even include effects such as multiple modes, which would represent multiple competing hypotheses about the true (best-fitting) value of $\boldsymbol{\theta}$. Summary information can be extracted from the posterior, including the mean (which is typically taken to be the the 'best guess' point estimate) and standard deviation (a representation of the amount of residual uncertainty). We can also extract one or two-dimensional marginal distributions, which simplify visualization of the features of the posterior.

However, as mentioned above, the posterior distribution can not usually be constructed analytically, and this will almost certainly not be possible when a complex simulation model appears inside the likelihood function. Markov chain Monte Carlo (MCMC) sampling is considered here, but this requires hundreds of thousands of evaluations of the likelihood function, which in the case of model calibration equates to hundreds of thousands of evaluations of the computer model $G(\cdot,\cdot)$. For most realistic models, this number of evaluations will not be feasible. In such situations, the analyst must usually resort to the use of a more inexpensive surrogate (a.k.a. response surface approximation) model. Such a surrogate might involve reduced order modeling (e.g. a coarser mesh) or data-fit techniques such as Gaussian process (a.k.a. kriging) modeling.

This work adopts the approach of using a Gaussian process surrogate to the true simulation. We find such an approach to be an attractive choice for use within the Bayesian calibration framework for several reasons. First, the Gaussian process model is very flexible, and it can be used to fit data associated with a wide variety of functional forms. Second the Gaussian process is stochastic, thus providing both an estimated response value and an uncertainty associated with that estimate.[1] Conveniently, the Bayesian framework allows us to take account of this uncertainty.

Our approach for accounting for the uncertainty introduced by the surrogate model is admittedly somewhat simpler than other approaches reported in the literature (specifically that of Kennedy and O'Hagan 2001), but we believe that we still address in a meaningful manner the uncertainty introduced by employing an approximation to the actual simulator. Additional details regarding the differences between this and other formulations are given by McFarland (2008, Sec. 5.4).

Through the assumptions used for Gaussian process modeling, the surrogate response conditional on a set of observed 'training points' follows a multivariate normal distribution. For a discrete set of new inputs, this response is characterized by a mean vector and a covariance matrix (see Equations (9.4) through (9.6)). Let us denote the mean vector and covariance matrix corresponding to the inputs $(\boldsymbol{\theta}, \boldsymbol{s}_1), \ldots, (\boldsymbol{\theta}, \boldsymbol{s}_n)$ as $\boldsymbol{\mu}_{GP}$ and $\boldsymbol{\Sigma}_{GP}$, respectively. It is easy to show that the likelihood function for $\boldsymbol{\theta}$ is then given by a multivariate normal probability density function (note that the likelihood

[1] We recognize that there are simpler models, such as polynomial regression models, which allow one to derive estimates of the uncertainty associated with predicted values. Most of these simpler models, however, do not have the desirable property that this uncertainty is zero at the training points.

function of Equation (9.13) can also be expressed as a multivariate normal probability density, with $\boldsymbol{\Sigma}$ diagonal):

$$L(\boldsymbol{\theta}) = (2\pi)^{-n/2} |\boldsymbol{\Sigma}|^{-1/2} \exp\left[-\frac{1}{2}(\boldsymbol{d} - \boldsymbol{\mu}_{GP})^T \boldsymbol{\Sigma}^{-1} (\boldsymbol{d} - \boldsymbol{\mu}_{GP})\right], \qquad (9.15)$$

where $\boldsymbol{\Sigma} = \sigma^2 \boldsymbol{I} + \boldsymbol{\Sigma}_{GP}$, so that both $\boldsymbol{\mu}_{GP}$ and $\boldsymbol{\Sigma}$ depend on $\boldsymbol{\theta}$. Simply put, the overall covariance matrix is equal to the sum of the covariance of the error terms ($\sigma^2 \boldsymbol{I}$) and the covariance of the Gaussian process predictions ($\boldsymbol{\Sigma}_{GP}$). Note that with this formulation, $\boldsymbol{\Sigma}$ has $n \times n$ elements, so that it is scaled by the number of experimental observations, not the number of training points used to construct the GP surrogate.

If the data $\boldsymbol{d}$ contain multiple derived features from a single system response, then the assumption that the residuals are independent is likely not valid. For example, if the data are discrete points taken from a single time-history response, then the assumption of independence will most likely not be appropriate. A more appropriate formulation requires the estimation of the covariance matrix associated with the residuals, $\boldsymbol{\varepsilon} = (\varepsilon_1, \ldots, \varepsilon_n)$.

There are several such approaches for constructing this covariance matrix, but a common approach is to formulate a parametric model that describes the error random process. If the data are temporally correlated, one of the simplest such models, described in more detail by Seber and Wild (2003), is the so-called first-order autoregressive, or AR(1), model, which has the form

$$\varepsilon_i = \varphi \varepsilon_{i-1} + \nu_i, \qquad (9.16)$$

where φ is a (scalar) correlation parameter that satisfies $|\varphi| < 1$, and the 'innovation errors' are independently and identically distributed as $\nu_i \sim N(0, \sigma_\nu^2)$. The pairwise correlation between any two residuals is given by

$$\rho_{i,j} = \varphi^{|i-j|}, \qquad (9.17)$$

and the variance of the residuals is given by

$$\mathrm{Var}[\varepsilon_i] = \frac{\sigma_\nu^2}{1 - \varphi^2}. \qquad (9.18)$$

Note that in order to use this model, the observations must be spaced evenly in time (although see Glasbey 1979, for an extension that allows for unequally spaced time intervals). The adoption of the AR(1) model introduces one additional parameter that must be estimated (φ), but its estimation can be handled nicely in the Bayesian framework by giving it a suitable prior distribution (perhaps uniform) and treating it as an additional object of Bayesian inference.

With the use of the AR(1) or other suitable model, the data covariance can be represented as $\boldsymbol{\Sigma} = \boldsymbol{\Sigma}_{AR} + \boldsymbol{\Sigma}_{GP}$, where the covariance matrix for the residuals ($\mathbf{\Sigma}_{AR}$) is constructed using Eqs. (9.17) and (9.18) as

$$\mathrm{Cov}\left[\varepsilon_i, \varepsilon_j\right] = \frac{\sigma_\nu^2}{1-\varphi^2}\varphi^{|i-j|}. \tag{9.19}$$

9.4 Case Study: Thermal Simulation of Decomposing Foam

A series of experiments have been conducted at Sandia National Laboratories in an effort to support the physical characterization and modeling of thermally decomposing foam (Erickson *et al.* 2004). An associated thermal model is described by Romero *et al.* (2006). The system considered here, often referred to as the 'foam in a can' system, consists of a canister containing a mock weapons component encapsulated by a foam insulation. A schematic of this system is shown in Figure 9.1.

The simulation model is a finite element model developed for simulating heat transfer through decomposing foam. The model contains roughly 81,000 hexahedral elements, and has been verified to give spatially and temporally converged temperature predictions. The heat transfer model is implemented using the massively parallel code CALORE (CAL 2005), which has been developed at Sandia National Laboratories under the ASC (Advanced Simulation and Computing) program of the NNSA (National Nuclear Security Administration).

The simulator has been configured to model the 'foam in a can' experiment, but several of the input parameters are still unknowns (either not measured or not measurable). In particular, we consider five calibration parameters: q_2, q_3, q_4, q_5, and *FPD*. The parameters q_2 through q_5 describe the applied heat flux boundary condition, which is not well-characterized in

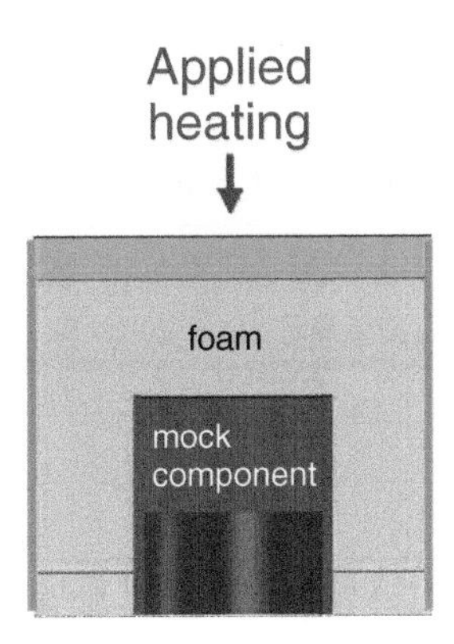

Figure 9.1 Schematic of the 'foam in a can' system.

the experiments. The last calibration parameter, *FPD*, represents the foam final pore diameter and is the parameter of most interest, because it will play a role in the ultimate modeling and prediction process. We want to consider the calibration of the simulator for the temperature response up to 2100 seconds, using temperature measurements made inside the mock component.

9.4.1 Preliminary Analysis

The first step is to collect a database of simulator runs for different values of the calibration parameters, from which the surrogate model will be constructed. Ideally, we would like our design of computer experiments to provide good coverage for the posterior distribution of the calibration inputs. However, since we don't know the form of the posterior beforehand, we have to begin with an initial guess for the appropriate bounds. Fortunately the Bayesian method provides feedback, so if our original bounds are not adequate, they can be revised appropriately. This type of sequential approach has previously been used for Bayesian model calibration and other studies (Aslett *et al.* 1998; Bernardo *et al.* 1992; Craig *et al.* 1996; Kennedy and O'Hagan 2001).

We make use of the DAKOTA (Eldred *et al.* 2006) software package for our design and collection of computer experiments. DAKOTA is an object-oriented framework for design optimization, parameter estimation, uncertainty quantification, and sensitivity analysis that can be configured to interface with the thermal simulator via external file input/output and a driver script. We use the DAKOTA software package to generate a Latin Hypercube (LH) sample of size 50 using the variable bounds listed in Table 9.1. For simplicity, we do not discuss a revision of these bounds here, although one is discussed by McFarland *et al.* (2008).

Using the results from the simulation runs, we can compare the ensemble of predicted time histories against the experimental time histories to see if the experimental data are 'enveloped' by the simulation data. Figure 9.2 compares the envelope of simulator outputs against the experimental data, which clearly shows that the experimental data are enveloped by the model predictions.

Table 9.1 Design of computer experiments.

Variable	Lower bound	Upper bound
FPD	2.0×10^{-3}	15.0×10^{-3}
q_2	25,000	150,000
q_3	100,000	220,000
q_4	150,000	300,000
q_5	50,000	220,000

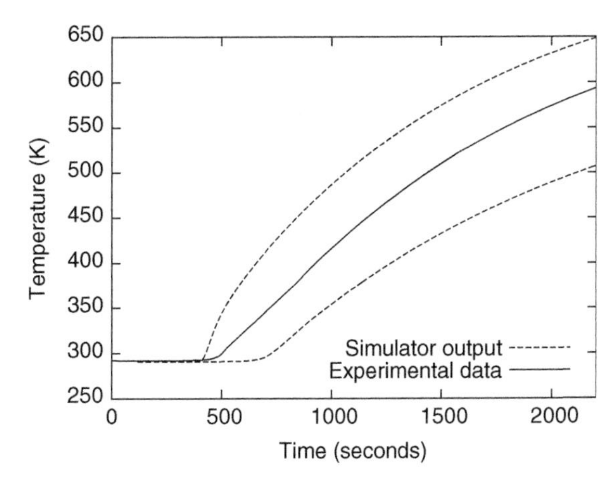

Figure 9.2 Temperature response comparison for envelope of 50 simulator outputs with observed data.

9.4.2 Bayesian Calibration Analysis

Here we present a Bayesian calibration of the CALORE simulator using temperature data obtained via thermocouple measurements from inside the mock component. The calibration data $\boldsymbol{d}$ contain 17 measurements of temperature evenly spaced at time intervals of 100 seconds from 500 to 2100 seconds (note from Figure 9.2 that the temperature response is trivial for $t < 500$ seconds).

We employ the statistical model of Equation (9.12) to relate the observations and model predictions. Because $\boldsymbol{d}$ contains measurements of the same quantity over time, the assumption of independence for the residuals, ε_i, is not appropriate, and we employ the first-order auto-regressive model described in Section 9.3 to account for serial correlation. We treat both parameters of the AR(1) model as objects of Bayesian inference, giving them independent vague reference priors (Lee 2004) in which the prior for σ_ν^2 is log-uniform and the prior for φ is uniform on $[-1, 1]$. For the prior distributions of the calibration parameters, we choose independent uniform distributions based on the bounds given in Table 9.1.

The FEM simulation, $G(\boldsymbol{\theta})$, is represented using a Gaussian process surrogate model that captures the output as a function of the calibration parameters *and* time. As discussed in Section 9.2.2, introducing time as an input tends to result in redundant training data for the surrogate, and we avoid this problem by employing the point selection algorithm proposed by McFarland *et al.* (2008). We use this algorithm to choose 100 points optimally from the 8,550 available points (171 time instants $\times$ 50 LH samples).

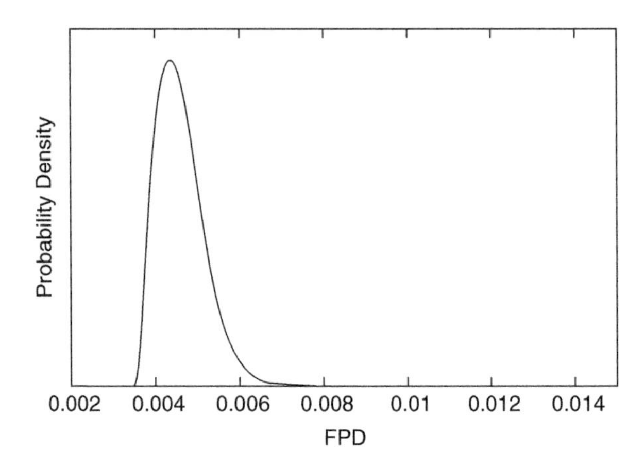

Figure 9.3 Posterior distribution of FPD (x-axis range represents prior bounds).

The MCMC simulation is adjusted appropriately and run for 100,000 iterations. While this seems like a very large number of iterations, it is computationally feasible because the primary cost of each iteration is the evaluation of the GP surrogate model, which can be performed very efficiently. The resulting marginal posterior distributions for the parameter of most interest, *FPD*, is shown in Figure 9.3, where the plotting ranges are representative of the bounds of the prior distribution. Recall that the prior distribution for $\boldsymbol{\theta}$ is independently uniform over the ranges listed in Table 9.1.

Some statistics of the marginal posteriors are given in Table 9.2 (note that although we report means and standard deviations, there is no assumption that the marginal posteriors are normally distributed), and the pairwise correlation coefficients are given in Table 9.3. The correlation coefficients indicate strong negative relationships between FPD and q_5 as well as FPD and q_3. For visualization, we use kernel density estimation (Silverman 1986) to plot the contour of the marginal density of *FPD* and q_5 corresponding to the 95 % confidence level in Figure 9.4.

Table 9.2 Posterior statistics based on the nominal calibration analysis.

Variable	Mean	Std. Dev.
FPD	4.58×10^{-3}	5.47×10^{-4}
q_2	73,526	40,234
q_3	177,330	24,142
q_4	162,400	12,430
q_5	159,180	12,888

Table 9.3 Pairwise correlation coefficients within the posterior distribution for nominal analysis.

	FPD	q_2	q_3	q_4	q_5
FPD	1.00	−0.52	−0.80	0.08	−0.98
q_2	−0.52	1.00	0.42	−0.17	0.54
q_3	−0.80	0.42	1.00	−0.41	0.82
q_4	0.08	−0.17	−0.41	1.00	−0.11
q_5	−0.98	0.54	0.82	−0.11	1.00

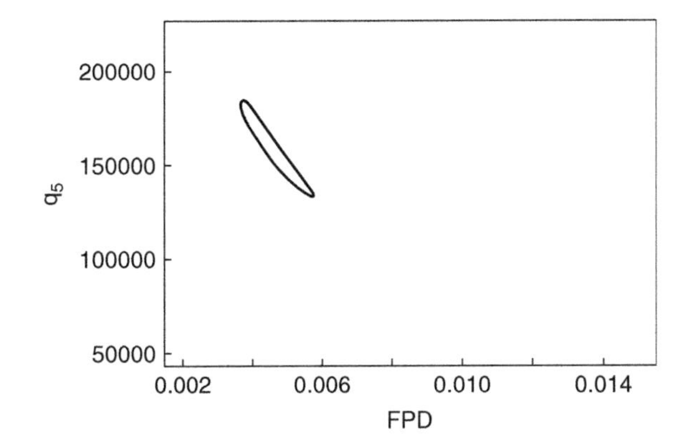

Figure 9.4 95 % confidence region for calibration parameters FPD and q_5. Plotting bounds represent prior bounds.

The predicted response (based on the GP surrogate model), after calibration, is compared with the experimental observations in Figure 9.5, where the calibration parameters are set to their posterior mean values. This type of plot serves as a check to ensure that the model output has been successfully calibrated to the experimental observations.

Finally, it is worth noting that this type of model calibration analysis could also be approached from a non-Bayesian standpoint, using classical nonlinear regression approaches, such as those discussed by Seber and Wild (2003). Specialized gradient-based algorithms are widely available for searching for a locally optimum point estimate of the calibration parameters; because these algorithms are fairly efficient, the development of a surrogate model for $G(\cdot)$ may not even be necessary. Classical approaches also exist for the quantification of uncertainty, but these methods are often based on approximations and can give poor results for nonlinear models. Even so, the graphical display of classical confidence regions is likely expensive enough to necessitate a surrogate for $G(\cdot)$, in which case no real computational savings would be gained as compared to a Bayesian approach (further, classical methods do not appear

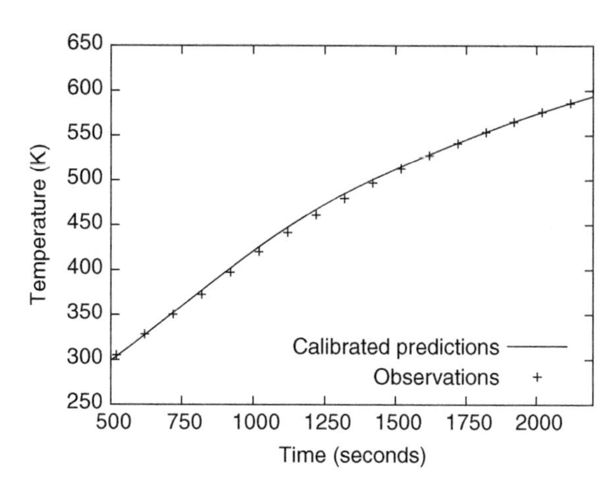

Figure 9.5 Comparison of experimental observations and calibrated surrogate model predictions (calibration parameters set to the posterior mean).

to provide a mechanism for accounting for uncertainty introduced by the use of a GP surrogate model, since this uncertainty is a function of the calibration parameters, $\boldsymbol{\theta}$). A more detailed comparison regarding classical and Bayesian approaches is given by McFarland (2008, Secs. 5.2, 6.4.4).

9.5 Conclusions

The important role that computational models play in prediction, design, and decision making necessitates appropriate methods for assessing the uncertainty in such predictions. Here we present the use of Bayesian model calibration as a tool for calibrating a computational simulation with experimental observations, while at the same time accounting for the uncertainty that is introduced in the process.

An important enabling tool for this analysis is Gaussian process surrogate modeling. We have shown how GP surrogate models can be used to represent the relationship between simulator inputs and outputs, even for time-dependent output. We have also shown how the uncertainty introduced by the use of a surrogate model can be accounted for in the model calibration process.

The example given here involves the calibration of five model parameters based on an experimentally observed temperature response. While the example does come from a real-world application, it may still lack some of the complexities seen in larger and less ideal systems. Nevertheless, we are confident that the above approach can be adapted to handle such cases. For example, some systems may require the estimation of significantly more model parameters. While Gaussian process surrogates and Markov chain Monte Carlo simulation have been shown to handle modest numbers of variables, systems requiring the estimation of hundreds or more parameters could certainly benefit from a preliminary sensitivity analysis to identify the important parameters. In other cases, experiments may yield a larger amount of data, perhaps as a result of the placement of sensors at multiple locations. These cases too can be handled using the above methodology, but more care must be taken in the formulation of appropriate statistical models to account for temporal and/or spatial dependence in the data.

References

Aslett, R., Buck, R.J., Duvall, S.G., Sacks, J. and Welch, W.J. 1998 Circuit optimization via sequential computer experiments: design of an output buffer. *Journal of the Royal Statistical Society: Series C (Applied Statistics)* **47**(1), 31–48.

Bernardo, M.C., Buck, R.J., Liu, L., Nazaret, W.A., Sacks, J. and Welch, W.J. 1992 Integrated circuit design optimization using a sequential strategy. *IEEE Transactions on Computer Aided Design of Integrated Circuits and Systems* **11**(3), 361–372.

CAL 2005 *Calore: A Computational Heat Transfer Program. Vol. 2 User Reference Manual for Version 4.1*.

Chib, S. and Greenberg, E. 1995 Understanding the Metropolis-Hastings algorithm. *American Statistician* **49**(4), 327–335.

Craig, P.S., Goldstein, M., Seheult, A.H. and Smith, J.A. 1996 Bayes linear strategies for matching hydrocarbon reservoir history. In *Bayesian Statistics 5* (ed. Bernardo, J.M., Berger, J.O., Dawid, A.P. and Smith, A.F.M.), Oxford University Press. Oxford, pp. 69–95.

Eldred, M.S., Brown, S.L., Adams, B.M., Dunlavy, D.M., Gay, D.M., Swiler, L.P., Giunta, A.A., Hart, W.E., Watson, J.P., Eddy, J.P., Griffin, J.D., Hough, P.D., Kolda, T.G., Martinez-Canales, M.L. and Williams, P.J. 2006 DAKOTA, a multilevel parallel object-oriented framework for design optimization, parameter estimation, uncertainty quantification, and sensitivity analysis: Version 4.0 reference manual. Technical Report SAND2006-4055, Sandia National Laboratories.

Erickson, K.L., Trujillo, S.M., Thompson, K.R., Sun, A.C., Hobbs, M.L. and Dowding, K.J. 2004 Liquefaction and flow behavior of a thermally decomposing removable epoxy foam. In *Computational Methods in Materials Characterisation* (ed. Mammoli, A.A. and Brebbia, C.A.). WIT press. Southampton-Boston, pp. 217–242.

Glasbey, C.A. 1979 Correlated residuals in non-linear regression applied to growth data. *Applied Statistics* **28**, 251–259.

Hastings, W.K. 1970 Monte Carlo sampling methods using Markov chains and their applications. *Biometrika* **57**(1), 97–109.

Kennedy, M.C. and O'Hagan, A. 2001 Bayesian calibration of computer models. *Journal of the Royal Statistical Society B* **63**(3), 425–464.

Lee, P. 2004 *Bayesian Statistics, an Introduction.* Oxford University Press, Inc., New York.

Martin, J. and Simpson, T. 2005 Use of kriging models to approximate deterministic computer models. *AIAA Journal* **43**(4), 853–863.

McFarland, J. 2008 *Uncertainty Analysis for Computer Simulations through Validation and Calibration.* PhD thesis. Vanderbilt University.

McFarland, J.M., Mahadevan, S., Romero, V.J. and Swiler, L.P. 2008 Calibration and uncertainty analysis for computer simulations with multivariate output. *AIAA Journal* **46**(5), 1253–1265.

Metropolis, N., Rosenbluth, A., Rosenbluth, M., Teller, A. and Teller, E. 1953 Equations of state calculations by fast computing machines. *Journal of Chemical Physics* **21**(6), 1087–1092.

Rasmussen, C. 1996 *Evaluation of Gaussian processes and other methods for non-linear regression.* PhD thesis. University of Toronto.

Ripley, B. 1981 *Spatial Statistics.* John Wiley & Sons, Inc., New York.

Romero, V.J., Shelton, J.W. and Sherman, M.P. 2006 Modeling boundary conditions and thermocouple response in a thermal experiment. *Proceedings of the 2006 International Mechanical Engineering Congress and Exposition* number IMECE2006-15046. ASME, Chicago, IL.

Santner, T.J., Williams, B.J. and Noltz, W.I. 2003 *The Design and Analysis of Computer Experiments.* Springer-Verlag, New York.

Seber, G.A.F. and Wild, C.J. 2003 *Nonlinear Regression.* John Wiley & Sons, Inc., Hoboken, New Jersey.

Silverman, B.W. 1986 *Density Estimation for Statistics and Data Analysis.* Chapman & Hall/CRC, New York.

Simpson, T., Peplinski, J., Koch, P. and Allen, J. 2001 Metamodels in computer-based engineering design: survey and recommendations. *Engineering with Computers* **17**(2), 129–150.

Stein, M.L. 1999 *Interpolation of Spatial Data: Some Theory for Kriging* Springer Series in Statistics. Springer-Verlag, New York.

Chapter 10

Bayesian Calibration of Expensive Multivariate Computer Experiments

R.D. Wilkinson

University of Nottingham, UK

This chapter is concerned with how to calibrate a computer model to observational data when the model produces multivariate output and is expensive to run. The significance of considering models with long run times is that they can be run only at a limited number of different inputs, ruling out a brute-force Monte Carlo approach. Consequently, all inference must be done with a limited ensemble of model runs. In this chapter we use this ensemble to train a meta-model of the computer *simulator*, which we refer to as an *emulator* (Sacks *et al.* 1989). The emulator provides a probabilistic description of our beliefs about the computer model and can be used as a cheap surrogate for the simulator in the calibration process. For any input configuration not in the original ensemble of model runs, the emulator provides a probability distribution describing our uncertainty about the model's output.

The Bayesian approach to calibration of computer experiments using emulators was described by Kennedy and O'Hagan (2001). Their approach was for univariate computer models, and in this chapter we show how those methods can be extended to deal with multivariate models. We use principal component analysis to project the multivariate model output onto a lower

Large-Scale Inverse Problems and Quantification of Uncertainty Edited by L. Biegler, G. Biros, O. Ghattas, M. Heinkenschloss, D. Keyes, B. Mallick, Y. Marzouk, L. Tenorio, B. van Bloemen Waanders and K. Willcox

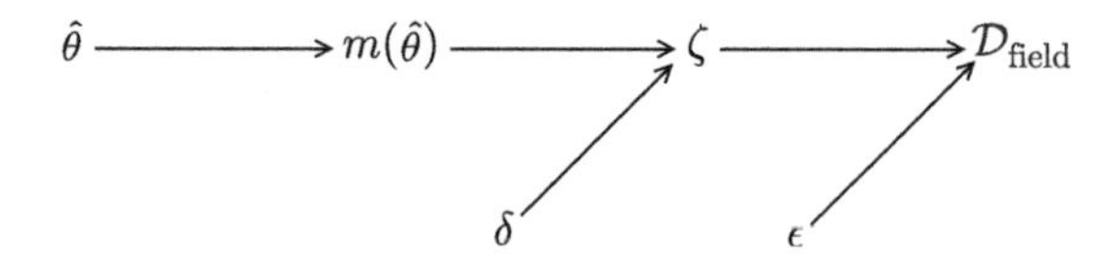

Figure 10.1 A Bayesian belief network showing the dependencies between the different components in the statistical model. Note that reality separates the model prediction $m(\hat{\theta})$ from the observations $\mathcal{D}_{\text{field}}$ and from the model discrepancy δ. See Pearl (2000) for an introduction to belief networks.

dimensional space, and then use Gaussian processes to emulate the map from the input space to the lower dimensional space. We can then reconstruct from the subspace to the original data space. This gives a way of bypassing the expensive multivariate computer model with a combination of dimension reduction and emulation in order to perform the calibration. Figure 10.4 shows a schematic diagram of the emulation process.

The result of a probabilistic calibration is a posterior distribution over the input space which represents our uncertainty about the best value of the model input given the observational data and the computer model. It is important when calibrating to distinguish between measurement error on the data and model error in the computer simulator predictions and in Section 10.1 we describe how both can and must be included in order to produce a fair analysis. A Bayesian belief network showing the statistical aspects of the calibration process is shown in Figure 10.1.

The layout of this chapter is as follows. In Section 10.1 we introduce the problem and describe the calibration framework. In Section 10.2 we introduce the idea of emulation and describe the principal component emulator and in Section 10.3 we give details of how to use this approach to calibrate multivariate models. To illustrate the methodology we use the University of Victoria intermediate complexity climate model, which we will calibrate to observational data collected throughout the latter half of the twentieth century. The model is introduced at the end of Section 10.1 and is returned to at the end of each subsequent section.

10.1 Calibration of Computer Experiments

In statistics, *calibration* is the term used to describe the inverse process of fitting a model to data, although it is also referred to as parameter estimation or simply as inference. Here, we consider the problem in which we have a computer model of a physical system along with observations of that system. The aim is to combine the science captured by the computer model with the physical observations to learn about parameter values and initial conditions for the model. We want to incorporate

1. the computer model, $m(\cdot)$,

2. field observations of the physical system, $\mathcal{D}_{\text{field}}$,

3. any other background information.

We consider each of these sources in turn.

The computer model

The computer model, $m(\cdot)$, is considered to be a map from the input space Θ, to the output space $\mathcal{Y} \subset \mathbb{R}^n$. The input parameters $\theta \in \Theta$ are the calibration parameters that we wish to estimate. They may be physical constants (although care is needed when identifying model parameters with physical constants; see Section 10.1.2), context-specific constants, or tuning parameters needed to make the model perform well. These are parameters that would not need to be specified if we were doing a physical experiment. Models also often have control parameters which act as context indicators or as switches between scenarios. For clarity, we ignore these inputs, but note that the approach presented here can be extended to deal with this case by considering the input space to be $\Theta \times \mathcal{T}$, where $\mathcal{T}$ is a space of control variables.

Later in the chapter we will be concerned with models which produce multivariate output, typically with values of a variable reported at spatial locations or over a range of times, or perhaps both. We could, if we chose to, introduce an index variable to the inputs in order to reduce these models to having scalar outputs, allowing us to use methods for univariate models. For example, a model predicting the temperature on a grid of locations could be considered as predicting the temperature at a single location, where that locations is specified by an index variable in the inputs. We choose not to do this here because this limits the complexity of the problem we can analyze (as well as presenting some conceptual challenges with the emulation) and because the dimension reduction techniques appear to work better when outputs are highly correlated as they usually are if the output is a spatial-temporal field. We write $m(\theta)$ for the multivariate model output run at θ, and refer to elements in the output through an index t, so that $m_t(\theta)$ is the model output at location/time t.

The focus in this chapter is on calibrating computer models which have long run times. A consequence of this cost will be that we will have only a limited ensemble of model runs available to us. In other words, there will be a set of N design points $D = \{\theta_i : i = 1, \ldots, N\}$ for which we know the output of the model $\mathcal{D}_{\text{sim}} = \{y_i = m(\theta_i) : i = 1, \ldots, N\}$; all of the information about the model will come solely from this ensemble. The question of how to choose design points D is discussed in Section 10.2.1, and we shall assume throughout that we are provided with a well designed ensemble $\mathcal{D}_{\text{sim}}$ for use in the analysis.

Field Observations

We assume that we have observations of the physical system, $\mathcal{D}_{\text{field}}$, that directly correspond to outputs from the computer model. We let ζ_t represent reality at t, where t is an index variable such as time or location, and assume that the field data is a measurement of reality at t with independent Gaussian error. That is

$$\mathcal{D}_{\text{field}}(t) = \zeta_t + \varepsilon_t \tag{10.1}$$

where $\varepsilon_t \sim N(\mu_t, \sigma_t^2)$. It will usually be the case that $\mu_t = 0$ for all t, and often the case that we have homoscedastic errors so that $\sigma_t^2 = \sigma^2$ for all t, however neither of these assumptions is necessary for the analysis. We treat model error as separate from measurement error for reasons specified in Section 10.1.2. One of the benefits of this is that ε then genuinely represents measurement error. As the error rate for most instrumentation is known, and will usually be reported with the measurements, we assume μ_t and σ_t are known constants throughout. If this is not the case, it is possible to learn these parameters along with the others.

Other Background Information

Calibration is primarily about combining the physics in the model with field observations of the system to produce estimates of parameter values. However, there will often be additional expert knowledge that has not been built into the model. Part of this knowledge will be prior information about the likely best input values, gained through previous experiments and reading the literature, and will be represented by prior distribution $\pi(\theta)$. The modellers may also know something about how accurately the simulator represents the system. As explained in more detail below, when calibrating a model it is important to account for any discrepancy between the model and reality. Model builders are often able to provide information about how and where the model may be wrong. They may, for example, have more confidence in some of the model outputs than others, or they may have more faith in the predictions in some contexts than in others. This information can all be built into the analysis. Ideally, information should be elicited from the experts before they observe either the ensemble of model runs or the field data, however, in practice this will often not be the case. Garthwaite *et al.* (2005) give an introduction to elicitation of expert beliefs.

10.1.1 Statistical Calibration Framework

The calibration method presented here is based on the approach given by Kennedy and O'Hagan (2001) and uses the concept of a *best-input* (Goldstein and Rougier 2009). The approach assumes that there is a single 'best' value of θ, which we label $\hat{\theta}$, such that the model run at $\hat{\theta}$ gives the most accurate representation of the system. Note that $\hat{\theta}$ is the best value here only in the sense of most accurately representing the data according to the specified error structure, and as commented later, the value found for $\hat{\theta}$ need not coincide with the true physical value of θ. A consequence of this assumption is that the model run at its best input is sufficient for the model in the calibration, in the sense that once we know $m(\hat{\theta})$ we can not learn anything further about reality from the simulator.

A common and incorrect assumption in calibration is to assume that we observe the simulator prediction plus independent random noise. If the computer simulator is not a perfect representation of reality, this assumption is wrong and may lead to serious errors in the analysis and in future predictions. In order to relate the simulator prediction to reality we must account for the existence of model error. We can do this with an additive error term and state that reality ζ is the best simulator prediction plus a model error δ:

$$\zeta = m(\hat{\theta}) + \delta. \tag{10.2}$$

Equations (10.1) and (10.2) completely describe the structural form assumed in the calibration. Combining the two equations gives

$$\mathcal{D}_{\text{field}} = m(\hat{\theta}) + \delta + \varepsilon \tag{10.3}$$

where all quantities in this equation are vectors. A consequence of the best input approach is that $m(\hat{\theta})$ is independent of δ allowing us to specify beliefs about each term without reference to the other. A schematic representation of the conditional independence structure assumed for the calibration is shown in Figure 10.1.

In later sections, distributional assumptions are discussed for all terms on the right-hand side of Equation (10.3), but before then we briefly consider the inferential process. Calibration is the process of judging which input values are consistent with the field data, the model and any prior beliefs. The Bayesian approach to calibration is to find the posterior distribution of the best input parameter given these three sources of information; namely, we aim to find

$$\pi(\hat{\theta}|\mathcal{D}_{\text{sim}}, \mathcal{D}_{\text{field}}, E),$$

where E represents the background information, $\mathcal{D}_{\text{sim}}$ the ensemble of model runs, and $\mathcal{D}_{\text{field}}$ the field observations. The posterior gives relative weights to all $\theta \in \Theta$, and represents our beliefs about the best input in the light of the available information.

To calculate the posterior distribution, we use Bayes theorem to find that the posterior of $\hat{\theta}$ is proportional to its likelihood multiplied by its prior distribution:

$$\pi(\hat{\theta}|\mathcal{D}_{\text{sim}}, \mathcal{D}_{\text{field}}, E) \propto \pi(\mathcal{D}_{\text{sim}}, \mathcal{D}_{\text{field}}|\hat{\theta}, E)\pi(\hat{\theta}|E). \tag{10.4}$$

Often, the hardest part of any calibration is specification of the likelihood $\pi(\mathcal{D}_{\text{sim}}, \mathcal{D}_{\text{field}}|\hat{\theta}, E)$, as once we have the prior and the likelihood, finding the posterior distribution is in theory just an integral calculation. In practice, however, this will usually require careful application of a numerical integration technique such as a Markov chain Monte Carlo (MCMC) algorithm. Once we have made distributional assumptions about δ, ε and possibly $m(\hat{\theta})$, the structure shown in Figure 10.1 allows us to calculate the likelihood function for the data.

10.1.2 Model Error

There are a large variety of reasons why simulators are nearly always imperfect representations of the physical system they were designed to predict. For example, modellers' understanding of the system may be flawed, or perhaps not all physical processes were included in the analysis, or perhaps there are numerical inaccuracies in the solver used on the underlying model equations, and so on. If we wish to make accurate predictions it is important to account for this error, as otherwise we may have an unrealistic level of confidence in our predictions.

If we choose not to use a model error term we make the assumption that observations are the best simulator prediction plus white noise, i.e.

$$\mathcal{D}_{\text{field}} = m(\hat{\theta}) + \varepsilon$$

where ε_t is independent of ε_s for $t \neq s$. It will often be found that the magnitude of the measurement error associated with the measurement instrument is insufficient to account for the variability observed in the system, leading to poor model fit in the calibration. One solution might be to inflate the error variance to account for this missing variability, however, this will also cause problems with the confidence level in the predictions. In contrast to the white error structure of ε, the model error term δ will usually have a much richer structure, usually with δ_t highly correlated with δ_s for t close to s. By using correlated errors a better degree of accuracy can be achieved.

A consequence of using an imperfect simulator in the calibration, is that model parameters may not correspond to their physical namesakes. For example, in a simulator of ocean circulation we may have a parameter called viscosity and it may be possible to conduct an experiment to measure the physical value of the viscosity in a laboratory. However, because the simulator is an imperfect representation of the system, we may find that using the physical value of the viscosity leads to poorer predictions than when using a value determined by a statistical calibration. For this reason, careful thought needs to be taken when considering which parameters to include in the calibration. The fact that model parameters may not be physical parameters should be strongly stressed to the experts when eliciting the prior distribution $\pi(\hat{\theta}|E)$ required in Equation (10.4).

The question of how to choose a suitable model for δ is an area of active statistical research, and the approach taken depends on the amount of data available. In situations where data is plentiful, data driven approaches can be used with an uninformative prior specification for δ. So for example, in weather prediction, each day a forecast is made and the following day data is collected which can be used to validate the previous days prediction. In situations such as this Kennedy and O'Hagan (2001) suggest the use of a Gaussian process for δ, with uninformative priors on any parameters in δ.

If the data available is limited, then expert judgement becomes important. Goldstein and Rougier (2009) introduce the idea of a reified model in a thought experiment designed to help elicit beliefs about the model error. The reified model is the version of the model we would run if we had unlimited computing resources. So for example, in global climate models the earth's surface is split into a grid of cells and the computation assumes each cell is homogeneous. If infinite computing resources were available we could let the grid size tend to zero, giving a continuum of points across the globe. While clearly an impossibility, thinking about the reified model helps us to break down the model error into more manageable chunks; we can consider the difference between the actual computer model and the reified model, and then the difference between the reified model and reality. This approach may provide a way to help the modellers think more carefully about $\delta(\cdot)$. Murphy *et al.* (2007) take a different approach and use an ensemble of models. They look at the calibrated predictions from a collection of different climate models and use these to assess what the model error might be for their model.

10.1.3 Code Uncertainty

If the computer model is quick to run, then we can essentially assume that its value is known for all possible input configurations, as in any inference procedure we can simply evaluate the model whenever its value is needed. In this case, calculation of the calibration posterior

$$\pi(\theta|\mathcal{D}_{\text{field}}, m, E) \propto \pi(\mathcal{D}_{\text{field}}, |\theta, m, E)\pi(\theta|E), \tag{10.5}$$

where m represents the computer model, is relatively easy as the calibration framework (10.3) gives that

$$\mathcal{D}_{\text{field}} - m(\hat{\theta}) = \delta + \varepsilon.$$

Given distributions for the model discrepancy δ and measurement error ε we can calculate the likelihood of the field data, and thus can find the posterior distribution. If the model is not quick running, then the model's value is unknown at all input values other than those in design D. This uncertainty about the model output at untried input configurations is commonly called *code uncertainty*. If we want to account for this source of uncertainty in the calibration then we need a statistical model that describes our beliefs about the output for all possible input values. We introduce the idea of emulation after the following example.

Example 10.1 (Uvic Climater Model)

In order to demonstrate the methodology we introduce an example from climate science which we present along with the theory. We use the University of Victoria Earth System Climate Model (UVic ESCM) coupled with a dynamic vegetation and terrestrial carbon cycle and an inorganic ocean carbon cycle (Meissner *et al.* 2003). The model was built in order to study potential feedbacks in the terrestrial carbon cycle and to see how these affect future climate predictions. We present a simplified analysis here, with full details available in Ricciuto *et al.* We consider the model to have just two inputs, Q_{10} and K_c, and to output a time-series of atmospheric CO_2 values. Input Q_{10} controls the temperature dependence of respiration and can be considered as controlling a carbon source, whereas K_c is the Michaelis–Menton constant for CO_2 and controls the sensitivity of photosynthesis and can be considered to control a carbon sink. The aim is calibrate these two parameters to the Keeling and Whorf (2005) sequence of atmospheric carbon dioxide measurements. Each model run takes approximately two weeks of computer time and we have an ensemble of 47 model runs with which to perform the analysis. The model output and the field observations are shown in Figure 10.2.

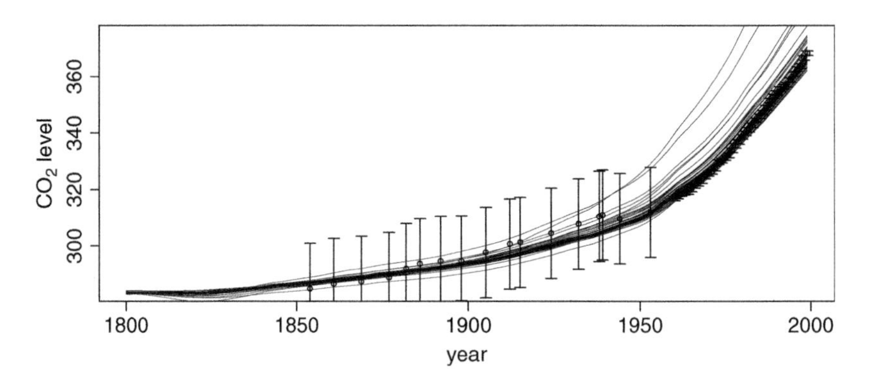

Figure 10.2 Ensemble of 47 model runs of the UVic climate model for a design on two inputs Q_{10} and K_c. The output (black lines) gives the atmospheric CO_2 predictions for 1800–1999, and the 57 field observations are shown as circles with error bars of two standard deviations.

10.2 Emulation

10.2.1 Bayesian

If the simulator, $m(\cdot)$, is expensive to evaluate, then its value is unknown at all input values except those in a small ensemble of model runs. We assume that the code has been run N times for all inputs in a space-filling design $D = \{\theta_i \in \Theta : i = 1, \ldots, N\}$ to produce output $\mathcal{D}_{\text{sim}} = \{m(\theta_i) \in \mathbb{R}^n : i = 1, \ldots, N\}$, and that further model runs are not available. The importance of the design D for computer experiments has been stressed by many authors (Morris and Mitchell 1995; Sacks *et al.* 1989) and there are numerous strategies available. A common aim is to find a space filling design that both sufficiently spans the input space and that does so in a way so that any possible θ in the input space will not be too far from a point in the design. Popular space filling designs include maximin Latin-hypercubes (Morris and Mitchell 1995) which maximize the minimum distance between points spreading them as far as possible, and low discrepancy sequences such as Sobol sequences (Morokoff and Caflisch 1994), which have an advantage over Latin-hypercubes of being generated sequentially so that extra points can be added as required. For the purpose of emulation, a Monte Carlo sample from the input space will nearly always perform more poorly than a carefully selected design of the same size.

For any $\theta \notin D$ we are uncertain about the value of the simulator for this input. However, if we believe that the model is a smooth continuous function of the inputs, then we can learn about $m(\theta)$ by looking at ensemble members with inputs close to θ. We could, for example, choose to predict $m(\theta)$ by linearly interpolating from the closest ensemble members. The function used to interpolate and extrapolate $\mathcal{D}_{\text{sim}}$ to other input values is commonly called an *emulator*, and there is extensive literature on emulation (sometimes called meta-modelling) for computer experiments (see Santner *et al.* (2003) for references).

We use a Bayesian approach to build an emulator which captures our beliefs about the model. We can elicit prior distributions about the shape of the function, e.g. do we expect linear, quadratic or sinusoidal output, and about the smoothness and variation of the output, e.g. over what kind of length scales do we expect the function to vary. A convenient and flexible semi-parametric family that is widely used to build emulators are Gaussian processes (Stein 1999). We say $f(\cdot)$ is a Gaussian process with mean function $g(\cdot)$ and covariance function $c(\cdot,\cdot)$ and write $f(\cdot) \sim GP(g(\cdot), c(\cdot,\cdot))$ if for any collection of inputs $(x_1, \ldots, x_n)$ the vector $(f(x_1), \ldots, f(x_n))$ has a multivariate Gaussian distribution with mean $(g(x_1), \ldots, g(x_n))$ and covariance matrix Σ where $\Sigma_{ij} = c(x_i, x_j)$. Gaussian process emulators can be used to predict the simulator's value at any input, giving predictions in the form of Gaussian probability distributions over the output space. They can incorporate prior beliefs about both the prior mean structure and the covariance structure, both of which affect the behaviour when interpolating and extrapolating from design points. They are a popular alternative to the use of neural networks (Rasmussen and Williams 2006) because they include with any estimate a measure of confidence in that prediction. The conditioning to update from the prior process to the posterior process after observing the ensemble of model runs is possible analytically, and both the prior and posterior process are simple to simulate from. For univariate computer models we write

$$m(\cdot)|\beta, \lambda, \sigma^2 \sim GP(g(\cdot), \sigma^2 c(\cdot, \cdot))$$

where $g(\theta) = \beta^T h(\theta)$ is a prior mean function which is usually taken to be a linear combination of a set of regressor functions, $h(\cdot)$, and where β represents a vector of coefficients. The prior variance is assumed here to be stationary across the input range and is written as the product of a prior at-a-point variance $\sigma^2 = \mathbb{V}\mathrm{ar}(m(\theta))$, and a correlation function $\mathbb{C}\mathrm{orr}(m(\theta_1), m(\theta_2)) = c(\theta_1, \theta_2)$. Common choices for the correlation function include the Matérn function and the exponential correlation functions (Abrahamsen 1997), such as the commonly used squared exponential family

$$c(\theta_1, \theta_2) = \exp\left[-(\theta_1 - \theta_2)^T \Lambda(\theta_1 - \theta_2)\right]. \tag{10.6}$$

Here, $\Lambda = \mathrm{diag}(\lambda_1, \ldots, \lambda_n)$ is a diagonal matrix containing the roughness parameters. The λ_i represent how quickly we believe the output varies as a function of the input, and can be thought of as a measure of the roughness of the function.

Once we observe the ensemble of model runs $\mathcal{D}_{\text{sim}}$, we update the prior beliefs to find the posterior distribution. If we choose a conjugate prior distribution for β such as an uninformative improper distribution $\pi(\beta) \propto 1$, or a Gaussian distribution, then we can integrate out β to find the posterior

$$m(\cdot)|\mathcal{D}_{\text{sim}}, \lambda, \sigma^2 \sim GP(g^*(\cdot), \sigma^2 c^*(\cdot, \cdot))$$

for modified functions $g^*(\cdot)$ and $c^*(\cdot, \cdot)$. Modified function g^* and c^* are the updated mean and covariance function of the Gaussian process after conditioning on observing the ensemble $\mathcal{D}_{\text{sim}}$. Expressions for g^* and c^* are given later by Equations (10.7) and (10.8) and details of the calculation can be found in Rasmussen and Williams (2006) and many other texts. It is not possible to find a conjugate prior distribution for the roughness parameters, so we take an empirical Bayes approach (Casella 1985) and give each λ_i a prior distribution and then find its maximum a posteriori value and fix λ_i at this value, approximating $\pi(m(\cdot)|\mathcal{D}_{\text{sim}}, \sigma^2)$ by $\pi(m(\cdot)|\mathcal{D}_{\text{sim}}, \hat{\lambda}, \sigma^2)$. If we give σ^2 an inverse chi-squared distribution it is possible to integrate it out analytically, however, this leads to a t-process distribution for $m(\cdot)$ which is inconvenient later, and so we leave σ^2 and use MCMC to integrate it out numerically later in the analysis. An example of how a simple univariate Gaussian process can be used as an emulator is shown in Figure 10.3.

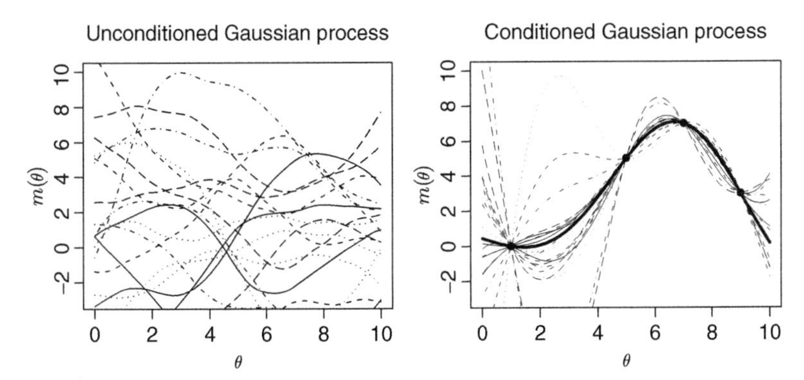

Figure 10.3 The left hand plots shows random draws from a Gaussian process with constant mean function ($g(\theta) = c$ for all θ) and covariance function given by Equation (10.6). These are realizations from our prior belief about the model behaviour. The right hand plot shows draws from the Gaussian process after having conditioned on four data points (shown as solid black points), i.e. draws from our posterior belief about $m(\cdot)$. The posterior mean of the Gaussian process $g^*(\cdot)$ is shown as a thick solid line. Note that at the data points, there is no uncertainty about the value of the process.

The Gaussian process emulator approach described above is for univariate models. For multivariate outputs we could build separate independent emulators for each output, although this ignores the correlations between the outputs and will generally perform poorly if the size of the ensemble is small (as we are throwing away valuable information). Conti and O'Hagan (2007) provide an extension of the above approach which allows us to model a small number of multivariate outputs capturing the correlations between them, and Rougier (2008) describes an outer product emulator which factorizes the covariance matrix in a way that allows computational efficiency and so can be used on a larger number of dimensions if we are prepared to make some fairly general assumptions about the form of the regressors and the correlations. Both of these approaches require careful thought about what correlations are expected between output dimensions. This can be difficult to think about, especially with modellers who may not have much experience with either probability or statistics. Both methods are also limited by the size of problem that can be tackled, although Rougier (2008) made advances on this front. For models with hundreds or thousands of outputs a direct emulation approach may not be feasible, and so here we use a data reduction method to reduce the size of the problem to something more manageable.

10.2.2 Principal Component

We take an approach here similar to Higdon *et al.* (2008), and use a dimension reduction technique to project the output from the computer model onto a

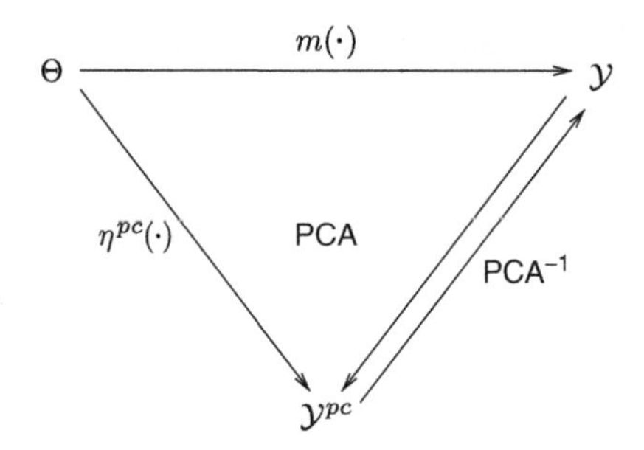

Figure 10.4 Schematic plot of the idea behind principal component emulation. Θ is the input space, $\mathcal{Y}$ the output space, and $m(\cdot)$ the computer model. We let $\eta^{pc}(\cdot)$ denote the Gaussian process emulator from Θ to principal subspace $\mathcal{Y}^{pc}$.

subspace with a smaller number of dimensions and then build emulators of the map from the input space to the reduced output space. The only requirement of the dimension reduction is that there is a method for reconstruction to the original output space. We use principal component analysis here (also known as the method of empirical orthogonal functions), as the projection is then guaranteed to be the optimal linear projection, in terms of minimizing the average reconstruction error, although we could use any other dimension reduction technique as long as there is a method of reconstruction. A schematic plot of this idea is shown in Figure 10.4. The computer model $m(\cdot)$ is a function from input space Θ to output space $\mathcal{Y}$. Principal component analysis provides a map from full output space $\mathcal{Y}$ to reduced space $\mathcal{Y}^{pc}$. We build Gaussian process emulators to map from Θ to $\mathcal{Y}^{pc}$ and then use the inverse of the original projection (also a linear projection) to move from $\mathcal{Y}^{pc}$ to $\mathcal{Y}$. This gives a computationally cheap map from the input space Θ to the output space $\mathcal{Y}$ which does not use the model $m(\cdot)$. This cheap surrogate, or emulator, approximately interpolates all the points in the ensemble (it is approximate due to the error in the principal component reconstruction) and gives probability distributions for the model output for any value of the input.

Principal component analysis is a linear projection of the data onto a lower dimensional subspace (the principal subspace) such that the variance of the projected data is maximized. It is commonly done via an eigenvalue decomposition of the correlation matrix, but for reasons of computational efficiency, we will use a singular value decomposition of the data here. Let Y denote an $N \times n$ matrix with row i the ith run of the computer model, $Y_{i\cdot} = m(\theta_i)$ (recall that the model output is n dimensional and that there are N runs in the ensemble $\mathcal{D}_{\text{sim}}$). The dimension reduction algorithm can then be described as follows:

1. Centre the matrix. Let μ denote the row vector of column means, let Y' be the matrix Y with μ subtracted from each row ($Y' = Y - \mu$) so that

the mean of each column of Y' is zero. We might also choose to scale the matrix, so that the variance of each column is one.

2. Calculate the singular value decomposition

$$Y' = U\Gamma V^*.$$

V is an $n \times n$ unitary matrix containing the principal components (the eigenvectors) and V^* denotes its complex conjugate transpose. Γ is an $N \times n$ diagonal matrix containing the principal values (the eigenvalues) and is ordered so that the magnitude of the diagonal entries decreases across the matrix. U is an $N \times N$ matrix containing the left singular values.

3. Decide on the dimension of the principal subspace, n^* say ($n^* < n$). An orthonormal basis for the principal subspace is then given by the first n^* columns of V (the leading n^* eigenvectors) which we denote as V_1 (an $n \times n^*$ matrix). Let V_2 denote the matrix containing the remaining columns of V.

4. Project Y' onto the principal subspace. The coordinates in the subspace (the factor scores) are found by projecting onto V_1:

$$Y^{pc} = Y'V_1.$$

The ith row of Y^{pc} then denotes the coordinate of the ith ensemble member in the space $\mathcal{Y}^{pc}$.

Some comments:

- Note that the principal component analysis is done across the columns of the matrix rather than across the rows as is usual. The result is that the eigenvalues are of the same dimension as the original output with the leading eigenvalue often taking the general form of the output.

- Although principal component analysis (PCA) is a linear projection, this method can be used on highly nonlinear models. The linear aspect of PCA is the projection of the output space $\mathcal{Y}$ onto a smaller output space $\mathcal{Y}^{pc}$. The map from the input space Θ to the reduced output $\mathcal{Y}^{pc}$ can still be nonlinear, and will be unaffected by any linear assumption used in the dimension reduction. The main requirement for the use of this method, is that there is a consistent covariance structure amongst the outputs. The projection on to the principal components is essentially assuming that the true number of degrees of freedom is less than n. If the outputs are all independent then PCA will not work as a dimension reduction technique.

- There is no established method for deciding on the dimension n^* of the principal subspace. The percentage of variance explained (sum of the corresponding eigenvalues in Γ) is often used as a heuristic, with the stated aim being to explain 95 % or 99 % of the variance. We must also decide which components to include in V_1. It may be found that components which only explain a small amount of the variance (small eigenvalues) are important predictively, as was found in principal component regression (Jolliffe 2002). One method of component selection is through the use of diagnostic plots as explained below.

This leaves us with the coordinates of the ensemble in the principal subspace $\mathcal{Y}^{pc}$, with each row corresponding to the same row in the original design D. Gaussian processes can now be used to emulate this map. Usually, we will have $n^* > 1$, and so we still need to use a multivariate emulator such as that proposed by Rougier (2008). However, emulating the reduced map with n^* independent Gaussian processes often performs as well as using a fully multivariate emulator, especially if the size of the ensemble N is large compared with n^*. Another computational aid that helps with the emulation is to scale the matrix of scores so that each column has variance one. This helps with tuning the MCMC sampler for the σ^2 parameters in the Gaussian process covariance function, as it makes the n^* dimensions comparable with each other.

To reconstruct from the subspace $\mathcal{Y}^{pc}$ to the full space $\mathcal{Y}$ is also a linear transformation. We can post-multiply the scores by V_1^T to give a deterministic reconstruction $Y'' = Y^{pc}V_1^T$. However, this does not account for the fact that by projecting into a n^*-dimensional subspace, we have discarded information in the dimension reduction. To account for this lost information we add random multiples of the eigenvectors which describe the discarded dimensions, namely V_2. We model these random multiples as zero-mean Gaussian distributions with variances corresponding to the relevant eigenvalues. This gives a stochastic rather than a deterministic reconstruction, which accounts for the error in the dimension reduction. In summary, we reconstruct as

$$Y'' = Y^{pc}V_1^T + \Phi V_2^T$$

where Φ is an $N \times (n - n^*)$ matrix with ith column containing N draws from a $N(0, \Gamma_{n^*+i,n^*+i})$ distribution. We then must add the column means of Y to each row of Y'' to complete the emulator.

A useful diagnostic tool when building emulators are leave-one-out cross validation plots. These are obtained by holding back one of the N training runs in the ensemble, training the emulator with the remaining $N - 1$ runs, and then predicting the held back values. Plotting the predicted values, with 95 % credibility intervals, against the true values for each output dimension gives valuable feedback on how the emulator is performing and allows us to validate the emulator. These plots can be used to choose the dimension of the

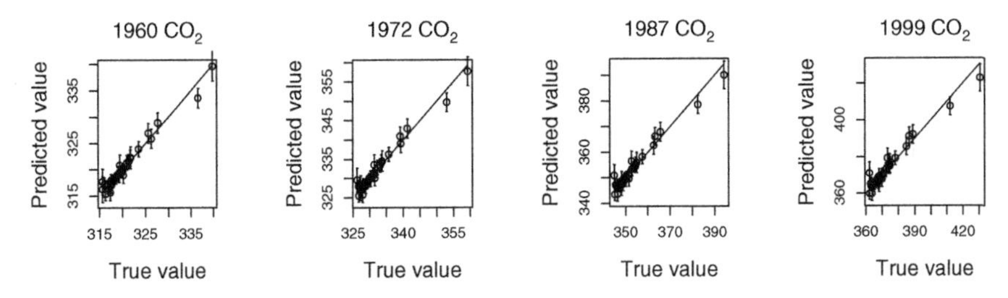

Figure 10.5 Leave-one-out cross validation plots for a selection of four of the 200 outputs. The error bars show 95 % credibility intervals on the predictions. The two outliers seen in each plots are for model runs with inputs on the edge of the design. These points are predictions where we extrapolate rather than interpolate from the other model runs.

principal subspace and which components to include. They are also useful for choosing which regressor functions to use in the specification of the mean structure. Once we have validated the emulator, we can then proceed to use it to calibrate the model.

Example 10.2 (Uvic continued)

We use principal component emulation to build a cheap surrogate for the UVic climate model introduced earlier. Recall that the output of the model is a time-series of 200 atmospheric CO_2 predictions. Figure 10.5 shows the leave-one-out cross-validation plots for a selection of four of the 200 output points. The emulation was done by projecting the time-series onto a 10-dimensional principal subspace and then emulating each map with independent Gaussian processes before reconstructing the data back up to the original space of 200 values. A quadratic prior mean structure was used, $h(\theta_1, \theta_2) = (1, \theta_1, \theta_2, \theta_1^2, \theta_2^2, \theta_1\theta_2)^T$, as the cross-validation plots showed that this gave superior performance over a linear or constant mean structure, with only negligible further gains possible by including higher order terms. The plots show that the emulator is accurately able to predict the held back runs and that the uncertainty in our predictions (shown by the 95 % credibility intervals) provide a reasonable measure of our uncertainty (with 91 % coverage on average).

10.3 Multivariate Calibration

Recall that our aim is to find the distribution of $\hat{\theta}$ given the observations and the model runs, namely

$$\begin{aligned}\pi(\hat{\theta}|\mathcal{D}_{\text{field}}, \mathcal{D}_{\text{sim}}) &\propto \pi(\mathcal{D}_{\text{field}}|\mathcal{D}_{\text{sim}}, \hat{\theta})\pi(\hat{\theta}|\mathcal{D}_{\text{sim}}) \\ &\propto \pi(\mathcal{D}_{\text{field}}|\mathcal{D}_{\text{sim}}, \hat{\theta})\pi(\hat{\theta})\end{aligned}$$

where we have noted that $\pi(\mathcal{D}_{\text{sim}}|\hat{\theta}) = \pi(\mathcal{D}_{\text{sim}})$ and so can be ignored in the posterior distribution of $\hat{\theta}$, leaving $\pi(\mathcal{D}_{\text{field}}|\mathcal{D}_{\text{sim}}, \hat{\theta})$ to be specified in order to find the posterior. The calibration framework given by Equation (10.3) contains three different terms, $m(\hat{\theta}), \delta$ and ε, which we need to model. The best input approach ensures that parameter $\hat{\theta}$ is chosen so as to make $m(\hat{\theta})$ and δ independent for all t (Kennedy and O'Hagan 2001), and the measurement error ε is also independent of both terms. This allows us to specify the distribution of each part of Equation (10.3) in turn, and then calculate the distribution of the sum of the three components. If all three parts have a Gaussian distribution, the sum will also be Gaussian. Distributional choices for ε and δ will be specific to each individual problem, but often measurement errors are assumed to be zero-mean Gaussian random variables, usually with variances reported with the data. Kennedy and O'Hagan recommend the use of Gaussian process priors for the discrepancy function δ. While this is convenient mathematically, sensible forms for the discrepancy will need to be decided with the modellers in each case separately. If some non-Gaussian form is used then there may be difficulty calculating the likelihood in Equation (10.5). For ease of exposition, we assume δ has a Gaussian process distribution here.

Finally, we must find the distribution of $m(\hat{\theta}, t)$ using the principal component emulator. Before considering the map from Θ to $\mathcal{Y}$, we must first consider the distribution of the emulator $\eta^{pc}(\cdot)$ from Θ to $\mathcal{Y}^{pc}$. Using independent Gaussian processes to model the map from the input space to each dimension of the principal subspace (i.e., $\eta^{pc} = (\eta_1^{pc}, \ldots, \eta_{n^*}^{pc})$), we have that the prior distribution for $\eta_i^{pc}(\cdot)$ is

$$\eta_i^{pc}(\cdot)|\beta_i, \sigma_i^2, \lambda_i \sim GP(g_i(\cdot), \sigma_i^2 c_i(\cdot, \cdot)).$$

If we give β_i a uniform improper prior $\pi(\beta_i) \propto 1$, we can then condition on $\mathcal{D}_{\text{sim}}$ and integrate out β_i to find

$$\eta_i^{pc}(\cdot)|\mathcal{D}_{\text{sim}}, \sigma_i^2, \lambda_i \sim GP(g_i^*(\cdot), \sigma_i^2 c_i^*(\cdot, \cdot))$$

where

$$g_i^*(\theta) = \hat{\beta}^T h(\theta) + t(\theta)^T A^{-1}(Y_{.i}^{pc} - H\hat{\beta}) \tag{10.7}$$

$$c_i^*(\theta, \theta') = c(\theta, \theta') - t(\theta)^T A^{-1} t(\theta') + (h(\theta)^T - t(\theta)^T A^{-1} H)(H^T A^{-1} H)^{-1} \times (h(\theta')^T - t(\theta')^T A^{-1} H)^T \tag{10.8}$$

and

$$\hat{\beta}_i = (H^T A^{-1} H)^{-1} H^T A^{-1} Y_{.i}^{pc}$$
$$t(\theta) = (c(\theta, \theta_1), \ldots, c(\theta, \theta_N))$$
$$\{A_i\}_{jk} = \{c_i(\theta_j, \theta_k)\}_{j,k=1,\ldots,N}$$
$$H^T = (h(\theta_1), \ldots, h(\theta_N))$$

assuming the regressors, $h(\cdot)$, are the same for each dimension. Here, $Y^{pc}_{\cdot i}$ denotes the ith column of matrix Y^{pc}, and $\theta_1, \ldots, \theta_N$ are the points in design D. The reconstruction to the full space, $\eta^e(\cdot) = \eta^{pc}(\cdot)V_1^T + \Phi V_2^T$, then has posterior distribution

$$\eta^e(\theta)|\mathcal{D}_{\text{sim}}, \sigma^2, \lambda \sim N(g^*(\theta)V_1^T, \sigma^2 c^*(\theta,\theta)V_1V_1^T + V_2\Gamma' V_2^T)$$

where $g^* = (g_1^*, \ldots, g_{n^*}^*)$ and $\Gamma' = \text{diag}(\Gamma_{n^*+1,n^*+1}, \ldots, \Gamma_{n,n})$. An empirical Bayes approach can be used for the roughness parameters by fixing them at their maximum likelihood estimates. We do not integrate σ^2 out analytically for reasons of tractability, but leave them in the calculation and use MCMC to integrate them out numerically later.

If all three parts of Equation (10.3) are Gaussian then we can write down the likelihood of the field data conditional on the parameters:

$$\pi(\mathcal{D}_{\text{field}}|\mathcal{D}_{\text{sim}}, \sigma^2, \theta, \gamma_\delta)$$

where γ_δ are parameters required for the discrepancy term $\delta(t)$. We elicit prior distributions for θ and γ_δ from the modellers and decide upon priors for σ^2 ourselves (emulators parameters are the responsibility of the person performing the emulation). We then use a Markov chain Monte Carlo algorithm to find the posterior distributions. It is possible to write down a Metropolis-within-Gibbs algorithm to speed up the MCMC calculations, although we do not give the details here.

■ **Example 10.3 (Uvic continued)**

Figure 10.6 shows the marginal posterior distributions from calibrating the UVic model to the Keeling and Whorf (2005) observations. We use an autoregressive process of order one for the discrepancy term with $\delta_t = \rho\delta_{t-1} + U$ where $U \sim N(0, \sigma_\delta^2)$. We give ρ a $\Gamma(5,1)$ prior truncated at one, and σ_δ^2 a $\Gamma(4, 0.6)$ prior distribution. The Markov chains were run for 1,000,000 iterations. The first 200,000 samples were discarded as burn-in and the remaining samples were thinned to every tenth value leaving 80,000 samples. Uniform prior distributions were used for Q_{10} and K_c ($Q_{10} \sim U[1,4]$ and $K_c \sim U[0.25, 1.75]$), and $\Gamma(1.5, 6)$ priors were used for each of the emulator variances σ^2. Tests were done to check the sensitivity of the results to choice of prior distribution, and the analysis was robust to changes in priors for σ^2 and γ_δ, but not to changes in the priors for Q_{10} and K_c.

This will not usually be the end of the calibration process. The results will be returned to the modellers, who may decide to use them to improve the model, before another calibration is performed. For details of this problem, and of the Metropolis-within-Gibbs algorithm, see Ricciuto *et al.*

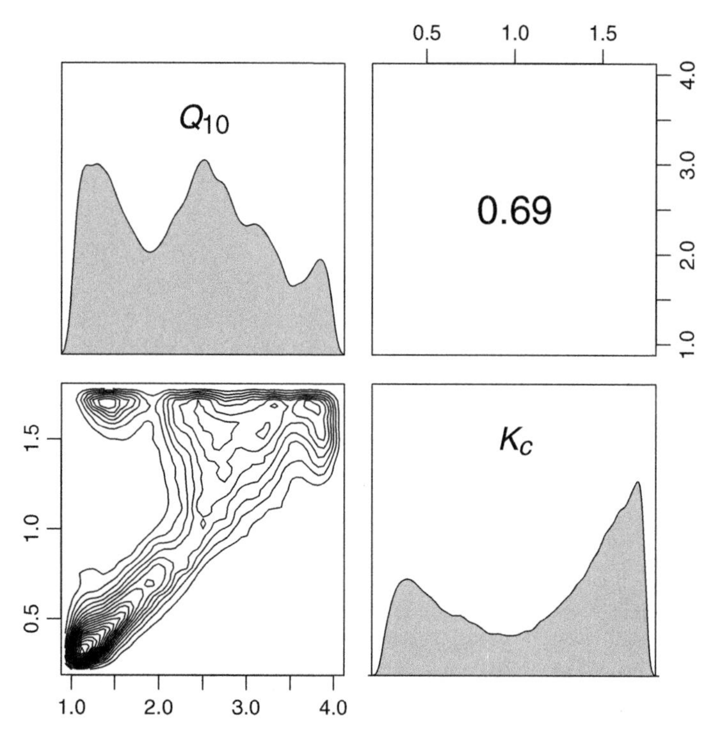

Figure 10.6 Marginal posterior distributions for two of the calibration parameters, Q_{10} and K_c, in the UVic climate model. The two plots on the leading diagonal show the individual marginal plots. The bottom left plot shows the pairwise marginal distribution, and the top right box shows the posterior correlation between Q_{10} and K_c.

10.4 Summary

In this chapter we have shown how to extend the calibration approach of Kennedy and O'Hagan (2001) to enable the calibration of computer models with multivariate outputs. The approach is based around the idea of emulating a reduced dimension version of the model, thus bypassing the need to repeatedly simulate from an expensive model. The method requires a well-designed ensemble of model evaluations $\mathcal{D}_{\text{sim}}$, measurements of the system $\mathcal{D}_{\text{field}}$, and expert beliefs about the measurement error ε and possibly also the model error δ. The method can then be broken down into three steps.

The first step is to build a cheap surrogate for the simulator to use in its place. We project the output from the true model onto a lower dimensional set of basis vectors using principal component analysis, before training Gaussian process emulators to map from the inputs to this new space. We then reconstruct to the original output space by using the inverse projection and account for data loss by using Gaussian multiples of the discarded basis vectors. The second stage is the specification of distributions in accordance with the conditional independence structure shown in Figure 10.1, with prior distributions needed for the calibration and error parameters. Decisions about how to model δ need to be based on each specific simulator, but Gaussian processes are a flexible form that have been found to be useful in previous applications. Finally, we must perform the inference using a Monte Carlo technique such as Markov chain Monte Carlo. This calibration methodology takes account of measurement error, code uncertainty, reconstruction error and model discrepancy, giving posterior distributions which incorporate expert knowledge as well as the model runs and field data.

The limitations of this method are primarily the limitations of the Gaussian process emulator technology. The use of a dimension reduction technique requires that the outputs are correlated (as they usually are in spatial-temporal fields) and some simulators will not be represented adequately in a lower dimensional space, but this will be detected when looking at the reconstruction error. Another potential limitation is that emulators are built on the premise that $m(x + h)$ will be close to $m(x)$ for small h. If this is not the case, as in chaotic systems, then emulators cannot be used. For such simulators, there is no alternative to repeatedly evaluating the model. Gaussian process emulators are also limited by the size of the ensemble they can handle, due to numerical instabilities in the inversion of the covariance matrix, although methods are being developed to improve this (Furrer *et al.* 2006). There exists a range of validation and diagnostic tools (Bastos and O'Hagan 2009) designed to detect and correct problems in emulators, and some form of validation should always be done when using emulators. Finally, it should be stressed that the resulting posteriors do not necessarily give estimates of the true value of physical parameters, but rather give values which lead the model to best explain the data. In order to estimate the true physical value of parameters, the model discrepancy δ must be very carefully specified. This is still a new area of research and much remains to be done in the area of modelling discrepancy functions.

References

Abrahamsen, P. 1997 A review of Gaussian random fields and correlation functions. *Technical Report* **917**, Norwegian Computing Center, Oslo.

Bastos, L.S. and O'Hagan, A. 2009 Diagnostics for Gaussian process emulators. *Technometrics*, to appear.

Casella, G. 1985 An introduction to empirical Bayes data analysis. *American Statistician*, **39(2)**: 83–87.

Conti, S. and O'Hagan, A. 2007 Bayesian emulation of complex multi-output and dynamic computer models. In submission. Available as Research Report No. 569/07, Department of Probability and Statistics, University of Sheffield.

Furrer, R., Genton, M.G. and Nychka, D. 2006 Covariance tapering for interpolation of large spatial datasets. *Journal of Computational and Graphical Statistics* **15(3)**, 502–523.

Garthwaite, P.H., Kadane, J.B. and O'Hagan, A. 2005 Statistical methods for eliciting probability distributions. *Journal of the American Statistical Association* **100**, 680–701.

Goldstein, M. and Rougier, J.C. 2009 Reified Bayesian modelling and inference for physical systems, with discussion and rejoinder. *Journal of Statistical Planning and Inference* **139(3)**, 1221–1239.

Higdon, D., Gattiker, J., Williams, B. and Rightley, M. 2008 Computer model calibration using high-dimensional output. *Journal of the American Statistical Association* **103**, 570–583.

Jolliffe, I.T. 2002 *Principal Component Analysis* 2nd edn. Springer.

Keeling, C.D. and Whorf, T.P. 2005 Atmospheric CO_2 records from sites in the SIO air sampling network. In *Trends: A Compendium of Data on Global Change*, Carbon Dioxide Information Analysis Center, Oak Ridge National Laboratory, US Department of Energy, Tenn., USA.

Kennedy, M. and O'Hagan, A. 2001 Bayesian calibration of computer models (with discussion). *Journal of the Royal Statistical Society, Series B* **63**, 425–464.

Meissner, K.J., Weaver, A.J., Matthews, H.D. and Cox, P.M. 2003 The role of land surface dynamics in glacial inception: a study with the UVic Earth System Model. *Climate Dynamics* **21**, 515–537.

Morokoff, W.J. and Caflisch, R.E. 1994 Quasi-random sequences and their discrepancies. *SIAM J. Sci. Comput* **15**, 1251–1279.

Morris, M.D. and Mitchell, T.J. 1995 Exploratory designs for computational experiments. *Journal of Statistical Planning and Inference* **43**, 161–174.

Murphy, J.M., Booth, B.B.B., Collins, M., Harris, G.R., Sexton, D.M.H. and Webb, M.J. 2007 A methodology for probabilitic predictions of regional climate change from perturbed physics ensembles. *Philosophical Transactions of the Royal Society A* **365** 1993–2028.

Pearl, J. 2000 *Causality: Models, Reasoning, and Inference*, Cambridge University Press.

Rasmussen, C.E. and Williams, C.K.I. 2006 *Gaussian Processes for Machine Learning*, MIT Press.

Ricciuto, D.M., Tonkonojenkov, R., Urban, N., Wilkinson, R.D., Matthews, D., Davis, K.J. and Keller, K. Assimilation of oceanic, atmospheric, and ice-core observations into an Earth system model of intermediate complexity. In submission.

Rougier, J.C. 2008 Efficient Emulators for Multivariate Deterministic Functions. *Journal of Computational and Graphical Statistics* **17(4)**, 827–843.

Sacks, J., Welch, W.J., Mitchell, T.J. and Wynn, H.P. 1989 Design and analysis of computer experiments. *Statistical Science* **4**, 409–423.

Sacks, J., Schiller, S.B. and Welch, W.J. 1989 Designs for computer experiments. *Technometrics*, **31**, 41–47.

Santner, T.J., Williams, B.J. and Notz, W. 2003 *The Design and Analysis of Computer Experiments*, Springer.

Stein, M.L. 1999 *Statistical Interpolation of Spatial Data: Some Theory for Kriging*, Springer.

Chapter 11

The Ensemble Kalman Filter and Related Filters

I. Myrseth and H. Omre

Norwegian University of Science and Technology, Norway

11.1 Introduction

Temporal phenomena are abundant in nature and in man-created activity. Traditionally, mathematicians have modeled these phenomena by differential equations while statisticians have relied on empirically based time series models. It is a natural challenge to combine these two modeling approaches, and hidden Markov models have proven efficient in doing so. The inherent local characteristics of differential equations justifies the Markov assumption while the empirical data is linked to the variables of interest through likelihood functions. Evaluation of the hidden Markov model can be done in a probabilistic setting by Bayesian inversion.

R. E. Kalmans' celebrated paper (Kalman 1960) was based on this line of thought. Under very specific assumptions about linearity and Gaussianity exact analytical solutions can be determined for the Bayesian inversion. Whenever deviations from these assumptions occur however, one has to rely on approximations. This gives room for a large variety of approaches including linearizations and simulation based inference.

The current chapter focuses on simulation based inference of hidden Markov models and on ensemble Kalman filters in particular. The ensemble

---

Large-Scale Inverse Problems and Quantification of Uncertainty Edited by L. Biegler, G. Biros, O. Ghattas, M. Heinkenschloss, D. Keyes, B. Mallick, Y. Marzouk, L. Tenorio, B. van Bloemen Waanders and K. Willcox

Kalman filter was introduced by G. Evensen in the papers Evensen (1994) and Burgers *et al.* (1998). The filter relies on simulation based inference and utilizes a linearization in the data conditioning. These approximations make the ensemble Kalman filter computationally efficient and well suited for high-dimensional hidden Markov models. Hence this filter has found widespread use in evaluation of spatio-temporal phenomena like ocean modeling, weather forecasting and petroleum reservoir evaluation, see Bertino *et al.* (2002), Houtekamer *et al.* (2005), Nævdal *et al.* (2005) and references therein.

The ensemble Kalman filter and its characteristics are the major theme of this chapter, but also related simulation based filters like the randomized maximum likelihood filter and the particle filter are briefly defined. The properties of the various filters are demonstrated on a small example.

In the closing remarks we emphasize the very different ideas on which ensemble Kalman filters and randomized likelihood filters on one side, and particle filters on the other side, are based. Moreover, we recommend that particle filters are used on small low-dimensional problems while ensemble Kalman filters are applied to high-dimensional computer demanding problems.

11.2 Model Assumptions

Consider an unknown, multivariate time series $[\boldsymbol{x}_0, \boldsymbol{x}_1, \ldots, \boldsymbol{x}_T, \boldsymbol{x}_{T+1}]$ with $\boldsymbol{x}_t \in \mathbb{R}^{p_x}; t = 0, \ldots, T+1$ containing the primary variable of interest and $\boldsymbol{x}_T$ being the current state. Assume that an associated time series of observations $[\mathbf{d}_0, \ldots, \mathbf{d}_T]$ with $\mathbf{d}_t \in \mathbb{R}^{p_d}$; $t = 0, \ldots, T$, is available. The primary objective of the study is to assess the forecasting problem, namely evaluate $\boldsymbol{x}_{T+1}$ given $[\mathbf{d}_0, \ldots, \mathbf{d}_T]$.

Define a prior stochastic model for $[\boldsymbol{x}_0, \ldots, \boldsymbol{x}_{T+1}]$ by assuming Markov properties:

$$\begin{aligned}
[\boldsymbol{x}_0, \ldots, \boldsymbol{x}_{T+1}] &\sim f(\boldsymbol{x}_0, \ldots, \boldsymbol{x}_{T+1}) \\
&= f(\boldsymbol{x}_0) \prod_{t=0}^{T} f(\boldsymbol{x}_{t+1} | \boldsymbol{x}_0, \ldots, \boldsymbol{x}_t) \\
&= f(\boldsymbol{x}_0) \prod_{t=0}^{T} f(\boldsymbol{x}_{t+1} | \boldsymbol{x}_t),
\end{aligned}$$

with $\boldsymbol{x} \sim f(\boldsymbol{x})$ reading 'the random variable $\boldsymbol{x}$ is distributed according to the probability density function (pdf) $f(\boldsymbol{x})$'. Let $f(\boldsymbol{x}_0)$ be a known pdf for the initial state, and $f(\boldsymbol{x}_{t+1}|\boldsymbol{x}_t)$ for $t = 0, \ldots, T$ be known forward pdfs. Hence the prior model for the time series of interest is Markovian with each state given the past, dependent on the previous state only.

Define the likelihood model for $[\mathbf{d}_0, \ldots, \mathbf{d}_T]$ given $[\boldsymbol{x}_0, \ldots, \boldsymbol{x}_{T+1}]$ by assuming conditional independence and single state dependence:

$$\begin{aligned}[\mathbf{d}_0, \ldots, \mathbf{d}_T | \boldsymbol{x}_0, \ldots, \boldsymbol{x}_{T+1}] &\sim f(\mathbf{d}_0, \ldots, \mathbf{d}_T | \boldsymbol{x}_0, \ldots, \boldsymbol{x}_{T+1}) \\ &= \prod_{t=0}^{T} f(\mathbf{d}_t | \boldsymbol{x}_0, \ldots, \boldsymbol{x}_{t+1}) \\ &= \prod_{t=0}^{T} f(\mathbf{d}_t | \boldsymbol{x}_t)\end{aligned}$$

where $f(\mathbf{d}_t|\boldsymbol{x}_t)$ for $t = 0, \ldots, T$ are known likelihood functions. Hence, the likelihood model entails that the observation at time t is a function of state $\boldsymbol{x}_t$ only and is independent of the other observations when $\boldsymbol{x}_t$ is given.

These prior and likelihood assumptions define a hidden Markov process as depicted by the diagram in Figure 11.1. The arrows in the graph represent causal, stochastic dependencies between nodes. The resulting posterior stochastic model is defined by Bayesian inversion:

$$\begin{aligned}[\boldsymbol{x}_0, \ldots, \boldsymbol{x}_{T+1} | \mathbf{d}_0, \ldots, \mathbf{d}_T] &\sim f(\boldsymbol{x}_0, \ldots, \boldsymbol{x}_{T+1} | \mathbf{d}_0, \ldots, \mathbf{d}_T) \\ &= \text{const} \times f(\mathbf{d}_0, \ldots, \mathbf{d}_T | \boldsymbol{x}_0, \ldots, \boldsymbol{x}_{T+1}) \\ &\quad \times f(\boldsymbol{x}_0, \ldots, \boldsymbol{x}_{T+1}) \\ &= \text{const} \times f(\boldsymbol{x}_0) f(\mathbf{d}_0 | \boldsymbol{x}_0) \\ &\quad \times \left[\prod_{t=0}^{T-1} f(\mathbf{d}_{t+1} | \mathbf{x}_{t+1}) f(\mathbf{x}_{t+1} | \mathbf{x}_t) \right] \\ &\quad \times f(\boldsymbol{x}_{T+1} | \boldsymbol{x}_T),\end{aligned}$$

with 'const' being a normalizing constant that is usually hard to assess. Hence the full posterior model is not easily available.

The forecasting problem is the major objective of this study. The forecasting pdf is:

$$\begin{aligned}[\boldsymbol{x}_{T+1} | \mathbf{d}_0, \ldots, \mathbf{d}_T] &\sim f(\boldsymbol{x}_{T+1} | \mathbf{d}_0, \ldots, \mathbf{d}_T) \\ &= \int \cdots \int f(\boldsymbol{x}_0, \ldots, \boldsymbol{x}_{T+1} | \mathbf{d}_0, \ldots, \mathbf{d}_T) \mathrm{d}\boldsymbol{x}_0 \ldots \mathrm{d}\boldsymbol{x}_T.\end{aligned}$$

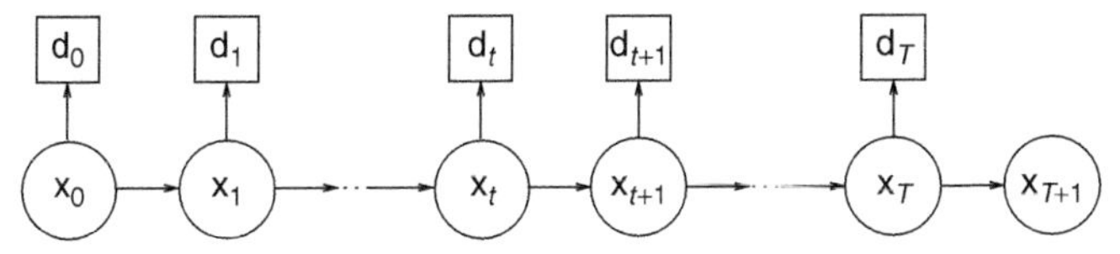

Figure 11.1 Hidden Markov process.

This forecasting pdf is computable by a recursive algorithm. In order to simplify notation, introduce:

$$\begin{aligned} \boldsymbol{x}_t^u &= [\boldsymbol{x}_t | \mathbf{d}_0, \ldots, \mathbf{d}_{t-1}] \\ \boldsymbol{x}_t^c &= [\boldsymbol{x}_t | \mathbf{d}_0, \ldots, \mathbf{d}_t], \end{aligned}$$

where indices u and c indicate unconditioned and conditioned on the observation at the current time, respectively. The recursive algorithm can be justified by the graph in Figure 11.1, and it is defined as:

Algorithm 11.1 *Recursive Forecasting*

- *Initiate:*

$$\boldsymbol{x}_0^u \sim f(\boldsymbol{x}_0^u) = f(\boldsymbol{x}_0)$$

- *Iterate* $t = 0, \ldots, T$

Conditioning:

$$\boldsymbol{x}_t^c \sim f(\boldsymbol{x}_t^c) = f(\boldsymbol{x}_t^u | \mathbf{d}_t) = \text{const} \times f(\mathbf{d}_t | \boldsymbol{x}_t^u) f(\boldsymbol{x}_t^u)$$

Forwarding:

$$\boldsymbol{x}_{t+1}^u \sim f(\boldsymbol{x}_{t+1}^u) = \int f(\boldsymbol{x}_{t+1}^u | \boldsymbol{x}_t^c) f(\boldsymbol{x}_t^c) \mathrm{d}\boldsymbol{x}_t^c$$

- *end iterate*

$$\boldsymbol{x}_{T+1}^u = [\boldsymbol{x}_{T+1} | \mathbf{d}_0, \ldots, \mathbf{d}_T] \sim f(\boldsymbol{x}_{T+1} | \mathbf{d}_0, \ldots, \mathbf{d}_T) = f(\boldsymbol{x}_{T+1}^u)$$

Hence the forecast pdf $f(\boldsymbol{x}_{T+1} | \mathbf{d}_0, \ldots, \mathbf{d}_T)$, which is the objective of the study, is obtained at time step $t = T + 1$. The recursion relies on a conditioning operation and a forwarding operation at each step. Note further that this recursive algorithm makes sequential conditioning on future observations possible.

Algorithm 11.1 relies on assumptions about independence made for the hidden Markov process, but beyond this no specific distributional assumptions are made. This entails that the prior model can be written as:

$$\begin{aligned} \boldsymbol{x}_0 &\sim f(\boldsymbol{x}_0) \\ [\boldsymbol{x}_{t+1} | \boldsymbol{x}_t] &= \omega_t(\boldsymbol{x}_t, \varepsilon_t^x) \sim f(\boldsymbol{x}_{t+1} | \boldsymbol{x}_t) \end{aligned} \tag{11.1}$$

where $\omega_t(.,.)$ is a known function $\mathbb{R}^{2p_x} \rightarrow \mathbb{R}^{p_x}$ and ε_t^x is a random variable from the normalized p_x-dimensional multivariate Gaussian distribution $N_{p_x}(\mathbf{0}, \mathbf{I}_{p_x})$ where $\mathbf{I}_{p_x}$ is a unit diagonal covariance matrix. This construction can generate a realization from an arbitrary $f(\boldsymbol{x}_{t+1} | \boldsymbol{x}_t)$. The likelihood model may be constructed in a similar manner:

$$[\mathbf{d}_t | \boldsymbol{x}_t] = \nu_t(\boldsymbol{x}_t, \varepsilon_t^d) \sim f(\mathbf{d}_t | \boldsymbol{x}_t), \tag{11.2}$$

where $\nu_t(.,.)$ is a known function $\mathbb{R}^{p_x+p_d} \rightarrow \mathbb{R}^{p_d}$ and $\boldsymbol{\varepsilon}_t^d$ is a normalized p_d-dimensional Gaussian random variable from $N_{p_d}(\mathbf{0}, \mathbf{I}_{p_d})$.

The associated Gauss-linear model is defined as:

$$\begin{aligned} \boldsymbol{x}_0 &\sim f(\boldsymbol{x}_0) = N_{p_x}(\boldsymbol{\mu}_0^x, \Sigma_0^x) \\ [\boldsymbol{x}_{t+1}|\boldsymbol{x}_t] &= \mathbf{A}_t\boldsymbol{x}_t + \boldsymbol{\varepsilon}_t^x \sim f(\boldsymbol{x}_{t+1}|\boldsymbol{x}_t) = N_{p_x}(\mathbf{A}_t\boldsymbol{x}_t, \Sigma_t^x) \\ [\mathbf{d}_t|\boldsymbol{x}_t] &= \mathbf{H}_t\boldsymbol{x}_t + \boldsymbol{\varepsilon}_t^d \sim f(\mathbf{d}_t|\boldsymbol{x}_t) = N_{p_d}(\mathbf{H}_t\boldsymbol{x}_t, \Sigma_t^d), \end{aligned} \tag{11.3}$$

where $\mathbf{A}_t$ and $\mathbf{H}_t$ are known matrices of proper dimensions and the error term is additive and Gaussian independent of $\boldsymbol{x}_t$.

Under the Gauss-linear assumptions the recursive Algorithm 11.1 is analytically tractable, and this analytical solution corresponds to the traditional Kalman filter (KF) which will be presented below. For other model assumptions the hidden Markov process must be assessed by some sort of approximation for which the recursive Algorithm 11.1 is well suited. In the current chapter we will focus on simulation based approximate solutions to the forecast problem.

Two Test Examples

In order to illustrate the basic characteristics of the various filters, we define two simple examples that will be used throughout this chapter. The first example is based on a Gauss-linear model, termed linear case, and it is subject to both analytical treatment and simulation based inference. The other case is based on a model with nonlinearities, termed nonlinear case, and it can only be evaluated by simulation based inference.

The variables of interest are $[\boldsymbol{x}_0, \ldots, \boldsymbol{x}_{11}]$, where $\boldsymbol{x}_t \in \mathbb{R}^{100}$; hence $\boldsymbol{x}_t$ is a 100- dimensional time series. Observations are available at $[\mathbf{d}_0, \ldots, \mathbf{d}_{10}]$. The current time is $T = 10$ and the objective is the forecast $[\boldsymbol{x}_{11}|\mathbf{d}_0, \ldots, \mathbf{d}_{10}]$. In Figure 11.2 the reference realizations of $[\boldsymbol{x}_{10}, \mathbf{d}_{10}]$ and $\boldsymbol{x}_{11}$ for both the linear and nonlinear cases are presented. The dimensions of $\boldsymbol{x}_t$ are denoted by nodes.

The linear case is defined as follows:

$$\begin{aligned} f(\boldsymbol{x}_0) &\sim N_{100}(\mathbf{0}, \Sigma_0^x) \\ [\boldsymbol{x}_{t+1}|\boldsymbol{x}_t] &= \mathbf{A}_t\boldsymbol{x}_t \\ [\mathbf{d}_t|\boldsymbol{x}_t] &= \mathbf{H}_t\boldsymbol{x}_t + \boldsymbol{\varepsilon}_t^d \sim f(\mathbf{d}_t|\boldsymbol{x}_t) = N_{13}(\mathbf{H}_t\boldsymbol{x}_t, \Sigma_t^d), \end{aligned}$$

where the initial covariance matrix Σ_0^x contains elements $\sigma_{i,j}^x = 20\exp(-3|i-j|/20)$ for $i,j = 1, \ldots, 100$. The forward model defined by $\mathbf{A}_t$ is a smoother that moves in steps of 5 from left to right for each time step. The matrix $\mathbf{A}_t$ is identical to an identity matrix where the 10×10 submatrix with leading

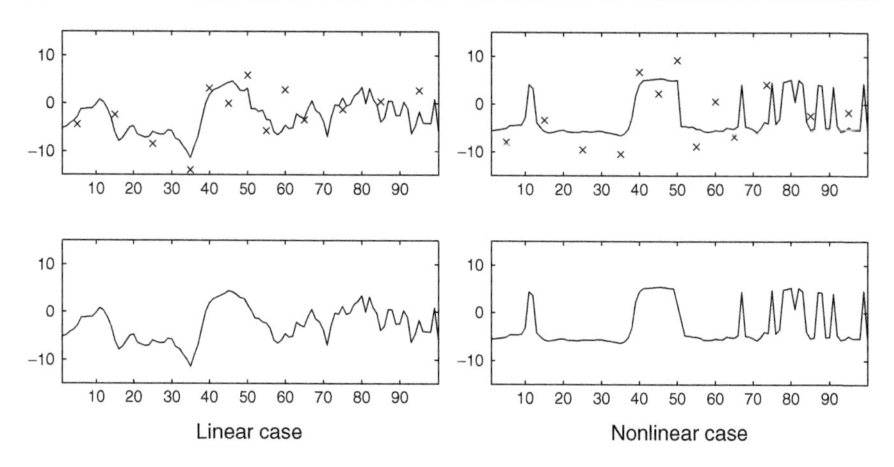

Figure 11.2 Reference realizations: $(\boldsymbol{x}_{10}, \mathbf{d}_{10})$ in the upper display and $\boldsymbol{x}_{11}$ in the lower display. Observations marked by (x). The linear case is determined by linear time dependence and linear likelihood model. The nonlinear model is determined by nonlinear time dependence and nonlinear likelihood model.

index $(5(t-1)+1, 5(t-1)+1)$ is replaced by the 10×10 matrix

$$\begin{bmatrix} \frac{1}{2} & \frac{1}{2} & 0 & & \cdots & & 0 \\ \frac{1}{3} & \frac{1}{3} & \frac{1}{3} & 0 & & \cdots & 0 \\ 0 & \frac{1}{3} & \frac{1}{3} & \frac{1}{3} & 0 & \cdots & 0 \\ \vdots & \ddots & \ddots & \ddots & \ddots & \ddots & \vdots \\ 0 & \cdots & 0 & \frac{1}{3} & \frac{1}{3} & \frac{1}{3} & 0 \\ 0 & & \cdots & 0 & \frac{1}{3} & \frac{1}{3} & \frac{1}{3} \\ 0 & & \cdots & & 0 & \frac{1}{2} & \frac{1}{2} \end{bmatrix}.$$

Consequently, the left part of $\boldsymbol{x}_{10}$ is smoother than the right part. The likelihood model defined by $(\mathbf{H}_t, \Sigma_t^d)$ returns 13 observations with independent errors of variance 10 at each time step, see Figure 11.2. The matrix $\mathbf{H}_t$ is independent of time and contains the elements $h_t(i,j)$ for $i = 1, \ldots, 13$ and $j = 1, \ldots, 100$:

$$h_t(i,j) = \begin{cases} 1 \text{ for } j = 5 + 10(i-1) & 1 \leq i \leq 4 \\ 1 \text{ for } j = 40 + 5(i-5) & 5 \leq i \leq 9 \\ 1 \text{ for } j = 65 + 10(i-10) & 10 \leq i \leq 13 \\ 0 \text{ otherwise} & \end{cases}.$$

The nonlinear case is defined as follows:

$$\begin{aligned}
f(\boldsymbol{x}_0) &\sim N_{100}(\mathbf{0}, \Sigma_0^x) \\
[\boldsymbol{x}_{t+1}|\boldsymbol{x}_t] &= c\mathbf{A}_t(\boldsymbol{x}_t + \arctan(\boldsymbol{x}_t)) \\
[\mathbf{d}_t|\boldsymbol{x}_t] &= \mathbf{H}_t(\boldsymbol{x}_t + \arctan(\boldsymbol{x}_t)) + \varepsilon_t^d \sim f(\mathbf{d}_t|\boldsymbol{x}_t) \\
&= N_{13}(\mathbf{H}_t(\boldsymbol{x}_t + \arctan(\boldsymbol{x}_t)), \Sigma_t^d),
\end{aligned}$$

where the model parameters are defined mostly as for the linear case, except for $c = .8$ being a scaling factor to align the variances in the linear and nonlinear case, and for arctan(.) being a functional that acts element-wise on $\boldsymbol{x}_t$. As Figure 11.2 shows, the nonlinear example appears more box-like than the linear one. The ability of the different algorithms to forecast $[\boldsymbol{x}_{11}|\mathbf{d}_0, \ldots, \mathbf{d}_{10}]$ will be used to measure their performance.

11.3 The Traditional Kalman Filter (KF)

The KF was introduced in Kalman (1960). The hidden Markov process was made analytically tractable by the recursive Algorithm 11.1 and the Gauss-linear model assumptions in Expression (11.3) with fully known parameter values.

The recursive algorithm will under these Gauss-linear assumptions reproduce Gaussianity from one step to the next, and the algorithm appears as:

Algorithm 11.2 *Kalman Filter*

- *Initiate:*

$$\begin{aligned}
&\boldsymbol{x}_0^u \sim f(\boldsymbol{x}_0^u) = N_{p_x}(\boldsymbol{\mu}_0^u, \Sigma_0^u) \\
&\boldsymbol{\mu}_0^u = \boldsymbol{\mu}_0^x \\
&\Sigma_0^u = \Sigma_0^x
\end{aligned}$$

- *Iterate* $t = 0, \ldots, T$

Conditioning:

$$\begin{aligned}
&\boldsymbol{x}_t^c \sim f(\boldsymbol{x}_t^c) = N_{p_x}(\boldsymbol{\mu}_t^c, \Sigma_t^c) \\
&\boldsymbol{\mu}_t^c = \boldsymbol{\mu}_t^u + \Sigma_t^u \mathbf{H}_t' [\mathbf{H}_t' \Sigma_t^u \mathbf{H}_t' + \Sigma_t^d]^{-1} (\mathbf{d}_t - \mathbf{H}_t \boldsymbol{\mu}_t^u) \\
&\Sigma_t^c = \Sigma_t^u - \Sigma_t^u \mathbf{H}_t' [\mathbf{H}_t \Sigma_t^u \mathbf{H}_t' + \Sigma_t^d]^{-1} \mathbf{H}_t \Sigma_t^u
\end{aligned}$$

Forwarding:

$$\boldsymbol{x}_{t+1}^u \sim f(\boldsymbol{x}_{t+1}^u) = N_{p_x}(\boldsymbol{\mu}_{t+1}^u, \Sigma_{t+1}^u)$$

$$\boldsymbol{\mu}_{t+1}^{u} = \mathbf{A}_t \boldsymbol{\mu}_t^{c}$$
$$\Sigma_{t+1}^{u} = \mathbf{A}_t \Sigma_t^{c} \mathbf{A}_t' + \Sigma_t^{x}$$

- *end iterate*

- $f(\boldsymbol{x}_{T+1}|\mathbf{d}_0, \ldots, \mathbf{d}_T) = N_{p_x}(\boldsymbol{\mu}_{T+1}^{u}, \Sigma_{T+1}^{u})$

Consequently, all relevant pdfs are analytically tractable and no approximations of the forecast problem is needed. Whenever deviations from these Gauss-linear assumptions arise however, the analytical tractability is lost. Approximate solutions can of course be made, and different approximations have defined a large family of Kalman Filter variations such as the extended Kalman filter (Jazwinski 1970), the unscented Kalman Filter (Julier and Uhlmann 1997) and others. These approximations follow the KF tradition by focusing on the two first moments of the pdf as they develop through time. Since traditional Kalman filters are not the focus of this presentation, we leave the reader with these references and the application to the test problem below.

Performance of the Test Examples

The traditional KF can only be applied to the linear case, and the solution is shown in Figure 11.3 together with the $\boldsymbol{x}_{11}$ reference realization. The KF prediction represented by the conditional expectation $\mathrm{E}\{\boldsymbol{x}_{11}|\mathbf{d}_0, \ldots, \mathbf{d}_{10}\}$ and associated .95 prediction intervals are displayed. This KF solution is the exact solution and can be computed analytically, and hence extremely computer efficient.

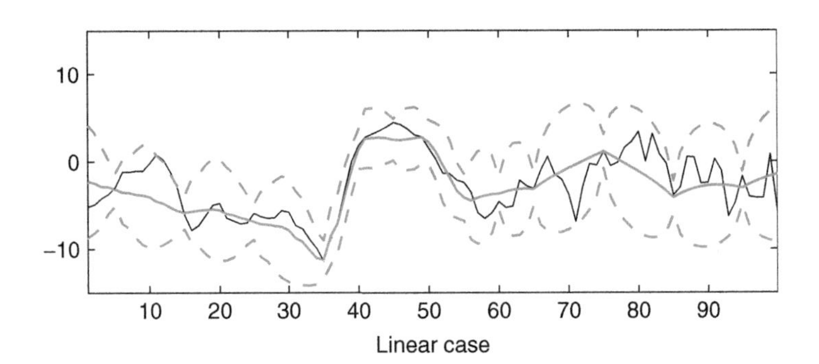

Figure 11.3 Results of the KF algorithm on the linear case: reference realization $\boldsymbol{x}_{11}$ (black), prediction (blue) and .95-prediction intervals (hatched blue).

11.4 The Ensemble Kalman Filter (EnKF)

The EnKF provides an approximate solution to the forecast problem both under Gauss-linear model assumptions and whenever deviations from these assumptions occur. The basic idea of the EnKF is to create a set of realizations, so called ensemble, from the initial model. These realizations are adjusted according to the likelihood model when an observation occurs and the adjusted realizations are then taken through the forward model to the next observation time. At time $t = T + 1$ a set of approximately independent realizations are available for empirical assessment of $f(\boldsymbol{x}_{T+1}|\mathbf{d}_0, \dots, \mathbf{d}_T)$. Hence, characteristics beyond the two first moments can be captured. Note also that the actual adjustment of realizations as observations appear makes it meaningful to visually inspect the ensemble members to seek for common features. Basic references for EnKF are Evensen (1994), Burgers *et al.* (1998), Evensen (2007) and references therein.

A time series of ensembles is defined by:

$$\boldsymbol{e}_t : \{(\boldsymbol{x}_t^u, \mathbf{d}_t)^{(i)}; i = 1, \dots, n_e\}; t = 0, \dots, T + 1,$$

where $\boldsymbol{x}_t^{u(i)} = [\boldsymbol{x}_t|\mathbf{d}_0, \dots, \mathbf{d}_{t-1}]^{(i)}$ are approximate realizations from $f(\boldsymbol{x}_t|\mathbf{d}_0, \dots, \mathbf{d}_{t-1})$ and $\mathbf{d}_t^{(i)}$ are associated realizations of the observation available at time t, i.e., $\mathbf{d}_t$. Note that at any step t - with t omitted in the notation - one has the expectation vector and covariance matrix:

$$\boldsymbol{\mu}_{xd} = \begin{bmatrix} \mathrm{E}\{\boldsymbol{x}^u\} \\ \mathrm{E}\{\mathbf{d}\} \end{bmatrix} = \begin{bmatrix} \boldsymbol{\mu}_x \\ \boldsymbol{\mu}_d \end{bmatrix}$$

$$\Sigma_{xd} = \begin{bmatrix} \mathrm{Cov}\{\boldsymbol{x}^u\} & \mathrm{Cov}\{\boldsymbol{x}^u, \mathbf{d}\} \\ \mathrm{Cov}\{\mathbf{d}, \boldsymbol{x}^u\} & \mathrm{Cov}\{\mathbf{d}\} \end{bmatrix} = \begin{bmatrix} \Sigma_x & \Gamma_{x,d} \\ \Gamma_{d,x} & \Sigma_d \end{bmatrix}.$$

Estimates of these model parameters based on the ensemble $\boldsymbol{e}_t$ are represented by a hat index. The EnKF algorithm in its general form is based on the recursive forecasting algorithm, Algorithm 11.1, and appears as:

Algorithm 11.3 *Ensemble Kalman Filter (General)*

- *Initiate*

$$\begin{aligned}
n_e &= \textit{no. of ensemble members} \\
&\boldsymbol{x}_0^{u(i)}; i = 1, \dots, n_e \ \textit{iid} \ f(\boldsymbol{x}_0) \\
&\boldsymbol{\varepsilon}_0^{d(i)} \sim N_{p_d}(\mathbf{0}, \mathbf{I}_{p_d}); i = 1, \dots, n_e \\
&\mathbf{d}_0^{(i)} = \nu_t(\boldsymbol{x}_0^{u(i)}, \boldsymbol{\varepsilon}_0^{d(i)}); i = 1, \dots, n_e \\
&\boldsymbol{e}_0 : \{(\boldsymbol{x}_0^u, \mathbf{d}_0)^{(i)}; i = 1, \dots, n_e\}
\end{aligned}$$

- *Iterate* $t = 0, \ldots, T$

  Conditioning:

  Estimate Σ_{xd} *from* $\mathbf{e}_t \rightarrow \widehat{\Sigma}_{xd}$

  $\boldsymbol{x}_t^{c(i)} = \boldsymbol{x}_t^{u(i)} + \widehat{\mathbf{\Gamma}}_{x,d}\widehat{\Sigma}_d^{-1}(\mathbf{d}_t - \mathbf{d}_t^{(i)})\,;\, i = 1, \ldots, n_e$

  Forwarding:

  $\boldsymbol{\varepsilon}_t^{x(i)} \sim N_{p_x}(\mathbf{0}, \mathbf{I}_{p_x})\,;\, i = 1, \ldots, n_e$

  $\boldsymbol{x}_{t+1}^{u(i)} = \omega_t(\boldsymbol{x}_t^{c(i)}, \boldsymbol{\varepsilon}_t^{x(i)})\,;\, i = 1, \ldots, n_e$

  $\boldsymbol{\varepsilon}_{t+1}^{d(i)} \sim N_{p_d}(\mathbf{0}, \mathbf{I}_{p_d})\,;\, i = 1, \ldots, n_e$

  $\mathbf{d}_{t+1}^{(i)} = \nu_{t+1}(\boldsymbol{x}_{t+1}^{u(i)}, \boldsymbol{\varepsilon}_{t+1}^{d(i)})\,;\, i = 1, \ldots, n_e$

  $\mathbf{e}_{t+1} \,:\, \{(\boldsymbol{x}_{t+1}^{u}, \mathbf{d}_{t+1})^{(i)}\,;\, i = 1, \ldots, n_e\}$

- *end iterate*

- *Assess*

  $f(\boldsymbol{x}_{T+1}|\mathbf{d}_0, \ldots, \mathbf{d}_T)$ *from* $\mathbf{e}_{T+1}$

The resulting ensemble $\mathbf{e}_{T+1}$ contains approximately independent realizations of

$$\boldsymbol{x}_{T+1}^{u} = [\boldsymbol{x}_{T+1}|\mathbf{d}_0, \ldots, \mathbf{d}_T]$$

from $f(\boldsymbol{x}_{T+1}|\mathbf{d}_0, \ldots, \mathbf{d}_T)$ which can be used to assess the forecast pdf. If the forecast expectation vector and covariance matrix are of interest, the standard estimators can be used:

$$\begin{aligned} \widehat{\boldsymbol{\mu}}_{T+1} &= \widehat{\mathrm{E}}\{\boldsymbol{x}_{T+1}|\mathbf{d}_0, \ldots, \mathbf{d}_T\} = \frac{1}{n_e}\sum_{i=1}^{n_e} \boldsymbol{x}_{T+1}^{u(i)} \\ \widehat{\Sigma}_{T+1} &= \widehat{\mathrm{Var}}\{\boldsymbol{x}_{T+1}|\mathbf{d}_0, \ldots, \mathbf{d}_T\} = \frac{1}{n_e - 1}\sum_{i=1}^{n_e}(\boldsymbol{x}_{T+1}^{u(i)} - \widehat{\boldsymbol{\mu}}_{T+1})(\boldsymbol{x}_{T+1}^{u(i)} - \widehat{\boldsymbol{\mu}}_{T+1})'. \end{aligned} \tag{11.4}$$

The EnKF algorithm, Algorithm 11.3, is recursive and each recursion consists of a conditioning operation and a forwarding operation. The conditioning expression is linear with weights estimated from the ensemble. The forwarding operation is defined by the forward pdf.

There are two implicit approximations in the EnKF:

1. Discretization of the sample space of $\boldsymbol{x}_t$. The initial ensemble of iid realizations is assumed to represent $f(\boldsymbol{x}_0)$. For high-dimensional problems a large number of ensemble members may be required to do so reliably.

2. The data conditioning expression is linearized. Moreover, the weights in the linearization are estimated from the ensemble. Note, however, that each ensemble member is conditioned individually and hence the linearization only applies to the conditioning not to the forward model. For highly non-Gaussian prior models and/or strongly nonlinear likelihood models this approximation may provide unreliable results.

Under these approximations, however, all types of models for the hidden Markov process can be evaluated. The EnKF is a consistent forecast procedure under a Gauss-linear model in the sense that the exact solution is obtained as $n_e \to \infty$ (see 11.8).

A graphical description of the EnKF is shown in Figure 11.4. At iteration t, the ensemble $\boldsymbol{e}_t$ with $n_e = 5$ is shown in display (a) together with the observed value $\mathbf{d}_t$. The regression slope of $\mathbf{d}_t$ on $\boldsymbol{x}_t$ is estimated by $\widehat{\boldsymbol{\Gamma}}_{x,d}\widehat{\Sigma}_d^{-1}$ and this corresponds to the regression slope in display (b). The EnKF-update is made on each ensemble member by using the estimated regression slope as presented in display (c). Note how the individual unconditioned members $\boldsymbol{x}_t^{u(i)}$ are moved to the conditioned ones $\boldsymbol{x}_t^{c(i)}$. To establish the ensemble $\boldsymbol{e}_{t+1}$ at iteration $t+1$, each $\boldsymbol{x}_t^{c(i)}$ is taken through the forward model to obtain $\boldsymbol{x}_{t+1}^{u(i)}$ and thereafter each $\boldsymbol{x}_{t+1}^{u(i)}$ is taken through the likelihood models to obtain $\mathbf{d}_{t+1}^{(i)}$. This brings us back to display (a) for iteration $t+1$.

The EnKF is a consistent forecast procedure for Gauss-linear models, but frequently deviations from Gaussianity in the prior and likelihood model appear. Consequences of these deviations from Gaussianity are discussed in Section 11.4.1. Other problems arise in the EnKF which are caused by the use of an estimate of Σ_{xd} based on $\boldsymbol{e}_t$ instead of the true covariance matrix. These problems include rank deficiency and estimation uncertainty due to the limited size of the ensemble, i.e., small values of n_e. This issue will be discussed further in Section 11.4.2.

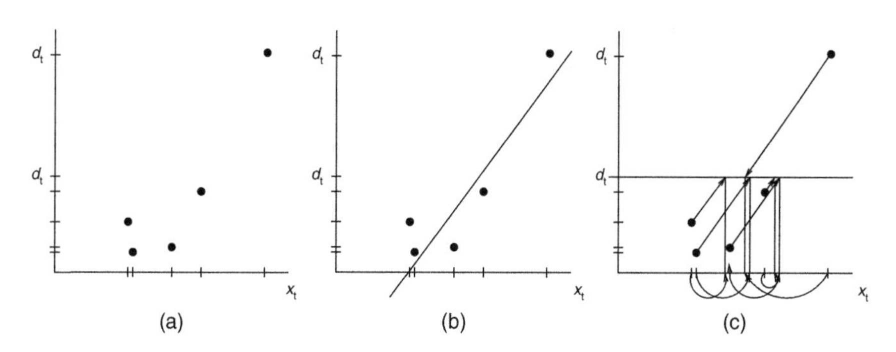

Figure 11.4 Graphical description of one step in the EnKF algorithm; ensemble (a), linearization (b) and ensemble conditioning (c). See explanation in text.

Performance of the Test Examples

The EnKF can be applied to both the linear and the nonlinear test cases. The solutions are presented in Figure 11.5 together with the $\boldsymbol{x}_{11}$ reference realization. The EnKF is based on the ensemble size $n_e = 100$, which is often used as rule of thumb. The exact motivation for this number is unclear, but Natvik and Evensen (2003a,b) present some kind of justification.

The upper displays in Figure 11.5 show the predictions $\widehat{\mathrm{E}}\{\boldsymbol{x}_{11}|\mathbf{d}_0, \ldots, \mathbf{d}_{10}\}$ and associated .95 empirical prediction intervals. The predictions are obtained as the average of the 100 ensemble members at each node. The empirical intervals are defined to be the range between the third smallest and the third largest ensemble member at any node, hence it covers the 94 central ensemble members.

The lower displays show the predictions from ten different independent EnKF runs, each of these with $n_e = 100$.

For the linear case, the correct solution is the KF solution shown in Figure 11.3. The EnKF solution with $n_e = 100$ does reproduce the main features of the reference realization $\boldsymbol{x}_{11}$, but as seen in the upper display, it seems that the prediction interval is underestimated. This is probably caused by estimation uncertainty in the weights in the conditioning step. Some of the variability appears to be shifted to variability in between predictions based on different independent ensembles as seen in the lower display. Recall that the EnKF algorithm is consistent in the sense that the KF solution is reproduced in the

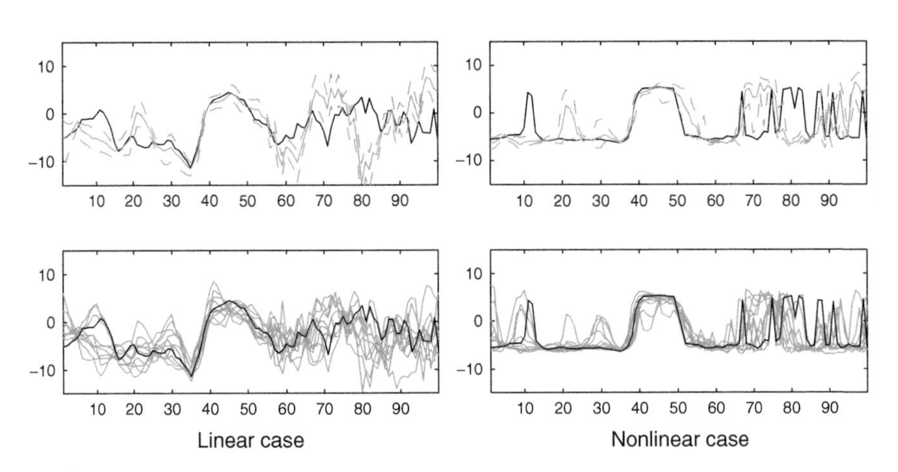

Figure 11.5 Results of the EnKF algorithm on the linear and nonlinear test cases: reference realization $\boldsymbol{x}_{11}$ (black), prediction (blue) and empirical .95-prediction intervals (hatched blue) for one run of the algorithm (top row); reference realization $\boldsymbol{x}_{11}$ (black) and predictions (blue) for 10 different runs of the algorithm (bottom row).

limit as $n_e \to \infty$. For the nonlinear case, the picture is basically the same. The results indicate that it is important to have a reasonably high number of ensemble members, to reduce estimation uncertainty in the model parameters.

11.4.1 Variable Characteristics

In this section the performance of the EnKF in cases where the true model deviates from Gaussianity will be discussed. The linear conditioning in the EnKF is correct if at each step the likelihood model $f(\mathbf{d}_t|\boldsymbol{x}_t^u)$ is Gauss-linear and the prior model $f(\boldsymbol{x}_t^u)$ is Gaussian. Often we can control the observation procedure or reformulate the model and ensure approximate Gauss-linearity in the likelihood, hence the discussion will focus on the characteristics of $\boldsymbol{x}_t^u$.

The EnKF is expected to be reliable as long as $f(\boldsymbol{x}_t^u)$ is unimodal and reasonably symmetric since all elliptically symmetric pdfs have conditional models that are linear in the variables (Kelker 1970). If, however, the $f(\boldsymbol{x}_t^u)$ is multimodal, the picture may be very different. Usually, multimodality indicates that the variable of interest $\boldsymbol{x}_t^u$ is a nested variable with a discrete underlying variable defining categorical classes and a continuous variable representing smooth variations within each class. The EnKF approach has no conditioning mechanism for shifting from one discrete outcome to the other, it fully relies on continuous adjustments through the linearized conditioning expression. Hence the EnKF solution, in cases with multimodal $f(\boldsymbol{x}_t^u)$, may deviate considerably from the correct solution $f(\boldsymbol{x}_t^u|\mathbf{d}_t)$. Several approaches to solving the multimodal problem have been suggested in the literature:

1. Mixtures of Gaussian pdfs may be used as model for $f(\boldsymbol{x}_t^u)$, and a nonlinear updating rule can be defined to obtain samples from $f(\boldsymbol{x}_t^u|\mathbf{d}_t)$ (Dovera and Della Rossa 2007). The nonlinear updating rule will be computationally demanding and may not be suitable for high-dimensional problems.

2. Reparametrization of the discrete underlying variable into continuous variables, preferably with unimodal pdfs. In spatial problems where the discrete process may be related to domains in space, the boundary of these domains may be characterized by continuous variables. The conditioning can then be made on these boundary variables (Bertino *et al.* 2003; Liu and Oliver 2005). The reparametrization is usually highly problem specific and the efficiency is hard to judge.

3. Basis-function expansion of the variables of interest $\boldsymbol{x}_t^u$. The associated coefficients are conditioned by the EnKF procedure (Jafarpour and McLaughlin 2007). The choice of basis functions is crucial for the success of the approach and no clear guideline seems to be available so far.

11.4.2 Parameter Estimates

In this section the impact on EnKF of using estimates of the model parameters will be discussed. The EnKF algorithm depends crucially on the term $\Gamma_{x,d}\Sigma_d^{-1}$ which must be inferred from the ensemble $\boldsymbol{e}_t$ of size n_e. Usually the classical covariance matrix estimators are applied:

$$\widehat{\boldsymbol{\Gamma}}_{x,d} = \frac{1}{n_e - 2}\sum_{i=1}^{n_e}(\boldsymbol{x}_t^{u(i)} - \widehat{\boldsymbol{\mu}}_x)(\mathbf{d}_t^{(i)} - \widehat{\boldsymbol{\mu}}_d)'$$

$$\widehat{\Sigma}_d = \frac{1}{n_e - 1}\sum_{i=1}^{n_e}(\mathbf{d}_t^{(i)} - \widehat{\boldsymbol{\mu}}_x)(\mathbf{d}_t^{(i)} - \widehat{\boldsymbol{\mu}}_x)',$$

with

$$\widehat{\boldsymbol{\mu}}_x = \frac{1}{n_e}\sum_{i=1}^{n_e}\boldsymbol{x}_t^{u(i)}$$

$$\widehat{\boldsymbol{\mu}}_d = \frac{1}{n_e}\sum_{i=1}^{n_e}\mathbf{d}_t^{(i)}.$$

If $\boldsymbol{e}_t$ contains independent members these estimators are unbiased and consistent, i.e., $\widehat{\boldsymbol{\Gamma}}_{x,d} \rightarrow \Gamma_{x,d}$ and $\widehat{\Sigma}_d \rightarrow \Sigma_d$ as $n_e \rightarrow \infty$, for all distributional models. For finite n_e, the estimates have minimum variance for Gauss-linear models. With only slight deviations from Gauss-linearity, however, the estimation variance may increase dramatically (Huber 1981). Moreover, the ranks of the estimated matrices $\widehat{\boldsymbol{\Gamma}}_{x,d}$ and $\widehat{\Sigma}_d$ can at most be $n_e - 1$. Hence the reliability of the estimates are highly dependent on the actual true model and the size of the ensemble. Two major problems may occur in using the EnKF:

1. Rank deficiency. Recall that $\Gamma_{x,d}$ and Σ_d are, respectively, $p_x \times p_d$ and $p_d \times p_d$ matrices. Hence $n_e \geq p_d + 1$ is required to ensure full rank of the estimates $\widehat{\boldsymbol{\Gamma}}_{x,d}$ and $\widehat{\Sigma}_d$. In simple time series applications this requirement is easy to meet. In spatio-temporal applications with computationally demanding forward models and many observations, as for example in petroleum reservoir engineering and meteorology, requiring $n_e \geq p_d + 1$ may be prohibited. With reduced rank, the expression $\widehat{\boldsymbol{\Gamma}}_{x,d}\widehat{\Sigma}_d^{-1}$ in the EnKF updating will be ill-defined.

2. Estimation uncertainty. Standard estimators for second order moments like the covariance matrices $\Gamma_{x,d}$ and Σ_d are notoriously unreliable due to extreme dependence on the tail-behavior of the underlying pdf. This sensitivity is caused by the second order term of the estimators. It is complicated to make wise bias/variance trade-offs without making definite distributional assumptions on which the estimator properties rely heavily. Lack of precision in $\widehat{\boldsymbol{\Gamma}}_{x,d}\widehat{\Sigma}_d^{-1}$ in the EnKF updating may cause

spurious values to appear in the conditioned $\boldsymbol{x}_t^c$. This effect may be worsened by nonlinear forward models. Moreover, since the same estimate $\widehat{\boldsymbol{\Gamma}}_{x,d}\widehat{\Sigma}_d^{-1}$ is used in the updating of all ensemble members in $\boldsymbol{e}_t$, unwanted dependencies between the members may be introduced. This may cause biased model parameter estimates at later times. Ideally, the estimation uncertainty in $\widehat{\boldsymbol{\Gamma}}_{x,d}$ and $\widehat{\Sigma}_d$ should be accounted for in the final assessment of $f(\boldsymbol{x}_{T+1}|\mathbf{d}_0, \ldots, \mathbf{d}_T)$, but this is not done in the current version of the EnKF.

Several approaches to solving the rank deficiency and estimation uncertainty problems have been suggested in the literature:

1. Singular value decomposition of Σ_d provides the linear combinations of the original variables that capture most of the variability in the observations. By retaining less than $n_e - 1$ of these combinations, the EnKF updating procedure is well defined in the reduced space (Skjervheim *et al.* 2007). Note that exact observations will not be reproduced if this dimensionality reduction is made. Moreover, the procedure contains no direct measure against effects of estimation uncertainty. The approach is expected to work whenever the observations $\mathbf{d}_t$ are highly correlated. Otherwise the information content in the observations can be severely reduced by the dimensionality reduction and the effect of the conditioning may be partly removed.

2. Localization is based on a moving neighborhood linearized conditioning concept. When conditioning a certain dimension of the variables of interest, $\boldsymbol{x}^u$, only observations in some predefined neighborhood of this dimension is included in the conditioning expression. The conditioning is performed by running sequentially through the dimensions of $\boldsymbol{x}^u$ with different neighborhood for each dimension (Evensen 2007; Houtekamer and Mitchell 1998). This reduction of number of conditioning observations for each dimension of $\boldsymbol{x}^u$ will also reduce the actual dimensions of the $\Gamma_{x,d}\Sigma_d^{-1}$-term and hence by adjusting the neighborhood definition full rank for a given ensemble size n_e can be ensured. The approach is expected to work whenever the observations in $\mathbf{d}_t$ are close to conditional independent given $\boldsymbol{x}_t$ and each observation has local influence on $\boldsymbol{x}_t$. Then reliable neighborhood rules can be defined. If this local dependence is not present, the information content in the observations may be severely reduced and the effect of the conditioning will be partly removed. Moreover, localization often introduces artifacts in the solutions since different sets of conditioning observations are used for different dimensions of $\boldsymbol{x}^u$.

3. Regularization approaches when estimating $\Gamma_{x,d}$ and Σ_d have been proposed. The estimates $\widehat{\boldsymbol{\Gamma}}_{x,d}$ and $\widehat{\Sigma}_d^{-1}$ have to be valid covariance matrices and hence non-negative definite. This constrains the class

of possible estimators. Non-negative definiteness is closed under addition and certain products, however, and these properties are used to introduce subjective shrinkage effects into the estimator (Hamill *et al.* 2001). Regularization can both ensure full rank and dampen the effect of large estimation uncertainty. It can of course be complicated to choose a suitable regularization factor and for many proposed regularization factors the consistency of the estimators as $n_e \to \infty$ is lost.

4. Experimental design approaches in generating the initial ensemble has been proposed to reduce the variance of $\widehat{\mathbf{\Gamma}}_{x,d}$ and $\widehat{\Sigma}_d^{-1}$ at later times (Evensen 2004). By designing the initial ensemble the iid properties are lost and it is unclear how this influences the bias of the estimates and the final inference of $f(\boldsymbol{x}_{T+1}|\mathbf{d}_0, \ldots, \mathbf{d}_T)$ based on the ensemble $\mathbf{e}_{T+1}$. Experimental design approaches will have no impact on the rank deficiency problem.

5. Ensemble splitting has been proposed in order to explore the estimation uncertainty and reduce the unwanted dependencies between the ensemble members (Houtekamer and Mitchell 1998). By splitting the ensemble the rank deficiency problem becomes even more acute and estimation uncertainty is even worse for each split ensemble. The effect of these features is unclear.

6. A hierarchical extension of the hidden Markov process such that the model parameters $\boldsymbol{\mu}_{xd}$ and Σ_{xd} are included in the probability space has been proposed to account for the parameter uncertainty in the EnKF results (Myrseth and Omre 2010). This approach also ensures full rank and provides robust estimates of $\Gamma_{x,d}$ and Σ_d. This hierarchical EnKF requires that prior pdfs be defined for $\boldsymbol{\mu}_{xd}$ and Σ_{xd}, which of course can be a challenge. Full consistency in the estimator as $n_e \to \infty$ is ensured, however.

Performance of the Test Examples

The EnKF with the ensemble size $n_e = 30$ is run on the linear test problem to illustrate the effect of estimation uncertainty. Note that rank deficiency is avoided by ensuring $n_e >= p_d + 1$. The solutions are presented in Figure 11.6 in a format similar to Figure 11.5. Figure 11.6 should be compared to Figure 11.5 where solutions from the same EnKF algorithm with $n_e = 100$ are presented. Note that the prediction interval based on a single ensemble is severely underestimated while almost all variability is shifted to differences between predictions of repeated EnKF runs. This is caused by increased estimation uncertainty in the weights in the conditioning step, which introduces dependence between ensemble members. Consequentially, empirical variance in the ensemble decreases and differences between averages of independent

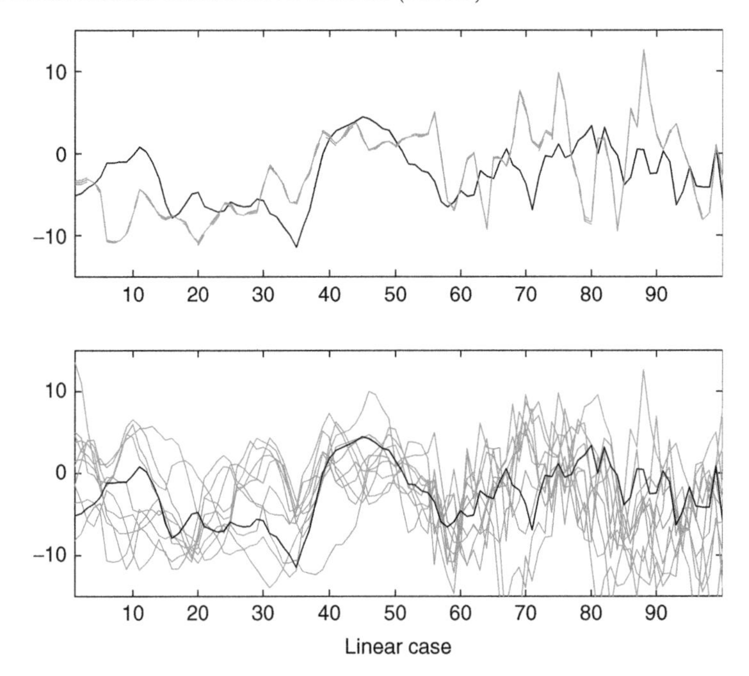

Figure 11.6 Results of the EnKF algorithm with $n_e = 30$ on the linear case: reference realization $\boldsymbol{x}_{11}$ (black), prediction (blue) and empirical .95-prediction intervals (hatched blue) (top row); reference realization $\boldsymbol{x}_{11}$ (black) and predictions (blue) for 10 different runs of the algorithm (bottom row).

EnKF runs increase. The results from the EnKF with a small number of ensemble members seem to be very unreliable.

11.4.3 A Special Case

We consider a hidden Markov processes with a nonlinear prior model and a Gauss-linear likelihood model in particular, since this case is frequently used in the literature (Evensen 2007). The model assumptions are:

$$\begin{aligned} \boldsymbol{x}_0 &\sim f(\boldsymbol{x}_0) \\ [\boldsymbol{x}_{t+1}|\boldsymbol{x}_t] &= \omega_t(\boldsymbol{x}_t, \varepsilon_t^x) \sim f(\boldsymbol{x}_{t+1}|\boldsymbol{x}_t) \\ [\mathbf{d}_t|\boldsymbol{x}_t] &= \mathbf{H}_t\boldsymbol{x}_t + \varepsilon_t^d \sim f(\mathbf{d}_t|\boldsymbol{x}_t) = N_{p_d}(\mathbf{H}_t\boldsymbol{x}_t, \Sigma_t^d), \end{aligned}$$

where the notation is identical to that in Expressions (11.1) and (11.3). The known forward model $\omega_t(\boldsymbol{x}_t, \varepsilon_t^x)$ may include a differential equation, and it may be extremely computationally demanding to evaluate. Under these model assumptions the EnKF algorithm requires the following ensembles:

$$\boldsymbol{e}_t : \{\boldsymbol{x}_t^{u(i)}; i = 1, \ldots, n_e\}; t = 0, \ldots, T+1,$$

where $\boldsymbol{x}_t^{u(i)} = [\boldsymbol{x}_t|\mathbf{d}_0,\ldots,\mathbf{d}_{t-1}]^{(i)}$ are approximately independent realizations from the conditional distribution $f(\boldsymbol{x}_t|\mathbf{d}_0,\ldots,\mathbf{d}_{t-1})$. At each step the associated expectation vector and covariance matrix are $\boldsymbol{\mu}_x$ and Σ_x with time reference t omitted. Note that the ensemble need not contain the realizations of observations since the associated $\Gamma_{x,d} = \Sigma_x\mathbf{H}'$ can be assessed from estimates of Σ_x. The actual EnKF algorithm is as follows:

Algorithm 11.4 *Ensemble Kalman Filter (Gauss-linear Likelihood)*

- *Initiate*

$$n_e = \textit{no. of ensemble members}$$
$$\boldsymbol{x}_0^{u(i)}\,;\, i = 1,\ldots,n_e \ \textit{iid}\ f(\boldsymbol{x}_0)$$
$$\mathbf{e}_0 \,:\, \{\boldsymbol{x}_0^{u(i)}\,;\, i = 1,\ldots,n_e\}$$

- *Iterate* $t = 0,\ldots,T$

Conditioning:

Estimate Σ_x *from* $\mathbf{e}_t \rightarrow \widehat{\Sigma}_x$

$$\mathbf{d}_t^{(i)} \sim N_{p_d}(\mathbf{H}_t\boldsymbol{x}_t^{u(i)}, \Sigma_t^d)\,;\, i = 1,\ldots,n_e$$
$$\boldsymbol{x}_t^{c(i)} = \boldsymbol{x}_t^{u(i)} + \widehat{\Sigma}_x\mathbf{H}_t'[\mathbf{H}_t\widehat{\Sigma}_x\mathbf{H}_t' + \Sigma_t^d]^{-1}(\mathbf{d}_t - \mathbf{d}_t^{(i)})\,;\, i = 1,\ldots,n_e$$

Forwarding:

$$\boldsymbol{\varepsilon}_t^{x(i)} \sim N_{p_x}(\mathbf{0}, \mathbf{I}_{p_x})\,;\, i = 1,\ldots,n_e$$
$$\boldsymbol{x}_{t+1}^{u(i)} = \omega_t(\boldsymbol{x}_t^{c(i)}, \boldsymbol{\varepsilon}_t^{x(i)})\,;\, i = 1,\ldots,n_e$$
$$\mathbf{e}_{t+1} \,:\, \{\boldsymbol{x}_{t+1}^{u(i)}\,;\, i = 1,\ldots,n_e\}$$

- *end iterate*
- *Assess*

$$f(\boldsymbol{x}_{T+1}|\mathbf{d}_0,\ldots,\mathbf{d}_T) \ \textit{from}\ \mathbf{e}_{T+1}$$

The resulting ensemble $\mathbf{e}_{T+1}$ contains approximately independent realizations of

$$\boldsymbol{x}_{T+1}^u = [\boldsymbol{x}_{T+1}|\mathbf{d}_0,\ldots,\mathbf{d}_T]$$

from $f(\boldsymbol{x}_{T+1}|\mathbf{d}_0,\ldots,\mathbf{d}_T)$; estimates of the forecast expectation vector and covariance matrix can be obtained as in Expression (11.4). Note that for linear conditioning to be exact, $f(\boldsymbol{x}_t^u)$ needs to be Gaussian and $f(\mathbf{d}_t|\boldsymbol{x}_t)$ Gauss-linear. The former need not be the case under the current model assumptions, hence only approximations are obtained.

Performance of the Test Examples

The EnKF-Special case requires the likelihood model to be Gauss-linear and known, hence it can only be applied to the linear case. The solutions are presented in Figure 11.7 together with the $\boldsymbol{x}_{11}$ reference realization. The EnKF-Special case is based on the ensemble size $n_e = 100$. The layout of the figure is similar to that in Figure 11.5. Figure 11.7 should be compared to Figure 11.3, which displays the correct KF-solution, and Figure 11.5, which displays the solution for the general EnKF algorithm. The latter does estimate the likelihood model. The EnKF-special case with the likelihood model given and $n_e = 100$ provides results that are very similar to the correct KF solution. For the linear case, the EnKF-Special case solution appears more reliable than that of the general algorithm. The former is based on the correct likelihood model and an estimate of Σ_x only. Hence, the estimation uncertainty in the weights in the conditioning is reduced and the EnKF results are improved. Note, however, that in the nonlinear case only the general EnKF can be used.

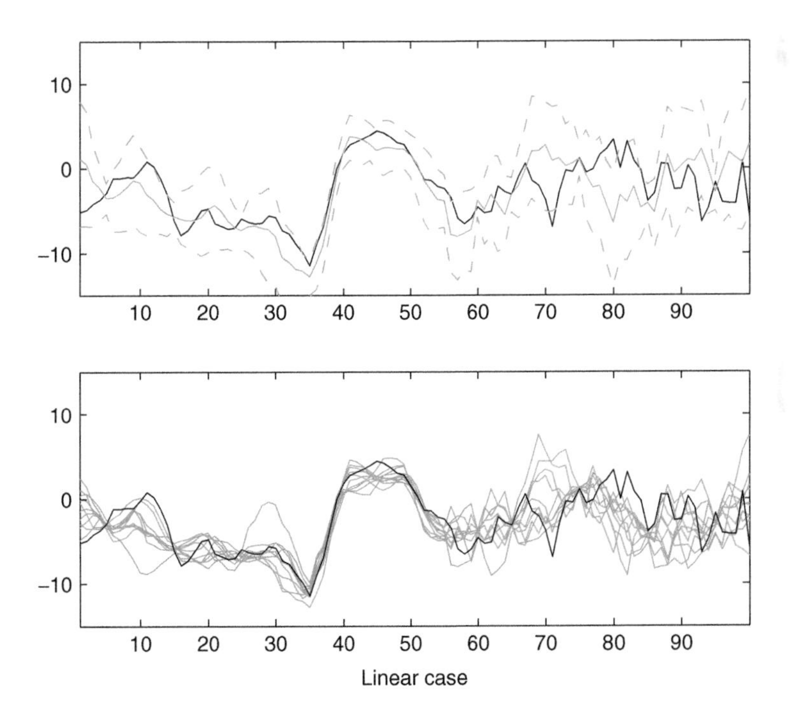

Figure 11.7 Results of the EnKF-special case algorithm applied to the linear case: reference realization $\boldsymbol{x}_{11}$ (black), prediction (blue) and empirical .95-prediction intervals (hatched blue) (top row); reference realization $\boldsymbol{x}_{11}$ (black) and predictions (blue) for 10 different runs of the algorithm (bottom row).

11.5 The Randomized Maximum Likelihood Filter (RMLF)

The basic idea of the RMLF is very similar to that of the EnKF, except for a different conditioning procedure for the observations. Realizations from the initial model, so called ensemble, are iteratively adjusted for observations and forwarded in time. At time $t = T + 1$, a set of approximately independent realizations is available for assessment of $f(\boldsymbol{x}_{T+1}|\mathbf{d}_0, \ldots, \mathbf{d}_T)$. Basic references for RMLF, also termed iterative EnKF, are Kitanidis (1986); Oliver (1996); Reynolds *et al.* (2006). The RMLF is well defined on a model for the hidden Markov process with a slightly constrained likelihood model:

$$\begin{aligned} \boldsymbol{x}_0 &\sim f(\boldsymbol{x}_0) \\ [\boldsymbol{x}_{t+1}|\boldsymbol{x}_t] &= \omega_t(\boldsymbol{x}_t, \boldsymbol{\varepsilon}_t^x) \sim f(\boldsymbol{x}_{t+1}|\boldsymbol{x}_t) \\ [\mathbf{d}_t|\boldsymbol{x}_t] &= \nu_t(\boldsymbol{x}_t) + \boldsymbol{\varepsilon}_t^d \sim f(\mathbf{d}_t|\boldsymbol{x}_t) = N_{p_d}(\nu_t(\boldsymbol{x}_t), \Sigma_t^d), \end{aligned}$$

where the notation is identical to that in Expressions (11.1) and (11.2). The likelihood model is still nonlinear in $\boldsymbol{x}_t$, but it is constrained to have an additive Gaussian error term independent of $\boldsymbol{x}_t$ with known covariance matrix.

0 A time series of ensembles is defined:

$$\mathbf{e}_t : \{\boldsymbol{x}_t^{u(i)} ; i = 1, \ldots, n_e\} ; t = 0, \ldots, T + 1,$$

where

$$\boldsymbol{x}_t^{u(i)} = [\boldsymbol{x}_t|\mathbf{d}_0, \ldots, \mathbf{d}_{t-1}]^{(i)}$$

are approximately independent realizations from $f(\boldsymbol{x}_t|\mathbf{d}_0, \ldots, \mathbf{d}_{t-1})$. At any time step the associated expectation vector and covariance matrix are $\boldsymbol{\mu}_x$ and Σ_x with time reference t omitted. The RMLF algorithm in its simplest form is included here to illustrate the basic idea of the approach:

Algorithm 11.5 *Randomized Maximum Likelihood Filter*

- *Initiate*

$$\begin{aligned} n_e &= \textit{no. of ensemble members} \\ &\boldsymbol{x}_0^{u(i)} ; i = 1, \ldots, n_e \ \textit{iid} \ f(\boldsymbol{x}_0) \\ &\mathbf{e}_0 : \{\boldsymbol{x}_0^{u(i)} ; i = 1, \ldots, n_e\} \end{aligned}$$

- *Iterate* $t = 0, \ldots, T$

Conditioning:

Estimate Σ_x *from* $\mathbf{e}_t \rightarrow \widehat{\Sigma}_x$

$$\mathbf{o}_t^{(i)} \sim N_{p_d}(\mathbf{d}_t, \Sigma_t^d) \,;\, i = 1, \ldots, n_e$$
$$\boldsymbol{x}_t^{c(i)} = \text{argmin}_{\boldsymbol{x}}\{(\mathbf{o}_t^{(i)} - \nu_t(\boldsymbol{x}))'\Sigma_t^{d^{-1}}(\mathbf{o}_t^{(i)} - \nu_t(\boldsymbol{x})) + (\boldsymbol{x} - \boldsymbol{x}_t^{u(i)})'\widehat{\Sigma}_x^{-1}(\boldsymbol{x} - \boldsymbol{x}_t^{u(i)})\} \,;\, i = 1, \ldots, n_e$$

Forwarding:

$$\boldsymbol{\varepsilon}_t^{x(i)} \sim N_{p_x}(\mathbf{0}, \mathbf{I}_{p_x}) \,;\, i = 1, \ldots, n_e$$
$$\boldsymbol{x}_{t+1}^{u(i)} = \omega_t(\boldsymbol{x}_t^{c(i)}, \boldsymbol{\varepsilon}_t^{x(i)}) \,;\, i = 1, \ldots, n_e$$
$$\boldsymbol{e}_{t+1} \,:\, \{\boldsymbol{x}_{t+1}^{u(i)} \,;\, i = 1, \ldots, n_e\}$$

- *end iterate*
- *Assess*

$f(\boldsymbol{x}_{T+1}|\mathbf{d}_0, \ldots, \mathbf{d}_T)$ *from* $\boldsymbol{e}_{T+1}$

The resulting ensemble $\boldsymbol{e}_{T+1}$ contains approximately independent realizations of

$$\boldsymbol{x}_{T+1}^u = [\boldsymbol{x}_{T+1}|\mathbf{d}_0, \ldots, \mathbf{d}_T]$$

from $f(\boldsymbol{x}_{T+1}|\mathbf{d}_0, \ldots, \mathbf{d}_T)$. Hence estimates of the forecast expectation vector and covariance matrix can be obtained with the usual estimators as in Expression (11.4). The RMLF algorithm, Algorithm 11.5, is also recursive with a conditioning operation and a forwarding operation in each step. Note that $\mathbf{o}_t^{(i)}$ is not identical to $\mathbf{d}_t^{(i)}$ used in EnKF, since the former is centered around the actual observation $\mathbf{d}_t$. There are two implicit approximations in the RMLF:

1. Discretization of the sample space of $\boldsymbol{x}_t$. The initial ensemble is assumed to represent $f(\boldsymbol{x}_0)$. This approximation is similar for all ensemble based approaches and it may require a large number of ensemble members in high-dimensional problems.

2. Conditioning on observations $\mathbf{d}_t$ is done by optimization with a criterion that relies on Gaussianity. This optimization approach is more flexible than the linearization used in the EnKF. Note that each ensemble member is conditioned individually, hence the approximation only applies to the conditioning, not to the forward model. For highly non-Gaussian prior models this approximation may provide unreliable solutions.

Under these approximations, however, all hidden Markov processes with nonlinear likelihood models and additive Gaussian error terms can be evaluated. The RMLF coincides with the EnKF whenever the likelihood model is Gauss-linear (see 11.9). This entails that it is a consistent forecast procedure under a full Gauss-linear model in the sense that the exact solution is obtained as $n_e \to \infty$ (see 11.8).

Two major problems that may arise with the RMLF are:

1. Complications caused by the use of an estimate of Σ_x based on the ensemble $\boldsymbol{e}_t$ instead of the true covariance matrix. These complications include rank deficiency and estimation uncertainty as in EnKF.

2. Computational cost of performing the optimization in each step; specially because the objective function may have several local optima. If the nonlinear likelihood function $\nu_t(.)$ is available in analytic form so that the derivatives are computable, an efficient optimization may be performed. For a general 'black box' $\nu_t(.)$ used in a computer code, the optimization may be prohibitively expensive to perform.

The RMLF is not widely used but several applications to reservoir evaluation problems demonstrate its characteristics (Gao *et al.* 2006; Gu and Oliver 2007). It is shown, however, that the conditioning procedure in the RMLF captures multimodality in the prior model and likelihood function better than the EnKF.

Performance of the Test Examples

The RMLF approach can be applied to both the linear and nonlinear cases, but due to the high computational cost of the optimization, only a subset of the $\boldsymbol{x}$-nodes, [40, 60], is evaluated, see Figure 11.8. For the linear case the RMLF and the EnKF-Special case solution in Figure 11.7 coincide since the likelihood model is known to be linear. All solutions are trustworthy with

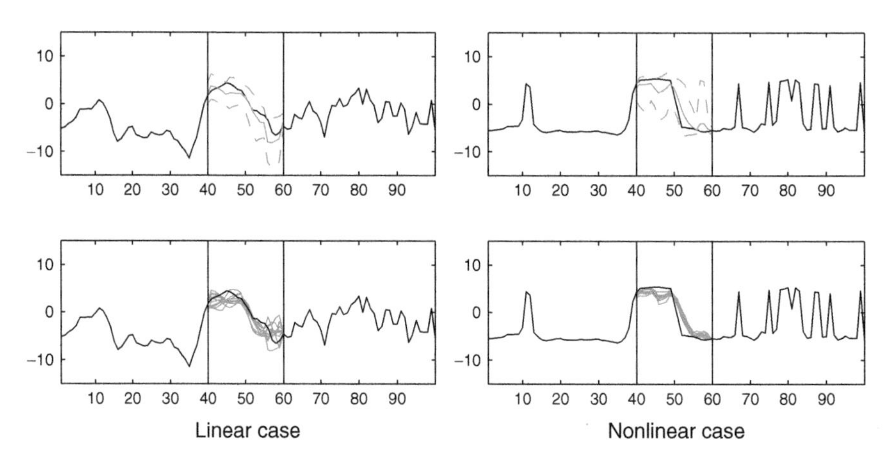

Figure 11.8 Results of the RMLF algorithm on the linear and nonlinear test cases: reference realization $\boldsymbol{x}_{11}$ (black), prediction (blue) and empirical .95-prediction intervals (hatched blue) (top row); reference realization $\boldsymbol{x}_{11}$ (black) and predictions (blue) for 10 different runs of the algorithm (bottom row).

the ensemble size $n_e = 100$. In the nonlinear case the RMLF solution can be compared to the EnKF solution in Figure 11.5, and the former appears more reliable since the prediction intervals look more trustworthy and there is less variability between repeated runs. This is caused by RMLF capturing some of the nonlinear effects in the likelihood, while EnKF is based on an empirical linearized approach. Computational demands may limit the use of the RMLF, however.

11.6 The Particle Filter (PF)

The basic idea of the particle filter is to create a set of iid realizations from the initial model, so called particles, and to take these realizations through the prior model. Each realization is assigned weights according to the likelihood model as observations occur through time. At time $t = T + 1$, a set of realizations with associated weights are available for assessment of $f(\boldsymbol{x}_{T+1}|\mathbf{d}_0, \ldots, \mathbf{d}_T)$. Note that the actual ensemble members of $\boldsymbol{e}_{T+1}$, without weighting, carry no information about the observations, hence visual inspection is of no use. Basic references for particle filters are Gordon *et al.* (1993) and Doucet *et al.* (2001).

A time series of ensembles is defined:

$$\boldsymbol{e}_t : \{(\boldsymbol{x}_t, w_t)^{(i)}; i = 1, \ldots, n_e\}; t = 0, \ldots, T + 1$$

where $\boldsymbol{x}_t^{(i)}$ are samples from the prior model of $\boldsymbol{x}_t$, and $w_t^{(i)}$ are associated weights that are updated as observations arise. The particle filter algorithm in its simplest form is as follows:

Algorithm 11.6 *Particle Filter*

- *Initiate:*

$$n_e = \textit{no. of ensemble members}$$
$$\boldsymbol{x}_0^{(i)} ; i = 1, \ldots, n_e \textit{ iid } f(\boldsymbol{x}_0)$$
$$w_0^{(i)} = 1 ; i = 1, \ldots, n_e$$
$$\boldsymbol{e}_0 : \{(\boldsymbol{x}_0, w_0)^{(i)} ; i = 1, \ldots, n_e\}$$

- *Iterate* $t = 0, \ldots, T$

Weight updating:

$$v^{(i)} = f(\mathbf{d}_t|\boldsymbol{x}_t^{(i)}) \times w_t^{(i)} ; i = 1, \ldots, n_e$$
$$w_{t+1}^{(i)} = v^{(i)} \times [\textstyle\sum_{i=1}^{n_e} v^{(i)}]^{-1} ; i = 1, \ldots, n_e$$

Forwarding:

$$\begin{aligned}
&\boldsymbol{\varepsilon}_t^{x(i)} \sim N_{p_x}(\mathbf{0}, \mathbf{I}_{p_x})\,;\; i = 1, \ldots, n_e \\
&\boldsymbol{x}_{t+1}^{u(i)} = \omega_t(\boldsymbol{x}_t^{c(i)}, \boldsymbol{\varepsilon}_t^{x(i)})\,;\; i = 1, \ldots, n_e \\
&\mathbf{e}_{t+1} \,:\, \{(\boldsymbol{x}_{t+1}, w_{t+1})^{(i)}\,;\; i = 1, \ldots, n_e\}
\end{aligned}$$

- *end iterate*
- *Assess*

$$f(\boldsymbol{x}_{T+1}|\mathbf{d}_0, \ldots, \mathbf{d}_T) \text{ from } \mathbf{e}_{T+1}$$

The resulting ensemble $\mathbf{e}_{T+1}$ contains samples from the prior model $f(\boldsymbol{x}_0)$ and associated normalized weights resulting from conditioning on $[\mathbf{d}_0, \ldots, \mathbf{d}_T]$ which can be used in assessing the forecast pdf $f(\boldsymbol{x}_{T+1}|\mathbf{d}_0, \ldots, \mathbf{d}_T)$. Estimates of the forecasting expectation vector and covariance matrix will be:

$$\begin{aligned}
\widehat{\boldsymbol{\mu}}_{T+1} =& \widehat{\mathrm{E}}\{\boldsymbol{x}_{T+1}|\mathbf{d}_0, \ldots, \mathbf{d}_T\} = \sum_{i=1}^{n_e} w_{T+1}^{(i)} \boldsymbol{x}_{T+1}^{(i)} \\
\widehat{\Sigma}_{T+1} =& \widehat{\mathrm{Var}}\{\boldsymbol{x}_{T+1}|\mathbf{d}_0, \ldots, \mathbf{d}_T\} = \sum_{i=1}^{n_e} w_{T+1}^{(i)} (\boldsymbol{x}_{T+1}^{(i)} - \widehat{\boldsymbol{\mu}}_{T+1})(\boldsymbol{x}_{T+1}^{(i)} - \widehat{\boldsymbol{\mu}}_{T+1})'.
\end{aligned}$$

The PF algorithm, Algorithm 11.6, is recursive with a weight updating operation and a forwarding operation in each step. The implicit approximation in the particle filter lies in the discretization of the sample space of $\boldsymbol{x}_t$. The initial ensemble is assumed to represent $f(\boldsymbol{x}_0)$. For high-dimensional problems a large number of ensemble members may be required to justify these assumptions. Under this approximation, however, all types of models for the hidden Markov process can be evaluated. The particle filter is a consistent forecast procedure for all hidden Markov models in the sense that the exact solution is obtained as $n_e \to \infty$.

The particle filter can be run efficiently but one major problem is that all weights often are assigned to one or few samples. This happens when the prior model is vague and/or the observations are very informative. Several ways of correcting this problem have been proposed (Cappé *et al.* 2007; Doucet *et al.* 2001). Most approaches are based on some sort of resampling from the weighted ensemble. Since particle filters are not the focus of this presentation we leave the reader with these references and the following example.

Performance of the Test Examples

The particle filter can be applied to both the linear and nonlinear cases, see Figure 11.9. The filter with a simple forward function is computationally efficient and several hundred thousand ensemble members were run. Even with so many ensemble members only a few members were assigned weights

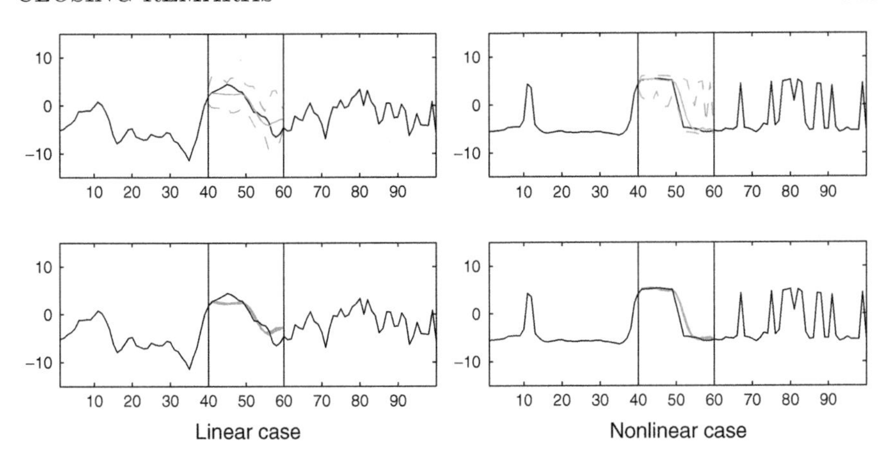

Figure 11.9 Results of the particle filter algorithm on the linear and nonlinear cases: reference realization $\boldsymbol{x}_{11}$ (black), prediction (blue) and empirical .95-prediction intervals (hatched blue) (top row); reference realization $\boldsymbol{x}_{11}$ (black) and predictions (blue) for 10 different runs of the algorithm (bottom row).

significantly different from zero for the full scale example, however. Hence we limited the study to a subset of the $\boldsymbol{x}$-nodes, $[40, 60]$, and used the ensemble size $n_e = 130\,000$, which corresponds to an effective number of ensemble members (Doucet *et al.* 2001) around 100. The particle filter results for the linear example should be compared to the exact solution obtained with the KF in Figure 11.3. It appears that the PF solution is close to the KF solution. For the nonlinear case the particle filter results should be compared to the EnKF and RMLF solutions in Figures 11.5 and 11.8, respectively. The PF solution appears reliable and the prediction interval seems to be close to the RMLF solution. The results indicate that the PF requires an extremely high number of ensemble members in order to provide reliable solutions. For computer demanding forward function such a high number of runs may be difficult to perform.

11.7 Closing Remarks

The forecasting of $[\boldsymbol{x}_{T+1}|\mathbf{d}_0, \ldots, \mathbf{d}_T]$ is based on a hidden Markov model. If this model is assumed to be Gauss-linear it is subject to analytical evaluation and traditional KF provides the exact solution. With deviations from Gauss-linearity approximate solutions must be sought, and a large variety of approaches are available. In the current chapter focus is on simulation based inference of the hidden Markov model, and the EnKF in particular. Also RMLF and PF are being introduced.

The content of this chapter can be summarized by:

- The EnKF and RMLF on one hand and PF on the other, are based on very different ideas, although both rely on simulation based inference. The EnKF and RMLF are based on an ensemble where each ensemble member is adjusted as observations appear. Hence the ensemble members are approximately independent realizations of $[\boldsymbol{x}_t|\mathbf{d}_0, \ldots, \mathbf{d}_{t-1}]$ at any time. The PF, on the other hand, is based on an ensemble where each ensemble member is fixed but weighted. The weights are sequentially adjusted as observations appear. Hence the weighted ensemble members represent $f(\boldsymbol{x}_t|\mathbf{d}_0, \ldots, \mathbf{d}_{t-1})$. In practice, one can visualize the ensemble members of the EnKF and the RMLF to illustrate the solution, but doing so for the PF ensemble has no meaning.

- The EnKF can be used for general hidden Markov models. It relies on a linearization of the conditioning expression where the weights are estimates from the ensemble. Hence EnKF will in the general case only provide an approximate solution even when the ensemble size tends to infinity. For high-dimensional problems the estimated weights may be unreliable if the ensemble size is too small. Further, this may cause unreliable predictions and underestimation of the prediction intervals. The EnKF is relatively computationally efficient.

- The RMLF can be used for general hidden Markov models, with minor constraints on the likelihood model. The conditioning expression is phrased as an optimization with an object function inferred from the ensemble. The RMLF will provide more reliable results than the EnKF whenever there are nonlinearities in the model. The solutions will only be approximate, however, even when the ensemble size tends to infinity. The RMLF may be very computationally demanding for high-dimensional problems with object functions with multiple optima.

- The PF can also be used for general hidden Markov models. The conditioning is made exactly, and the correct solution will be obtained whenever the ensemble size tends to infinity. Moreover, the PF is extremely computationally efficient as long as the forward function is simple. For high-dimensional problems, however, the convergence towards the correct solution is extremely slow. In practice, only very few ensemble members are assigned weight which are significantly different from zero. If the forward model is computer demanding, the PF may be prohibited to use in practice.

The authors, general recommendations are: For very low-dimensional problem the PF is to be recommended due to its favorable asymptotic properties even for nonlinear models. For intermediate- and high-dimensional problems the EnKF is expected to perform better than the PF given comparable computer resources. The adjustment of the individual ensemble members in the EnKF provides better coverage in high probability areas of the sample space. Attention must be given to the reliability of the EnKF solution, however, due to possible lack of precision in the predictions and too narrow prediction intervals. The RMLF is expected to find its use in special intermediate-dimensional problems with nonlinear relations.

References

Bertino, L., Evensen, G. and Wackernagel, H. 2002 Combining geostatistics and Kalman filtering for data assimilation in an estuarine system. *Inverse Problems* **18**(1), 1–23.

Bertino, L., Evensen, G. and Wackernagel, H. 2003 Sequential data assimilation techniques in oceanography. *International Statistical Review* **71**, 223–241.

Burgers, G., van Leeuwen, P.J. and Evensen, G. 1998 Analysis scheme in the ensemble Kalman filter. *Monthly Weather Review* **126**(6), 1719–1724.

Cappé, O., Godsill, S.J. and Moulines, E. 2007 An overview of existing methods and recent advances in sequential Monte Carlo. *Proceedings of the IEEE* **95**, 899–924.

Doucet, A., de Freitas, N. and Gordon, N. (eds) 2001 *Sequential Monte Carlo Methods in Practice*. Springer-Verlag.

Dovera, L. and Della Rossa, L. 2007 Ensemble Kalman filters for Gaussian mixture models. In *Extended Abstracts book of Petroleum Geostatistics 2007*, Cascais, Portugal.

Evensen, G. 1994 Sequential data assimilation with nonlinear quasi-geostrophic model using Monte Carlo methods to forecast error statistics. *Journal of Geophysical Research* **99**(C5), 10,143–10,162.

Evensen, G. 2004 Sampling strategies and square root analysis schemes for the EnKF. *Ocean Dynamics* **54**(6), 539–560.

Evensen, G. 2007 *Data Assimilation; The Ensemble Kalman Filter*. Springer.

Gao, G., Zafari, M. and Reynolds, A.C. 2006 Quantifying uncertainty for the PUNQ-S3 problem in a Bayesian setting with RML and EnKF. *SPE Journal* **11**(4), 506–515.

Gordon, N.J., Salmond, D.J. and Smith, A.F.M. 1993 Novel approach to nonlinear/non-Gaussian Bayesian state estimation. *In IEE Proceedings-F*, vol. 140, pp. 107–113.

Gu, Y. and Oliver, D.S. 2007 An iterative ensemble Kalman filter for multiphase fluid flow data assimilation. *SPE Journal* **12**(4), 438–446.

Hamill, T.M., Whitaker, J.S. and Snyder, C. 2001 Distance-dependent filtering of background error covariance estimates in an ensemble Kalman filter. *Monthly Weather Review* **129**(11), 2776–2790.

Houtekamer, P. and Mitchell, H.L. 1998 Data assimilation using an ensemble Kalman filter technique. *Monthly Weather Review* **126**(3), 796–811.

Houtekamer, P., Mitchell, H.L., Pellerin, G., Buehner, M., Charron, M., Spacek, L. and Hansen, B. 2005 Atmospheric data assimilation with an ensemble Kalman filter: Results with real observations. *Monthly Weather Review* **133**(3), 604–620.

Huber, P.J. 1981 *Robust Statistics*. John Wiley & Sons, Ltd.

Jafarpour, B. and McLaughlin 2007 History matching with an ensemble Kalman filter and discrete cosine parameterization. In *SPE Annual Technical Conference and Exhibition, 11-14 November 2007*, Anaheim, California, USA.

Jazwinski, A.H. 1970 *Stochastic Processes for Filtering Theory*. Academic Press.

Julier, S.J. and Uhlmann, J.K. 1997 A new extension of the Kalman filter to nonlinear systems In *Proceedings of the 12th International Symposium on Aerospace/Defense Sensing, Simulation and Controls*, pp. 182–193.

Kalman, R.E. 1960 A new approach to linear filtering and prediction problems. *Journal of Basic Engineering* **82**(1), 35–45.

Kelker, D. 1970 Distribution theory of spherical distributions and a location-scale parameter generalization. *Sankhya: The Indian Journal of Statistics* **32**, 419–430.

Kitanidis, P.K. 1986 Parameter uncertainty in estimation of spatial functions: Bayesian estimation. *Water resources research* **22**(4), 499–507.

Liu, N. and Oliver, D.S. 2005 Critical evaluation of the ensemble Kalman filter on history matching of geologic facies. *SPE Reservoir Evaluation & Engineering* **8**, 470–477.

Myrseth, I. and Omre, H. 2010 Hierarchical ensmble Kalman filter. *SPE Journal* **15**(2), 569–580.

Nævdal, G., Johnsen, L.M., Aanonsen, S.I. and Vefring, E.H. 2005 Reservoir monitoring and continuous model updating using ensemble Kalman filter. *SPE Journal* **10**(1), 66–74.

Natvik, L.J. and Evensen, G. 2003a Assimilation of ocean colour data into a biochemical model of the North Atlantic. Part 1. Data assimilation experiments. *Journal of Marine Systems* **40–41**, 127–153.

Natvik, L.J. and Evensen, G. 2003b Assimilation of ocean colour data into a biochemical model of the North Atlantic. Part 2. Statistical analysis. *Journal of Marine Systems* **40–41**, 155–159.

Oliver, D.S. 1996 On conditional simulation to inaccurate data. *Journal of Mathematical Geology* **28**(6), 811–817.

Reynolds, A.C., Zafari, M. and Li, G. 2006 Iterative forms of the ensemble Kalman filter. In *Proceedings of the 10th European conference on the Mathematics of Oil Recovery*, pp. 182–193.

Skjervheim, J., Evensen, G., Aanonsen, S., Ruud, B. and Johansen, T. 2007 Incorporating 4D seismic data in reservoir simulation models using ensemble Kalman filter. *SPE Journal* **12**(3), 282–292.

Appendix A: Properties of the EnKF Algorithm

We show that the EnKF algorithm is exact in the limit $n_e \to \infty$ under a Gauss-linear model.

The Gauss-linear model is

$$\begin{aligned} f(\boldsymbol{x}_0) &\sim N_{p_x}(\boldsymbol{\mu}_0^u, \Sigma_0^x) \\ f(\boldsymbol{x}_{t+1}|\boldsymbol{x}_t) &\sim N_{p_x}(\mathbf{A}_t\boldsymbol{x}_t, \Sigma_t^x) \\ f(\mathbf{d}_t|\boldsymbol{x}_t) &\sim N_{p_d}(\mathbf{H}_t\boldsymbol{x}_t, \Sigma_t^d). \end{aligned}$$

Note that for consistent estimators and $n_e \to \infty$:

$$\widehat{\Sigma}_{xd} \to \Sigma_{xd} = \begin{bmatrix} \Sigma_x & \boldsymbol{\Gamma}_{x,d} \\ \Gamma_{d,x} & \Sigma_d \end{bmatrix} = \begin{bmatrix} \boldsymbol{\Sigma}_t^u & \boldsymbol{\Sigma}_t^u\mathbf{H}_t' \\ \mathbf{H}_t\boldsymbol{\Sigma}_t^u & \mathbf{H}_t\boldsymbol{\Sigma}_t^u\mathbf{H}_t' + \Sigma_t^d \end{bmatrix}.$$

The proof is done by induction: Assume that

$$\begin{aligned} \boldsymbol{x}_t^{u(i)} &\sim N_{p_x}(\boldsymbol{\mu}_t^u, \boldsymbol{\Sigma}_t^u) \\ [\mathbf{d}_t^{(i)}|\boldsymbol{x}_t^{u(i)}] &\sim N_{p_d}(\mathbf{H}_t\boldsymbol{x}_t^{u(i)}, \Sigma_t^d). \end{aligned}$$

The conditioning rule is:

$$[\boldsymbol{x}_t^{c(i)}|\mathbf{d}_t^{(i)}, \boldsymbol{x}_t^{u(i)}] = \boldsymbol{x}_t^{u(i)} + \boldsymbol{\Gamma}_{x,d}\Sigma_d^{-1}(\mathbf{d}_t - \mathbf{d}_t^{(i)}).$$

Due to linearity and Gaussianity in $[\mathbf{d}_t^{(i)}, \boldsymbol{x}_t^{u(i)}]$ one has:

$$\boldsymbol{x}_t^{c(i)} \sim N_{p_x}(\boldsymbol{\mu}_t^c, \Sigma_t^c),$$

with

$$\begin{aligned} \boldsymbol{\mu}_t^c = \mathrm{E}\{\boldsymbol{x}_t^{c(i)}\} &= \boldsymbol{\mu}_t^u + \boldsymbol{\Gamma}_{x,d}\Sigma_d^{-1}(\mathbf{d}_t - \mathbf{H}_t\boldsymbol{\mu}_t^u) \\ &= \boldsymbol{\mu}_t^u + \boldsymbol{\Sigma}_t^u\mathbf{H}_t'[\mathbf{H}_t\boldsymbol{\Sigma}_t^u\mathbf{H}_t' + \Sigma_t^d]^{-1}(\mathbf{d}_t - \mathbf{H}_t\boldsymbol{\mu}_t^u) \\ \Sigma_t^c = \mathrm{Var}\{\boldsymbol{x}_t^{c(i)}\} &= \boldsymbol{\Sigma}_t^u + \boldsymbol{\Gamma}_{x,d}\Sigma_d^{-1}[\mathbf{H}_t\boldsymbol{\Sigma}_t^u\mathbf{H}_t' + \Sigma_t^d]\Sigma_d^{-1}\Gamma_{d,x} \\ &\quad - 2\boldsymbol{\Gamma}_{x,d}\Sigma_d^{-1}\mathbf{H}_t'\boldsymbol{\Sigma}_t^u \\ &= \boldsymbol{\Sigma}_t^u + \boldsymbol{\Sigma}_t^u\mathbf{H}_t'[\mathbf{H}_t\boldsymbol{\Sigma}_t^u\mathbf{H}_t' + \Sigma_t^d]^{-1}\mathbf{H}_t\boldsymbol{\Sigma}_t^u \\ &\quad - 2\boldsymbol{\Sigma}_t^u\mathbf{H}_t'[\mathbf{H}_t\boldsymbol{\Sigma}_t^u\mathbf{H}_t' + \Sigma_t^d]^{-1}\mathbf{H}_t\boldsymbol{\Sigma}_t^u \\ &= \boldsymbol{\Sigma}_t^u - \boldsymbol{\Sigma}_t^u\mathbf{H}_t'[\mathbf{H}_t\boldsymbol{\Sigma}_t^u\mathbf{H}_t' + \Sigma_t^d]^{-1}\mathbf{H}_t\boldsymbol{\Sigma}_t^u, \end{aligned}$$

which corresponds to the KF solution in Algorithm 11.2.

Moreover:

$$\begin{aligned} [\boldsymbol{x}_{t+1}^{u(i)}|\boldsymbol{x}_t^{c(i)}] &\sim N_{p_x}(\mathbf{A}_t\boldsymbol{x}_t^{c(i)}, \Sigma_t^x) \\ \boldsymbol{x}_{t+1}^{u(i)} &\sim N_{p_x}(\mathbf{A}_t\boldsymbol{\mu}_t^c, \mathbf{A}_t\Sigma_t^c\mathbf{A}_t' + \Sigma_t^x), \end{aligned}$$

which corresponds to the KF solution.

Since Gauss-linearity entails that for $t = 0$:

$$\boldsymbol{x}_0^{u(i)} \sim N_{p_x}(\boldsymbol{\mu}_0^x, \Sigma_0^x)$$
$$[\mathbf{d}_0^{(i)} | \boldsymbol{x}_0^{u(i)}] \sim N_{p_d}(\mathbf{H}_0 \boldsymbol{x}_0^{u(i)}, \Sigma_0^d)$$

it is concluded by induction that all ensemble members in EnKF under a Gauss-linear model are sampled from the correct pdf. QED.

Appendix B: Properties of the RMLF Algorithm

We show that the RMLF algorithm coincides with the EnKF algorithm whenever the likelihood model is Gauss-linear.

The Gauss-linear likelihood model:

$$\mathbf{o}_t^{(i)} \sim N_{p_d}(\mathbf{d}_t, \Sigma_t^d)$$
$$f_r^{(i)}(\boldsymbol{x}) = N_{p_x}(\boldsymbol{x}_t^{u(i)}, \widehat{\Sigma}_x)$$
$$f_r^{(i)}(\mathbf{o}_t^{(i)} | \boldsymbol{x}) = N_{p_d}(\mathbf{H}_t \boldsymbol{x}, \Sigma_t^d).$$

The conditioning rule with $f^{(i)}(\boldsymbol{x}|\mathbf{o}_t^{(i)})$ being Gaussian is:

$$\begin{aligned}
\boldsymbol{x}_t^{c(i)} &= \text{argmin}_{\boldsymbol{x}}\{(\mathbf{o}_t^{(i)} - \mathbf{H}_t\boldsymbol{x})\Sigma_t^{d^{-1}}(\mathbf{o}_t^{(i)} - \mathbf{H}_t\boldsymbol{x})' + (\boldsymbol{x} - \boldsymbol{x}_t^{u(i)})\widehat{\Sigma}_x^{-1}(\boldsymbol{x} - \boldsymbol{x}_t^{u(i)})'\} \\
&= \text{argmax}_{\boldsymbol{x}}\{f_r^{(i)}(\mathbf{o}_t^{(i)}|\boldsymbol{x}) f_r^{(i)}(\boldsymbol{x})\} \\
&= \text{argmax}_{\boldsymbol{x}}\{f^{(i)}(\boldsymbol{x}|\mathbf{o}_t^{(i)})\} \\
&= \text{MAP}\{\boldsymbol{x}|\mathbf{o}_t^{(i)}\} \\
&= \text{E}\{\boldsymbol{x}|\mathbf{o}_t^{(i)}\}.
\end{aligned}$$

Hence a better name would have been randomized maximum aposteriori filter (RMAPF).

Hereby

$$\begin{aligned}
\boldsymbol{x}_t^{c(i)} &= \boldsymbol{x}_t^{u(i)} + \widehat{\Sigma}_x \mathbf{H}_t' \Sigma_o^{-1}[\mathbf{o}_t^{(i)} - \mathbf{H}_t \boldsymbol{x}_t^{u(i)}] \\
&= \boldsymbol{x}_t^{u(i)} + \widehat{\Sigma}_x \mathbf{H}_t' [\mathbf{H}_t \widehat{\Sigma}_x \mathbf{H}_t' + \Sigma_t^d]^{-1}(\mathbf{d}_t - \mathbf{d}_t^{(i)}),
\end{aligned}$$

with

$$\begin{aligned}
\mathbf{d}_t &= \mathbf{o}_t^{(i)} - \boldsymbol{\varepsilon}_t^d \\
\mathbf{d}_t^{(i)} &= \mathbf{H}_t \boldsymbol{x}_t^{u(i)} + \boldsymbol{\varepsilon}_t^d,
\end{aligned}$$

which corresponds to the EnKF conditioning rule in Algorithm 11.4. QED.

Chapter 12

Using the Ensemble Kalman Filter for History Matching and Uncertainty Quantification of Complex Reservoir Models

A. Seiler,[1,2] G. Evensen,[1,2] J.-A. Skjervheim,[1]
J. Hove[1] and J.G. Vabø[1]

[1]*Statoil Research Centre, Bergen, Norway*

[2]*Mohn Sverdrup Center at Nansen Environmental and Remote Sensing Center, Bergen, Norway*

12.1 Introduction

The traditional inverse problem of estimating poorly known parameters in a reservoir simulation model, named *history matching* by reservoir engineers, can be formulated as a combined parameter and state estimation problem using a Bayesian framework. The general Bayesian formulation leads to the standard minimization problem where Gaussian priors allow the problem to be solved by minimizing a cost function. On the other hand, the general

Large-Scale Inverse Problems and Quantification of Uncertainty Edited by L. Biegler, G. Biros, O. Ghattas, M. Heinkenschloss, D. Keyes, B. Mallick, Y. Marzouk, L. Tenorio, B. van Bloemen Waanders and K. Willcox

Bayesian problem can also be written as a recursion where measured data are processed sequentially in time, and this recursion is well suited to be solved using ensemble methods.

The solution workflow of the combined parameter and state estimation problem using the ensemble Kalman filter (EnKF) is described in some detail. The EnKF uses a Monte Carlo approach for representing and evolving the joint probability density function (pdf) for the model state and parameters, and it computes the recursive update steps by introducing an approximation where only the first and second order moments of the predicted pdf are used to compute the update increments. Each model realization is updated individually by adding the update increments as in the traditional Kalman filter.

The recursive Bayesian formulation can in principle be solved using the particle filter (Doucet *et al.* 2001). However, for high dimensional systems the computational cost becomes overwhelming and approximate methods must be used. The recursive formulation can be solved using the EnKF under the assumption that predicted error statistics is nearly Gaussian. This assumption is not true in general for nonlinear dynamical systems, but it is claimed that the recursive formulation leads to a better posed problem when approximate methods such as the EnKF are used. In particular, the recursive updating with observations may keep the model realizations on track and consistent with the measurements, and nonlinear contributions are not freely allowed to develop (Evensen and van Leeuwen 2000) as in a pure unconstrained ensemble prediction.

The EnKF was introduced by Evensen (1994) for updating non-linear ocean models. The method is now used operationally in both oceanography and meteorology and it is being used in a large number of applications where dynamical models are conditioned to additional measured data. There exist an extensive literature on the EnKF and we refer to Evensen (2009), which explains the EnKF and its implementation in detail, and references therein.

The proposed method and workflow turn out to work well when applied with a reservoir simulation model. Contrary to traditional history matching methods, which minimize a cost function, the EnKF can handle huge parameter spaces, it avoids getting trapped in local minima, and the recursive formulation is ideal for use in an advanced reservoir management workflow.

Several publications have discussed the use of the EnKF for parameter estimation in oil reservoirs, and have shown promising results. Nevertheless, most published papers present synthetic cases (see e.g. Gu and Oliver 2005; Nævdal *et al.* 2002), while real field applications have only recently been considered. Previous works that demonstrate the capability to use the EnKF for history matching real reservoir models are Skjervheim *et al.* (2007), Haugen *et al.* (2006), Bianco *et al.* (2007), and Evensen *et al.* (2007). All these studies conclude that the EnKF is able to significantly improve the history match of a reservoir simulation model. Previously, the focus has mainly been on the estimation of porosity and permeability fields in the simulation models.

In Evensen *et al.* (2007) parameters such as initial fluid contacts, and fault and vertical transmissibility multipliers, are included as additional uncertain parameters to be estimated.

In the following section the EnKF is presented as a method for history matching reservoir simulation models and discussed in relation to traditional methods where a cost function is minimized. The properties of the EnKF are demonstrated in a real field application where it is illustrated how a large number of poorly known parameters can be updated and where the uncertainty is reduced and quantified throughout the assimilation procedure. It is shown that the introduction of additional model parameters such as the relative permeability leads to a significant improvement of the results when compared to previous studies.

12.2 Formulation and Solution of the Inverse Problem

Methods for history-matching that are currently used in the oil industry, are normally based on the minimization of a cost function. Conditioning reservoir stochastic realizations to production data is generally described as finding the optimal set of model parameters that minimizes the misfit between a set of measurements and the corresponding responses calculated on the realization of the stochastic model.

However, methods for assisted history matching can also be formulated and derived from Bayes' theorem, which states that the *posterior pdf* of a random variable, conditional on some data, is proportional to the *prior pdf* for the random variable times the *likelihood function* describing the uncertainty distribution of the data. The use of Gaussian priors allows for Bayes theorem to be expressed as a quadratic cost function, whose solution is a fixed sample defined by the minimum of the cost function that corresponds to the Maximum a posteriori (MAP) solution. The minimization of the cost function solves the inverse problem but does not provide an uncertainty estimate.

This Bayesian formulation is the common starting point for traditional minimization methods and sequential data assimilation methods. Figure 12.1 illustrates how different fundamental assumptions lead to the two routes for solving the inverse problem.

12.2.1 Traditional Minimization Methods

Starting from Bayes' theorem a general cost function can be defined by adopting Gaussian distributions for the prior random variables and the measurements (Figure 12.1 A.1). One then obtains a so-called weak constraint variational cost function. The notion 'weak constraint' was introduced by Sasaki (1970), and denotes that the cost function allows for the dynamical

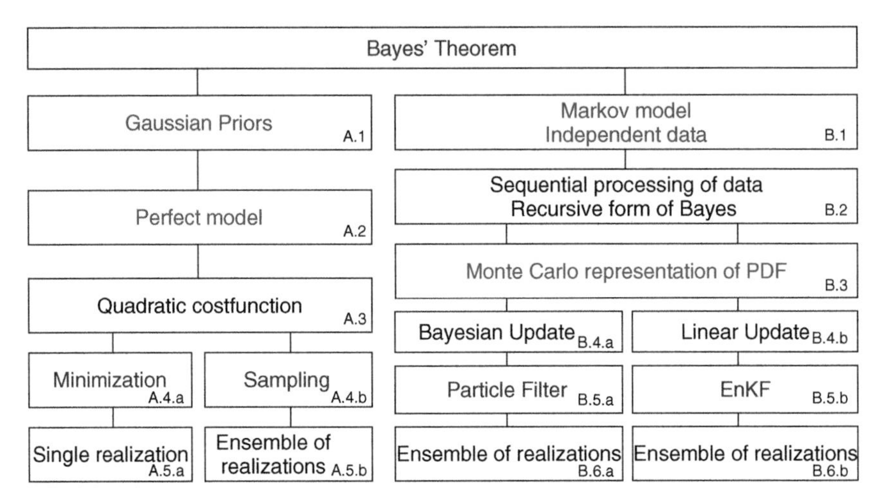

Figure 12.1 Solution of the inverse problem starting from Bayes' rule. The left route illustrates how the assumptions of Gaussian priors and a perfect model are needed to arrive at a cost function that can be minimized to solve the inverse problem. The right route illustrates how the assumption of the model being a Markov process and the measurement errors being uncorrelated in time leads to a sequence of inverse problems that are well suited for ensemble methods. In a linear case the methods derived using the two routes will give identical results.

model to contain errors, in addition to the errors in the parameters, the initial conditions, and the data. Such variational problems have been solved using the representer method[1] by Bennett (1992, 2002); Eknes and Evensen (1997); Muccino and Bennett (2001).

An additional assumption is often imposed where the dynamical model is treated as perfect and no model errors are accounted for. Furthermore the errors in the initial conditions are often neglected as well. These assumptions lead to a cost function that only measures the distance between the parameter estimate and its prior, plus the distance between the model solution and the measurements (Figure 12.1 A.3). Thus, only the parameters are solved for, and a given set of parameters defines a unique model solution. However, it is

[1]The representer method solves the Euler Lagrange equations resulting from a weak constraint variational formulation where a linear dynamical model is included as a soft constraint. An iteration may be used with nonlinear dynamical models. The representer method searches for the solution as the first guess unconstrained model solution plus a linear combination of influence functions, or representer functions, one for each measurement. The method is identical to Gauss-Markov smoothing in space and time, and the representers are space-time covariance functions between the predicted measurements and the model state vector in space and time. The beauty of the method is that the solution can be found very efficiently without actually computing the representer functions.

important to realize that we have now neglected other possible errors in the model and in the initial conditions.

The resulting 'strong constraint' cost function (Sasaki 1970) is normally minimized using different kinds of descent methods. Typical methods involve the use of gradients where the gradient may be evaluated by solving the adjoint model (see e.g. Evensen 2009). The solution obtained is a single realization that hopefully represents the global minimum of the cost function (Figure 12.1 A.5.a). With a single realization representing the solution there is no error estimate available.

A serious problem is that the cost function is highly nonlinear, and the parameter-estimation problem becomes extremely challenging even with relatively few parameters to estimate. Using minimization methods the solution is searched for in a space with dimensions equal to the number of poorly known parameters. The nonlinearities of the inverse problem lead to a cost function with multiple local minima. Thus, the probability of converging to a local minimum is high and in most realistic applications the global minimum is never found. It is also clear that the inverse problem becomes harder to solve for long time intervals since non-Gaussian contributions have more time to develop.

Thus, as an alternative to standard minimization, various sampling methods, have been proposed (Figure 12.1 A.4.b), which applies guided Monte Carlo sampling of the solution space. Sampling methods increase the likelihood for finding the global minimum but to an, often, unacceptable high numerical cost associated with the huge number of model simulations required when the dimension of the parameter space becomes large. If the global minimum can be found, the sampling methods, such as genetic algorithms (Goldberg 1989), also provide an estimate of the posterior pdf around the global minimum and an error estimate can be derived.

12.2.2 Sequential Processing of Measurements

An alternative route is taken in Evensen (2009). Starting from Bayes' theorem two assumptions are made (Figure 12.1 B.1). First the simulator or dynamical model is assumed to be a Markov model, which means that the solution at one time-instant is only dependent on the solution at the previous time-instant. This property is normally satisfied for most time-dependent models. The second assumption is that the measurement errors are uncorrelated in time. This assumption is often implicitly used in methods that minimize a cost function by the use of a diagonal error covariance matrix for the measurement errors, and also in the Kalman filter by the sequential processing of measurements.

As is discussed in Evensen (2009), the assumption of independent data in time allows for Bayes' theorem to be written as a recursion where data are processed sequentially in time and we end up with a sequence of inverse problems (Figure 12.1 B.2). Each of the inverse problems can be solved using any minimization method. However, to proceed in the recursion, access is needed to both the prior estimate and its uncertainty.

The particle filter by Doucet *et al.* (2001) is proposed as a general method for solving the sequence of inverse problems (Figure 12.1 B.4.a to B.6.a). The pdf for the solution is approximated by a large number of particles, or model states. Each of the model states are integrated forward in time according to the model equations, which include stochastic terms to represent model errors. At each time when data are available a Bayesian update is computed by resampling of the posterior distribution. The solution is a large ensemble of model realizations which represents the posterior pdf. The particle filter requires a large number of realizations to converge, and is so far only applicable to rather low dimensional problems.

The Ensemble Kalman Filter (EnKF) is to some extent similar to the particle filter, except that a simplification is imposed in the update step (Figure 12.1 B.4.b to B.6.b). It is assumed that the predicted prior pdf is well approximated using only the first and second order moments of the pdf. It is then possible to efficiently compute a linear update using the procedure outlined by Evensen (1994, 2009), where each realization is updated according to the standard Kalman filter update equation, but using measurements contaminated with simulated noise representing the measurement errors.

It is shown by Evensen (2009) that for linear systems the ensemble Kalman smoother leads to the same result as the representer method which solves the weak constraint inverse problem, as long as measurements are independent in time and Gaussian priors are used.

In the EnKF the sequential updating has the nice property of introducing Gaussianity into the pdf at each update step (see Evensen 2009). The model solution is kept on track and is consistent with the data and true solution, and non-Gaussian contributions will not develop to the same extent as if the ensemble is integrated forward in time without being conditioned on measurements. In fact the assimilation of measurements introduces Gaussianity into the predicted pdf, and this benefits the EnKF by reducing the severity of the neglected non-Gaussian contributions in the EnKF updates. The implication is that the linear update step used in the EnKF in many cases is a reasonable and valid approximation. This fact does to a large extent explain the success of the EnKF in many nonlinear inverse problems.

12.3 EnKF History Matching Workflow

The history matching workflow, independent of the method used, involves three major steps; first a *parameterization* where the parameters that are uncertain and at the same time characterize the major uncertainty of the model solution are identified, thereafter a *prior error model* is specified for the selected parameters based on an initial uncertainty analysis, and finally a *solution method* needs to be selected. All three steps may be equally important and the selections and choices made will depend on the problem at hand.

Traditional methods for assisted history matching are constrained to include a low number of model parameters in the optimization process. The history matching process is then performed using only the most influential parameters, typically identified from a sensitivity study. Often aggregated parameters are used to reduce the number of parameters in the estimation, and an example can be a multiplier of a field variable.

The EnKF is, on the other hand, not limited by the number of model parameters. The reason is that the dimension of the inverse problem is reduced to the number of realizations included in the ensemble. Thus, the solution is searched for in the space spanned by the ensemble members rather than the high dimensional parameter space. It is however important that the major variability of the parameters can be represented by a number of modes that are of the same order as the number of realizations.

The typical uncertain elements to consider in a reservoir characterization study are: the structural model which is based on seismic time interpretation and time-to-depth conversion; petrophysical evaluation of wells and property mapping; depth of fluid contacts (i.e. the interfaces between different fluids in the reservoir); horizontal and vertical barriers to fluid flow including vertical and horizontal permeability, and fault transmissibility.

An initial uncertainty analysis leads to a quantification of the prior uncertainties of the parameters, which is in turn represented using pdfs. The specified pdfs then represent our prior belief concerning the uncertainty of each particular parameter. The priors must be defined to obtain a realistic relative weighting on the first guesses of parameters, the model dynamics, and the measured data.

A reservoir and its uncertainties are normally represented by a so called geological model that is built using a reservoir modelling software. The geological model integrates all available prior information from exploration seismic surveys, well-log analysis from drilled test wells, and a general understanding of the geology and depositional environment of the reservoir.

Stochastic simulations are used to produce multiple realizations of the porosity and permeability fields in the reservoir model. The multiple stochastic realizations then represent the uncertainty in the property model. The realizations are conditioned to the well observations and honour the statistical properties such as trends and spatial correlation of the well-log data. It is also possible to condition the properties on seismic data. The petrophysical realizations are then up-scaled or interpolated to the simulation grid.

There is a large number of additional parameters that need to be specified in the simulation model, and most of these will also have an uncertainty associated with them, which targets them for further tuning and estimation. Some other parameters previously estimated using the EnKF include fault transmissibility, horizontal and vertical barriers to fluid flow, and the initial

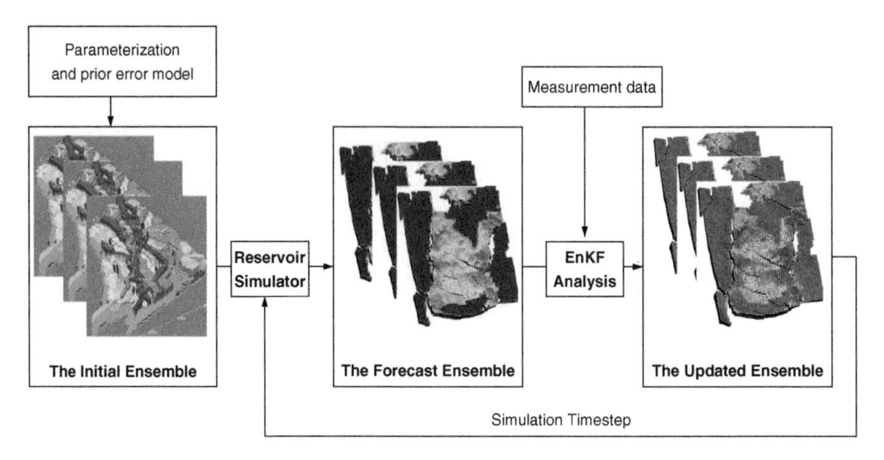

Figure 12.2 The general EnKF workflow for oil reservoir applications: The initial ensemble that expresses explicitly the model uncertainty is the starting point for the Ensemble Kalman filter. Forward integration of each ensemble member by using the reservoir simulator leads to the forecast ensemble. Updates are performed at each time when measurements of production data are available. These two processes, the forecast and the analysis, comprise the main EnKF loop that results in the updated ensemble.

vertical distribution of fluids in the model through the specification of the initial fluid contacts.

The uncertain parameters are normally characterized by a Gaussian distribution with mean equal to the best estimate and a standard deviation reflecting the uncertainty. Parameter values for the different realizations are then generated by random sampling from the prescribed distributions.

The dynamic variables, pressure and saturation grid-cell values, are included in the initial ensemble through an initialization using the flow simulator.

Thus, when using the EnKF one first creates an ensemble of reservoir models expressing explicitly the model uncertainty (Figure 12.2 The Initial Ensemble). The ensemble mean is considered as the best estimate and the spreading of the ensemble realizations around the mean reflects the uncertainty in the estimate.

12.3.1 Estimation of Relative Permeability

In this chapter we also, for the first time, include the relative permeability as an uncertain parameter to be estimated in the EnKF. Relative permeability is defined as the ratio of effective permeability of a particular fluid at a particular saturation (i.e. in presence of another fluid) to the absolute permeability of the porous medium at single phase saturation.

In a heterogeneous medium, where relative permeability properties are derived from core-samples and may not be representative at the reservoir scale, history matching of up-scaled relative permeability curves may lead to improved results. The relative permeability curves obtained from the core laboratory experiments should not be directly used in the reservoir simulation model, but need to be up-scaled to compensate for fluid forces, numerical dispersion, and geological heterogeneity effects. Unfortunately, the up-scaling techniques require large computational time and may not be robust. Therefore, our approach is to obtain relative permeability properties directly at the coarse scale by data assimilation. Hence, estimating the shape of relative permeability curves at the coarse scale are closely related to the up-scaling issues and the lack of information of the fine scale permeability heterogeneity.

Several recent publications show the potential of estimating relative permeability properties on a coarse scale reservoir simulation model. In Okano *et al.* (2005) the authors adjust the relative permeability curves during history matching using a stochastic sampling algorithm, called the Neighbourhood Approximation algorithm. The method is applied on synthetic cases, and the results show that they are able to match the production data by adjusting the relative permeability curves. Eydinov *et al.* (2007) use an adjoint method to perform a simultaneous estimation of absolute and relative permeability by automatic history matching of three-phase flow production data.

In this chapter, we employ Corey functions (Brooks and Corey 1964) to parameterize coarse scale relative permeability curves. The Corey parameterization is flexible and is often used in the petroleum industry. According to this model the relative permeability in an oil-water system is given by

$$k_{\mathrm{rw}} = k_{\mathrm{rw}}^{\star} \left(\frac{S_{\mathrm{w}} - S_{\mathrm{wc}}}{1 - S_{\mathrm{wc}} - S_{\mathrm{orw}}} \right)^{e_{\mathrm{w}}}, \tag{12.1}$$

$$k_{\mathrm{row}} = k_{\mathrm{row}}^{\star} \left(\frac{1 - S_{\mathrm{w}} - S_{\mathrm{orw}}}{1 - S_{\mathrm{wc}} - S_{\mathrm{orw}}} \right)^{e_{\mathrm{ow}}}, \tag{12.2}$$

where k_{rw} is the water relative permeability and k_{row} is the oil relative permeability. The relative permeabilities in an oil-gas system are given as

$$k_{\mathrm{rg}} = k_{\mathrm{rg}}^{\star} \left(\frac{S_{\mathrm{g}} - S_{\mathrm{gc}}}{1 - S_{\mathrm{wc}} - S_{\mathrm{gc}} - S_{\mathrm{org}}} \right)^{e_{\mathrm{g}}}, \tag{12.3}$$

$$k_{\mathrm{rog}} = k_{\mathrm{rog}}^{\star} \left(\frac{1 - S_{\mathrm{g}} - S_{\mathrm{org}}}{1 - S_{\mathrm{wc}} - S_{\mathrm{gc}} - S_{\mathrm{org}}} \right)^{e_{\mathrm{og}}}. \tag{12.4}$$

where k_{rg} is the gas relative permeability and k_{rog} is the oil relative permeability.

In a three phase system, the three phase oil relative permeability at a particular water and gas saturations is extrapolated from the input two phase relative permeability model.

Large flexibility in the relative permeability parameterizations is important, and the following parameters in the Corey function can be updated in the data assimilation; the relative permeability end points $k^{\star}_{\mathrm{rg}}$, $k^{\star}_{\mathrm{rog}}$, $k^{\star}_{\mathrm{row}}$, $k^{\star}_{\mathrm{rw}}$, the Corey exponents e_{w}, e_{ow}, e_{g}, e_{og}, the connate water and gas saturation S_{wc}, S_{gc}, and the residual oil and gas saturation S_{orw}, S_{org}. In the current implementation the connate gas saturation is equal to the critical gas saturation, and the connate water saturation is equal to the critical water saturation.

12.3.2 Transformed Fault Transmissibility Multipliers

Faults can act as both barriers and conduits to fluid flow, and are normally included in reservoir-simulation models as grid offset and using 2D transmissibility multipliers. The fault transmissibility multipliers should be limited to the interval $[0, 1]$, where a numerical value of 0 reflects a complete flow barrier and a value of 1 characterizes an open fault. The upper bound can be relaxed in many simulators without much influence on the flow properties. Anything in between 0 and 1 corresponds to a partial barrier to fluid flow. Experience shows that a large change in the fault transmissibility is needed to obtain significant changes in the flow behaviour. Note also that the transmissibility needs to be almost identically zero to maintain a difference in the pressure across the fault.

There are generally large uncertainties associated with the fault fluid-flow properties and faults transmissibility multipliers have already been included as parameters to be estimated using the EnKF in Evensen *et al.* (2007). In their field application, the prior guess for the uncertain multipliers is set to 1.0, with a standard deviation of 0.2. Results from the EnKF assimilation show that the uncertainty of the multipliers is not significantly reduced and it is concluded that the production is not very sensitive to this parameter. Similarly, we estimated the fault multipliers using a Gaussian distribution to reflect the uncertainty. However, only small updates of the order 0.20 to 0.35 were obtained and it is impossible to determine whether a fault is closed or open, which is often the level of uncertainty. The use of appropriate transformations to some extent enables us to overcome the bottleneck linked to updating non-Gaussian variables in the EnKF. With this concept, the EnKF updates a Gaussian variable, which is transformed before it is used in the reservoir simulator. The following transformation gives satisfactory results and is used in our field studies for updating faults as well as vertical transmissibility multipliers,

$$y = \frac{1}{\sqrt{2\pi}\sigma} \int_{-\infty}^{x} \mathrm{d}t\, \mathrm{e}^{-\frac{(t-\mu)^2}{2\sigma^2}}, \text{ with } x \sim N(0, 1). \tag{12.5}$$

The transformation (12.5) ensures that the output variable y is in the range $(0, 1)$ and, depending on the values of the variables σ and μ, it is possible to get either a reasonably uniform distribution, a bimodal distribution with peaks close to 0 and 1, or a uni-modal distribution peaked around a value in $(0, 1)$.

12.3.3 State Vector

The state vector $\boldsymbol{\Psi}$ is a high-dimensional vector consisting of static parameters (all the uncertain parameters we want to estimate), dynamic variables (pressure and saturations), simulated production data and/or seismic data, and it is written as

$$\boldsymbol{\Psi} = \begin{pmatrix} \boldsymbol{\psi} \text{ dynamic state variables} \\ \boldsymbol{\alpha} \text{ static parameters} \\ \boldsymbol{d} \text{ predicted data} \end{pmatrix}. \tag{12.6}$$

The 3D dynamic state variables consist of the reservoir pressure P, the water saturation S_{W}, the gas saturation S_{G}, the initial solution gas-oil ratio R_{S}, and the initial vapor oil-gas ratio R_{V}.

The static parameters include the 3D fields of porosity and horizontal and vertical permeability, fault transmissibility multipliers for each fault in the model, vertical transmissibility multipliers that regulate flow between various zones in the reservoir, the relative permeability parameterization as discussed above, and the depth of initial water-oil contacts and gas-oil contacts.

The predicted measurements are included in the state vector, since they are nonlinearly related to the model state and their inclusion in the state vector simplifies the comparison with the measured data in the EnKF update scheme. The predicted measurements are typically the oil, gas and water rates for each well. The gas and water rates are often represented by the gas-oil ratio and the water cut, which therefore are added to the state vector as well. When 4D seismic data is used to condition the model dynamics, the predicted seismic response is also included in the state vector in some form (Skjervheim *et al.* 2007).

The total number of variables and parameters to be estimated is then 8 times the number of active grid cells in the simulation model plus the additional parameters describing the initial contacts, the relative permeability parameters, and fault and vertical transmissivities.

12.3.4 Updating Realizations

Once an initial ensemble of reservoir models is generated, the EnKF is used to update the ensemble sequentially in time to honour the new observations at the time they arrive (Figure 12.2). Sequential data assimilation is a particularity of the EnKF and is a major strength compared to traditional history matching methods. It allows for real-time applications and fast model updating. The model can be updated whenever new data become available without the need to re-integrate the complete history period.

The EnKF consists of a forward integration to generate *the forecast* followed by the updating of state variables to generate *the analysis.* In the forecast step, the ensemble of reservoir models is integrated forward in time using the dynamical model. Each ensemble member is integrated until the next time when production measurements are available, leading to the forecast ensemble.

The assimilated observations $\boldsymbol{d}$ are considered as random variables having a distribution with the mean equal to the observed value and an error covariance $\boldsymbol{C}_{\varepsilon\varepsilon}$ reflecting the accuracy of the measurement. Thus, following Burgers *et al.* (1998) we generate an ensemble of observations $\boldsymbol{d}_j = \boldsymbol{d} + \boldsymbol{\varepsilon}_j$ where $\boldsymbol{\varepsilon}_j$ represents the measurement error and $\boldsymbol{C}_{\varepsilon\varepsilon} = \overline{\boldsymbol{\varepsilon}\boldsymbol{\varepsilon}^T}$ where the overline denotes the average over the ensemble.

In the analysis, the following updates are computed for each of the ensemble members,

$$\boldsymbol{\Psi}_j^{\mathrm{a}} = \boldsymbol{\Psi}_j^{\mathrm{f}} + \boldsymbol{C}_{\psi\psi}\boldsymbol{M}^{\mathrm{T}}(\boldsymbol{M}\boldsymbol{C}_{\psi\psi}\boldsymbol{M}^{\mathrm{T}} + \boldsymbol{C}_{\varepsilon\varepsilon})^{-1}\,(\boldsymbol{d}_j - \boldsymbol{M}\boldsymbol{\Psi}_j^{\mathrm{f}}), \tag{12.7}$$

where $\boldsymbol{\Psi}_j^{\mathrm{f}}$ represents the state vector for realization j after the forward integration to the time when the data assimilation is performed, while $\boldsymbol{\Psi}_j^{\mathrm{a}}$ is the corresponding state vector after assimilation. The ensemble covariance matrix is defined as $\boldsymbol{C}_{\psi\psi} = \overline{(\boldsymbol{\psi} - \overline{\boldsymbol{\psi}})(\boldsymbol{\psi} - \overline{\boldsymbol{\psi}})^{\mathrm{T}}}$, where $\overline{\boldsymbol{\psi}}$ denotes the average over the ensemble. The matrix $\boldsymbol{M}$ is an operator that relates the state vector to the production data and $\boldsymbol{M}\boldsymbol{\Psi}_j^{\mathrm{f}}$ extracts the predicted or simulated measurement value from the state vector $\boldsymbol{\Psi}_j^{\mathrm{f}}$. The result is an updated ensemble of realizations, conditioned on all previous production data, that properly characterize the estimate and its uncertainty. The updated ensemble is then integrated until the next update-time t_{k+1}.

12.4 Field Case

The proposed workflow based on the EnKF has been applied to history match several North Sea simulation models. The results from a complex real field case is presented and discussed below.

12.4.1 Reservoir Presentation

The Omega field is part of a larger N-S elongated fault block, and has a length of approximately 8 km and a width between 2.5 and 3.5 km (Figure 12.3). The structure represents an open syncline and is bounded by faults, which are assumed to be sealing. The main reservoir consists of shallow-marine deposits and associated near-shore, deltaic sediments. High spatial and temporal variations are the consequences of a succession

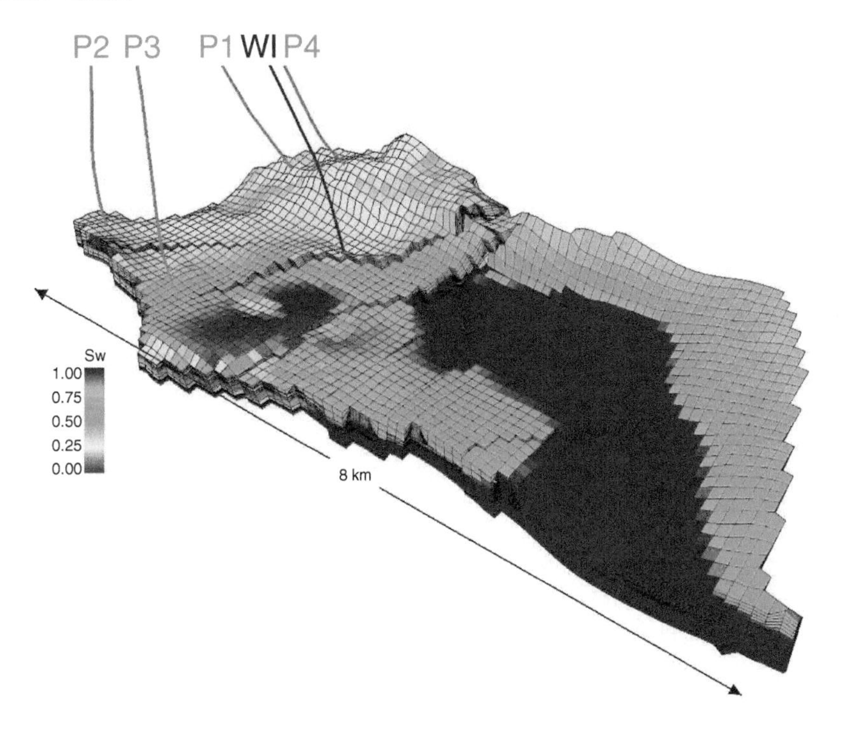

Figure 12.3 The Omega field initial water saturation. Four horizontal producers (P1 to P4) drain the northern part of the field, while a vertical water injector (WI) ensures pressure support.

of transgressive-regressive events combined with intense fault activity. The architectural style is complex, flow properties are heterogeneous and vertical communication is relatively poor.

Numerous faults lead to considerable structural complexity. The understanding of the fault network is to some extent guided by interpretations of water-oil contacts (WOC), although original observations of WOC's are rare, thus primarily based on pressure gradient analysis. In the Omega field, the controlling mechanisms of the fluid contacts are poorly understood and fault properties are highly uncertain.

The reservoir was initially at saturated conditions. The presence of a small gas cap in the north of the structure is interpreted as resulting from gas migration from the west. The field was set in production in 2000 and is drained by four horizontal producers and one water injector ensuring pressure support. Water injection started in 2004.

The simulation grid consists of 37×80 cells, with a lateral spacing of 100×100 m, and 40 layers with varying thickness. A total of 60 000 cells were active in the simulation.

12.4.2 The Initial Ensemble

The method described in the section 'EnKF history matching workflow' was used to build the initial ensemble. The focus was set on the target reservoir, located between layers 17 and 40. The main uncertain parameters identified in the Omega model are porosity and permeability fields, depths of initial fluid contacts, relative permeability and fault multipliers.

The porosity and permeability distributions are derived from analysis of well-log data. Due to spurious permeability measurements in the horizontal producing wells, it is decided not to constrain the Omega model to the well-log data from any of the four producing wells. Only the well-log data from the appraisal wells and the water injector are used. The geo-statistics from the initial ensemble are summarized in Table 12.1. Note that the upper part of the model (layers 1 to 16) is not assigned any uncertainty. Sequential Gaussian simulation is used to generate multiple stochastic realizations of the porosity and permeability fields. A deterministic correlation coefficient of 0.7 between porosity and permeability and a vertical to horizontal permeability ratio of 0.1 are assumed. The fields resulting from the geostatistical simulation are then up-scaled to the simulation grid.

The Omega structure is divided into five different compartments each having individual water-oil and gas-oil contacts. Direct measurements of the depths of the initial fluid contacts in the wells are rare, and thus have been estimated by pressure gradient analysis. The depths of the contacts are assumed to be Gaussian, with the prior mean defined as the best guess interpretation and a standard deviation of 5 meters.

Relative permeability properties have been derived from Special Core Analysis (SCAL) laboratory measurements and adapted to Corey curves. For simplification, only one set of relative permeability properties is defined for the

Table 12.1 Geomodel statistics properties.

Layers	Porosity mean (std)	Permeability mean (std)	Azimut	Correlation X(Azimut),Y	Vert. corr.
17-26	0.16 (0.03)	65 (86)	0	3000,3000	5
27-28	0.20 (0.03)	910 (1085)	60	2000,600	5
29-30	0.23 (0.03)	292 (397)	160	1500,1000	25
31	0.14 (0.04)	4.6 (4)	340	1000,500	5
32-33	0.16 (0.05)	10 (40)	340	1000,500	25
34-35	0.23 (0.03)	300 (350)	340	1500,1000	5
36	0.06 (0.03)	6 (30)	0	1000,1000	5
37	0.05 (0.03)	39 (20)	0	1000,1000	5
38-39	0.19 (0.04)	47 (190)	60	2000,600	25
40	0.10 (0.01)	8 (22)	60	2000,1000	5

Table 12.2 Relative permeability prior statistics.

Parameter	Mean	Std.dev
$k^{\star}_{rw}$	0.6	0.05
e_w	3	0.5
e_{ow}	4	0.5
e_g	1.8	0.5
e_{og}	5.4	0.5
S_{wc}	0.25	0.05
S_{gc}	0	0.02
S_{orw}	0.18	0.05
S_{org}	0.05	0.03

whole field in the current experiment. The measured curves are used as the prior mean in the EnKF and the critical saturations are the best guess averaged field values. We assume a Gaussian prior for the relative permeability parameters. Table 12.2 presents the parameters in the Corey function that were updated in the data assimilation and associated statistics. It should be mentioned that the relative permeability in the Omega model has previously not been considered as a history matching parameter.

Interpretation of the fault transmissibility properties has to some extent been guided by pressure measurements and inferred from sensitivity studies. A large uncertainty nevertheless remains. For faults without any prior knowledge about the behavior with respect to flow, we have used values $\sigma = 1$ and $\mu = 0$ leading to a prior fault transmissibility multiplier that is uniformly distributed between 0 and 1. A few faults are known to be pressure barriers, thus, the value of μ is increased to lead to a distribution of fault multipliers skewed towards zero. A total of 7 fault multipliers is estimated and the total number of variables to update is over 200 000.

The production data considered in the history matching are monthly averaged oil-production rates, water rates and gas rates from each well. The history-matching period covers 6 years. The measurement uncertainty is specified as a percentage of the measured rate. We have used 10 % of the measured value for the oil rate, 15 % for the gas rate and 20 % for the water rate. In the simulation, all wells are controlled by specifying a target for the reservoir-fluid volume-rate.

The computational cost of the EnKF increases linearly with the size of the ensemble. The major computation is connected to the forward integration of the ensemble of realizations, which can be done in parallel, while the computation of the update only contributes with a negligible fraction of the total cost. The overall run time for the current field case where the EnKF is used with 100 realizations is less than one day on our 80 nodes Linux cluster.

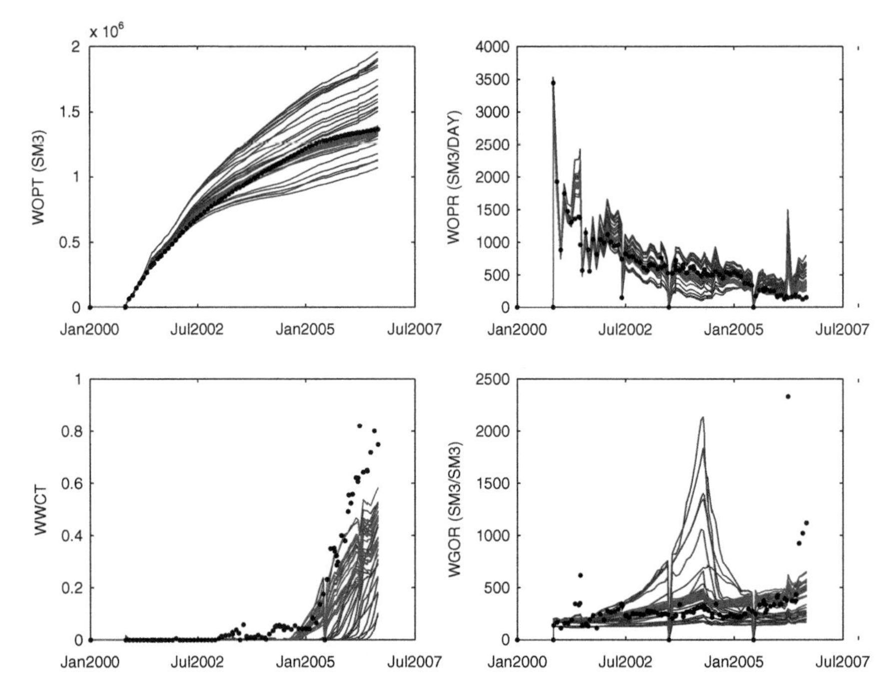

Figure 12.4 Well P1 oil production total (upper left), oil production rate (upper right), water cut (lower left) and gas-oil ratio (lower right). Ensemble prediction based on prior ensemble of parameters (blue lines) and EnKF estimated parameters (red lines). Black dots are measurements.

12.4.3 Results

Validation of Updated Models

The results from the EnKF history match are shown in Figures 12.4 to 12.7, where we have plotted for each well the oil production total (WOPT), the oil production rate (WOPR), the gas-oil ratio (WGOR) and the water cut (WWCT) for each well. The blue curves represent 20 realizations of the prior ensemble, unconditioned to production data. The red curves are from a rerun of the 20 ensemble members when initialized with the EnKF updated parameters. The black dots represent the observations.

The initial uncertainty on the model parameters leads to a significant uncertainty in the simulated productions, which is illustrated by the prior spread in cumulative oil production for each well. For all the producers the EnKF updated ensemble manages to reproduce the observed data much better than the prior ensemble and the uncertainty is significantly reduced. It has previously been a challenge to obtain a satisfactory match of the water

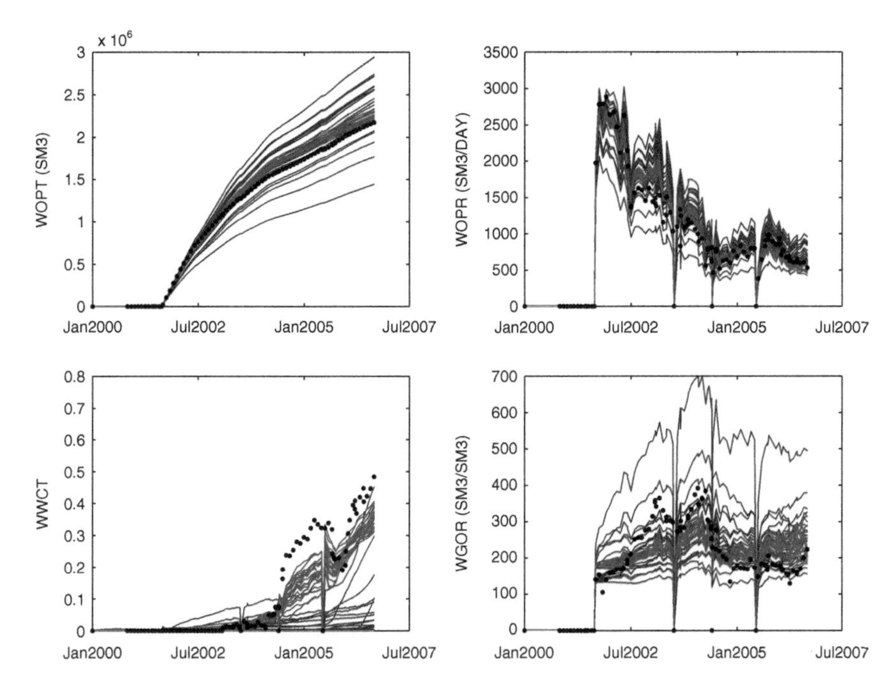

Figure 12.5 Well P2 oil production total (upper left), oil production rate (upper right), water cut (lower left) and gas-oil ratio (lower right). Ensemble prediction based on prior ensemble of parameters (blue lines) and EnKF estimated parameters (red lines). Black dots are measurements.

production in the Omega structure, and the reservoir models have had a poor prediction of the water breakthrough. The results from the EnKF illustrate that constraining the prior model on production data enables to capture the water production more accurately in all the wells. The improvements in wells P2 and P3 are satisfactory, as these have been very difficult to match manually. GOR observations are also reasonably reproduced.

Porosity and Permeability Updates

It can be challenging to get an overall picture of the updates of the porosity and permeability fields by scrolling through the different realizations, layer by layer. The applied modifications are maybe best analysed by comparing the initial and updated average fields, as well as the standard-deviation fields.

In Figure 12.8 we have plotted the initial and updated average porosity field for one layer, and in the lower part the corresponding standard deviation. The initial standard-deviation field shows that the northern part of the model domain, with low standard deviation (0.015), has been constrained to

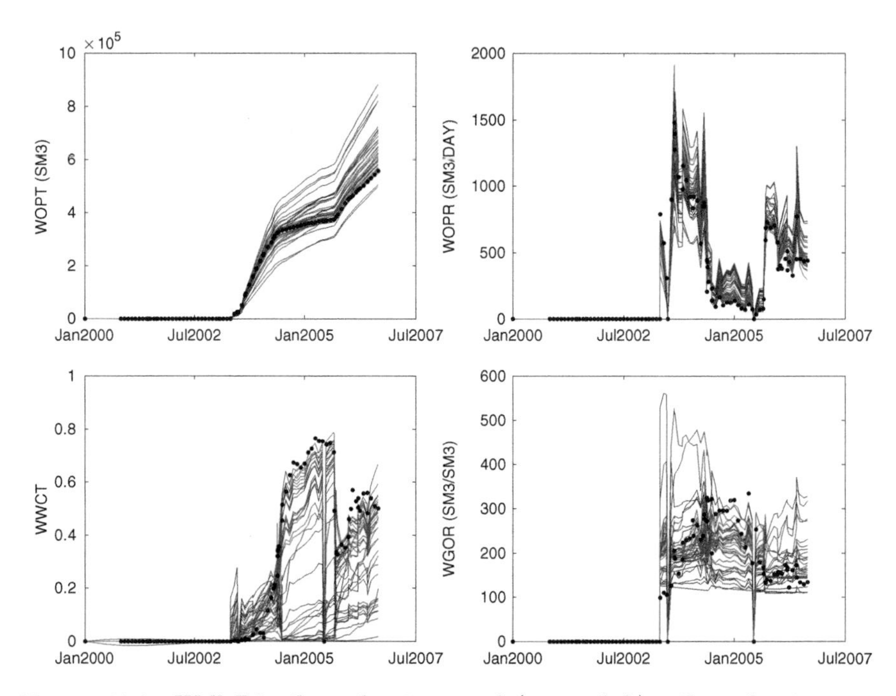

Figure 12.6 Well P3 oil production total (upper left), oil production rate (upper right), water cut (lower left) and gas-oil ratio (lower right). Ensemble prediction based on prior ensemble of parameters (blue lines) and EnKF estimated parameters (red lines). Black dots are measurements.

well-log data, whereas the southern part of the reservoir, with higher standard deviation (around 0.03), has not been constrained by well-log data. The updated porosity field shows only a small modification in the northern part, but we notice an important decrease in the average porosity at the toe of well P3. Furthermore, the final standard deviation is greatly reduced in this region, indicating that the updates can be interpreted with confidence.

The average field updates clearly show that an increase in porosity corresponds to an increase in permeability. This is reflecting the correlation coefficient (0.77) used when generating the prior realizations. Thus, updates in the petrophysical parameters will mainly impact the reservoir volumes and only to a minor extent the water or gas breakthrough time at the wells. The relative-permeability updates have the major impact on controlling the breakthroughs.

Relative Permeability Updates

The strong updates in the parameters demonstrate that the relative permeability properties are important parameters to consider in the Omega model.

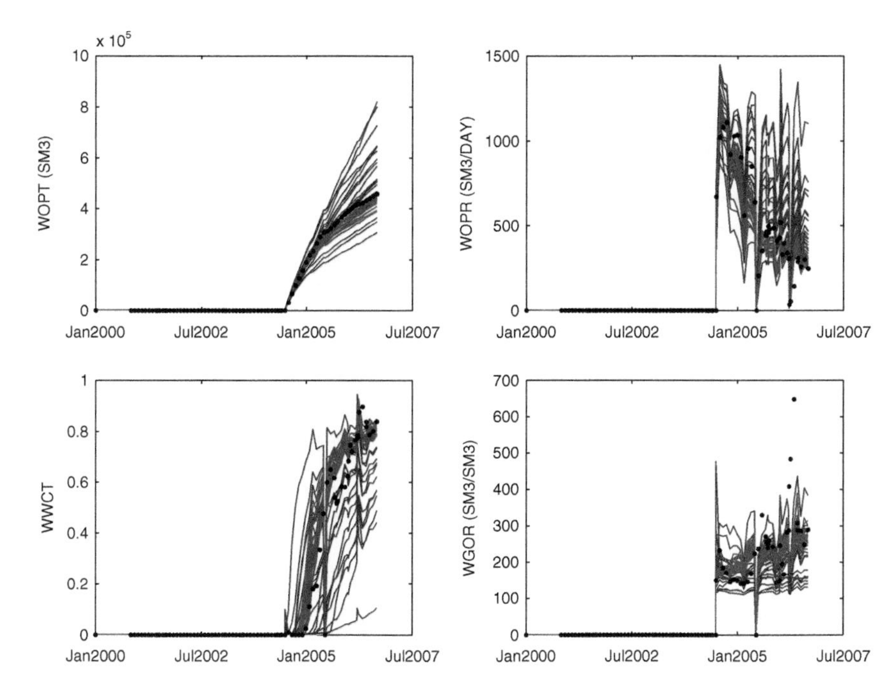

Figure 12.7 Well P4 oil production total (upper left), oil production rate (upper right), water cut (lower left) and gas-oil ratio (lower right). Ensemble prediction based on prior ensemble of parameters (blue lines) and EnKF estimated parameters (red lines). Black dots are measurements.

The left plot in Figure 12.9 shows the initial (blue) and updated (red) water/oil relative permeability curves. The results indicate clearly that for high water saturation ($S_w > 0.5$) the water mobility is increased. As a result, water breakthrough will occur faster. On the contrary the mobility at lower saturation is reduced, due to an increase in the critical water saturation. Thus, the initial water in the oil zone will be less mobile. The oil curve shows similar behaviour. The mobility at high oil saturation is increased, but more oil is left behind as a result of a higher residual oil saturation. The updated estimate of the residual oil saturation (0.36) is in accordance with the fractional flow curve obtained from core flooding experiments, which indicates that the residual oil after water flood is minimum 35 %.

The right plot in Figure 12.9 shows the initial (blue) and updated (red) oil/gas relative permeability curves. The gas exponent is very sensitive to the well observations and the uncertainty is reduced. The gas mobility is increased. The residual oil in presence of gas (S_{org}) is updated from 0 to 0.05 and the oil to gas exponent is decreased from 5.4 to 4.25. The EnKF updates are in line with conclusions drawn by the asset team where it is reported that this exponent is probably too high in the original model.

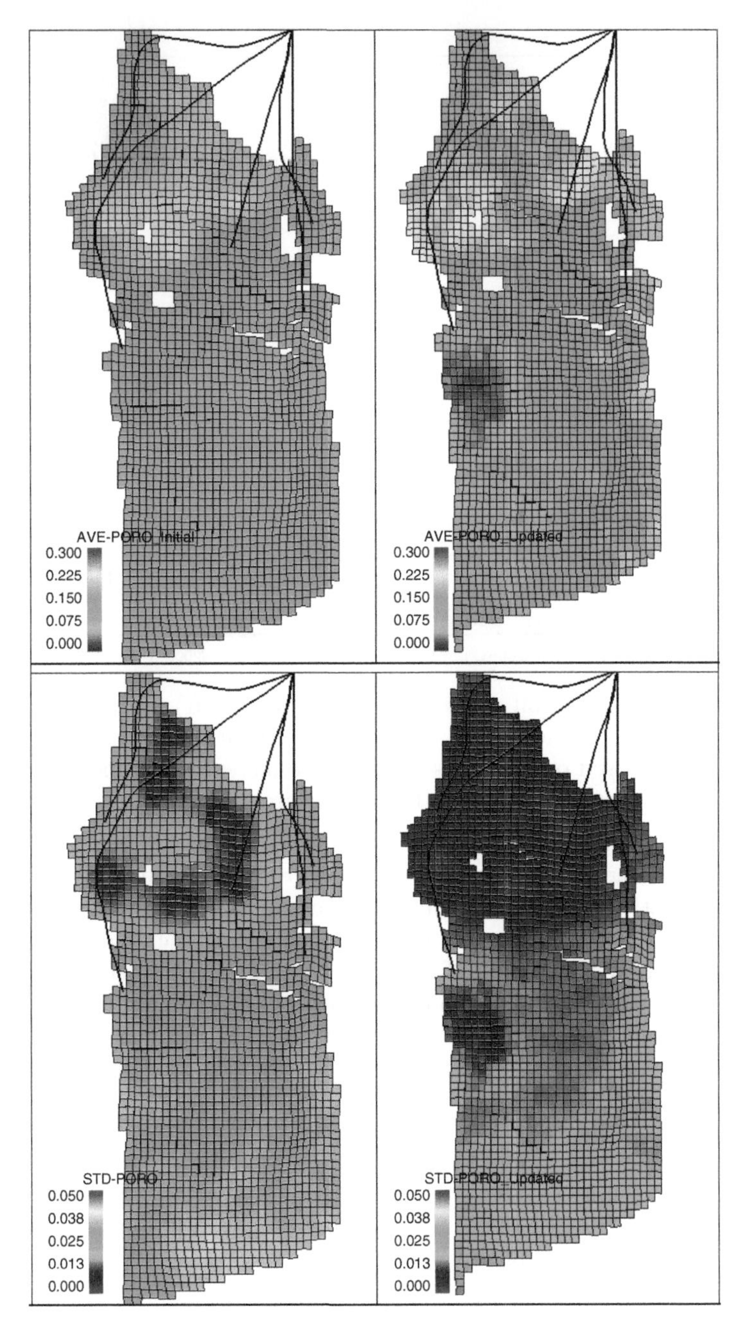

Figure 12.8 EnKF updates of porosity in a representative layer. Upper row; initial (left) and updated (right) average porosity in layer 19. Lower row; corresponding standard deviation, initial (left) and updated (right). Black lines show well paths of producers and injector.

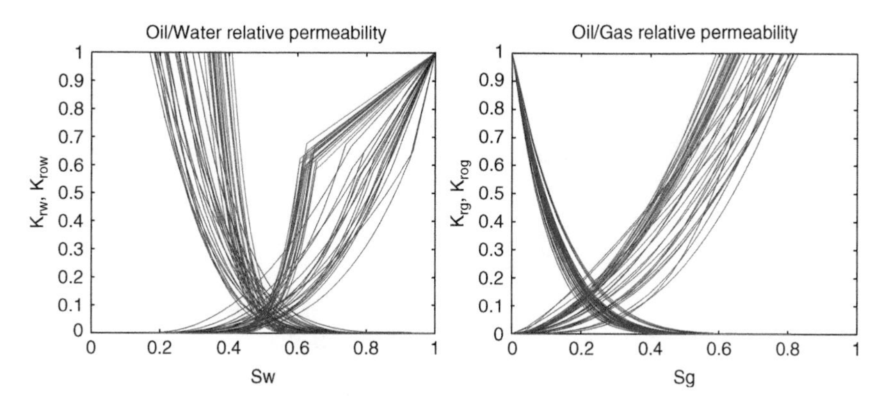

Figure 12.9 EnKF updates of relative permeability curves. Left; oil/water relative permeability. Right; oil/gas relative permeability. The blue lines show the initial uncertainty and the red lines the updated uncertainty.

It is stressed that the updated curves should be considered as up-scaled curves, accounting for numerical dispersion and geological heterogeneity effects and thus, they will naturally differ from the measured properties on core samples, particularly in a heterogeneous medium.

The permeability and relative permeability are indirectly linked, both controlling in some way the mobility of the fluids. In the current experiment we obtained an increase in both the relative permeability curves and the permeability field in the model. We also have some indications that the prior values of the permeability field may be too low and this may then be compensated by too high values of the relative permeability curves. Thus, the prior permeability field should be revised.

In an earlier experiment where the relative permeability was not updated, the initial ensemble could not capture the water breakthrough and the updated ensemble was not able to correct for this initial bias. The updated water cut profile indicated that the injection water did not reach the well and in order to match the high WCT values observed near the end of the simulation, more formation water was mobilized by reducing the depth of the WOC. It is worth mentioning that similar bias in the simulation performances were obtained by the asset team, working with a new model, based on an improved structural interpretation as well as a detailed facies model. Consequently, the additional updating of the relative permeability properties as well as the fault transmissibility multipliers appears to be a crucial element in order to capture the reservoir flow behaviour.

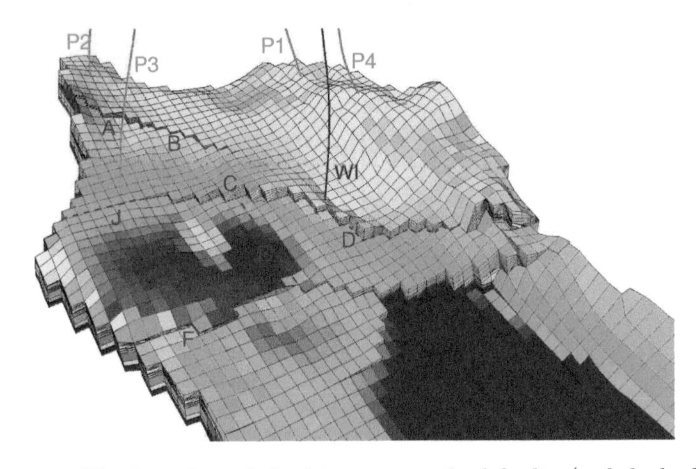

Figure 12.10 The location of the history-matched faults (red-dashed lines).

Fault Transmissibility Multiplier Updates

The locations of the history-matched faults are illustrated in Figure 12.10. The sequential updating of the fault transmissibility multiplier is plotted in Figure 12.11 for the faults D, E and F.

For fault D, the initial distribution the fault transmissibility multiplier is skewed towards zero. During the assimilation the spread of the prior ensemble is significantly reduced, indicating that the assimilated data are strongly correlated with the transmissibility multiplier, and furthermore, that the updates can be interpreted with confidence. The updated ensemble clearly indicates a closed fault, which is in line with an observed pressure difference that is maintained across the fault.

The simulated production data are not significantly sensitive to the fault multipliers for faults C, B, and E, and for these faults the ensemble spread remains throughout the data assimilation, possibly due to the larger distance to the production wells. Thus, one should be careful in interpreting the results from these faults.

The variance is reduced for the fault F, which is located in the toe region of P3, and fault A, crossing well P2, and the updated ensemble suggests that these faults are not completely sealing but nevertheless act as a barrier to flow.

12.5 Conclusion

A thorough workflow for updating reservoir simulation models using the EnKF is presented and demonstrated through a successful North Sea field case application.

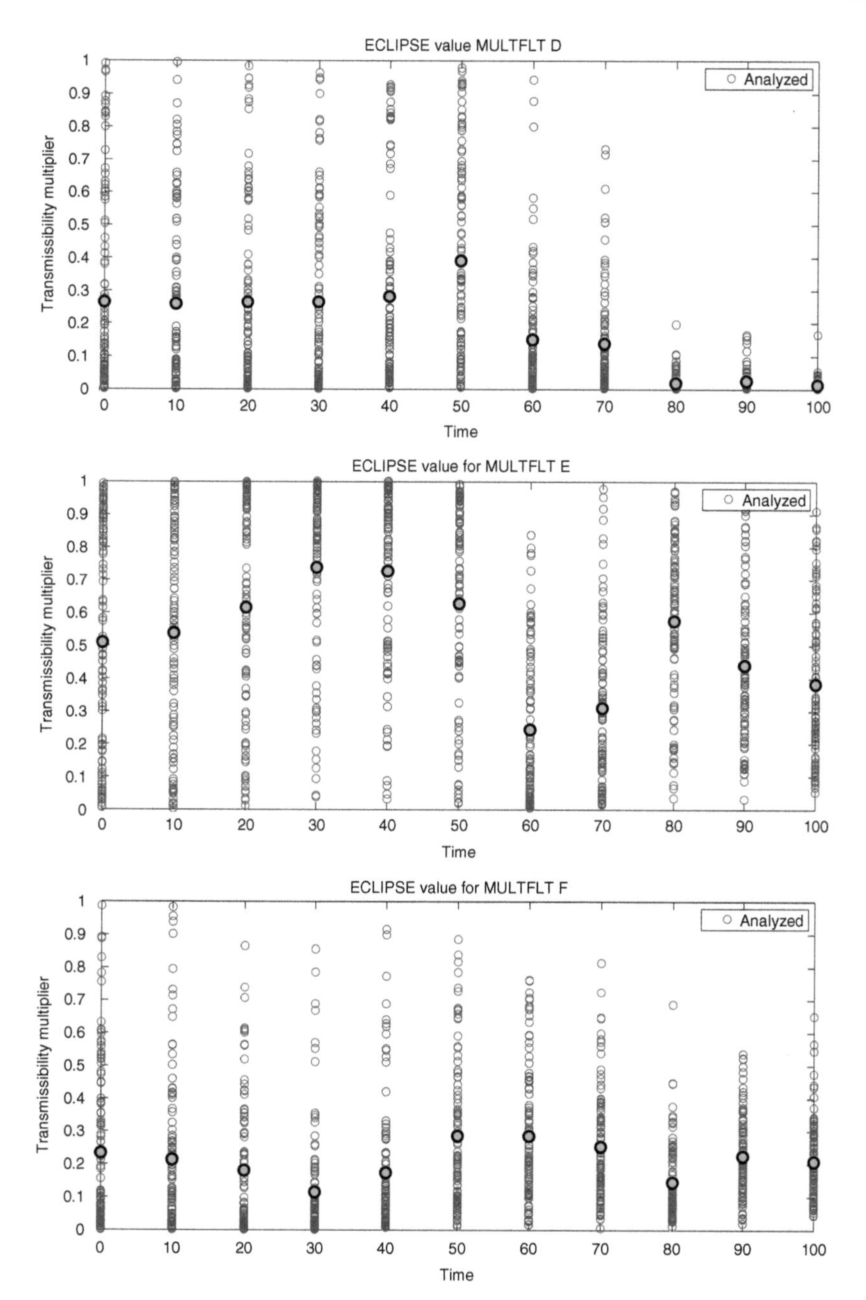

Figure 12.11 Sequential update of the fault transmissibility multiplier for faults D (upper plot), E (middle plot), and F (lower plot). The green circle indicates the ensemble mean.

The potential and advantages of the EnKF as an assisted history matching tool is demonstrated based on its capability of handling large parameter spaces, its sequential processing of measurements, and on the fact that it solves the combined state and parameter estimation problem as derived from a fundamental Bayesian formulation.

The EnKF provides an ensemble of updated reservoir realizations conditioned on production data, as well as improved estimates of the model parameters, the state variables, and their uncertainty. It forms an optimal starting point for computing predictions with uncertainty estimates.

It is demonstrated that the coarse-scale relative permeability properties can be estimated using the EnKF and that is an important parameter to consider in real field case applications. Assimilated production data are sensitive to relative permeability properties and estimating coarse-scale relative permeabilities can be crucial to obtain a satisfactory history match. Updating relative permeability properties by production data assimilation allows for the accounting of fluid forces, numerical dispersion and geological heterogeneity effects. In the presented field case, only one set of relative permeability curves is defined for the entire model. In the future, it might be valuable to differentiate between different regions, as well as horizontal versus vertical curves.

An improved parameterization for updating the fault transmissibility multipliers based on the use of transformations is presented. It is shown how the transformation enables us to successfully determine if a specific fault is open or closed with respect to flow.

References

Bennett, A.F. 1992 *Inverse Methods in Physical Oceanography*. Cambridge University Press.

Bennett A.F. 2002 *Inverse Modeling of the Ocean and Atmosphere*. Cambridge University Press.

Bianco A., Cominelli, A., Dovera, L., Naevdal, G. and Valles, B. 2007 History matching and production forecast uncertainty by means of the ensemble Kalman filter: a real field application. Paper SPE 107161 presented at the SPE Europec/EAGE Annual Conference and Exhibition, London, 11–14 June.

Brooks, R.H. and Corey, A.T. 1964 Hydraulic properties of porous media. *Hydrolo. Pap. 3*. Civ. Eng. Dep., Colo. State Univ., Fort Collins.

Burgers, G., van Leeuwen, P.J. and Evensen, G. 1998 Analysis scheme in the ensemble Kalman filter. *Mon. Weather Rev.* **126**, 1719–1724.

Doucet, A., de Freitas, N. and Gordon, N. (eds) 2001 *Sequential Monte Carlo Methods in Practice* Statistics for Engineering and Information Science. Springer-Verlag New York.

Eknes, M. and Evensen, G. 1997 Parameter estimation solving a weak constraint variational formulation for an Ekman model. *J. Geophys. Res.* **102**(C6), 12,479–12,491.

Evensen, G. 1994 Sequential data assimilation with a nonlinear quasi-geostrophic model using Monte Carlo methods to forecast error statistics. *J. Geophys. Res.* **99**(C5), 10,143–10,162.

Evensen, G. 2009 *Data Assimilation: The Ensemble Kalman Filter* (2nd eds). Springer-Verlag: Berlin.

Evensen, G. and van Leeuwen, P.J. 2000 An ensemble Kalman smoother for nonlinear dynamics. *Mon. Weather Rev.* **128**, 1852–1867.

Evensen, G., Hove, J., Meisingset, H., Reiso, E., Seim, K. and Espelid, Ø. 2007 Using the EnKF for assisted history matching of a North Sea reservoir model. Paper SPE-106184-MS presented at the SPE Reservoir Simulation Symposium, Woodlands, 26–28 February.

Eydinov, D., Gao, G., Li, G. and Reynolds, A.C. 2007 Simultaneous estimation of relative permeability and porosity/permeability fields by history matching prediction data. Petroleum Society's 8th Canadian International Petroleum Conference.

Goldberg, D.E. 1989 *Genetic Algorithms in Search, Optimization and Machine Learning*. Addison-Wesley Longman Publishing Co., Inc., Boston, MA, USA.

Gu, Y. and Oliver, D.S. 2005 History matching of the PUNQ-S3 reservoir model using the ensemble Kalman filter. *SPE Journal*, **10**(2), 217–224. SPE 89942-PA.

Haugen, V., Natvik, L.-J., Evensen, G., Berg, A., Flornes, K. and Nævdal, G. 2006 History matching using the ensemble Kalman filter on a North Sea field case. Paper SPE 102430 presented at the SPE Annual Technical Conference and Exhibition, San Antonio, 24–27 September.

Muccino, J.C. and Bennett, A.F. 2001 Generalized inversion of the Korteweg–de Vries equation. *Dyn. Atmos. Oceans* **35**, 227–263.

Nævdal, G., Mannseth, T. and Vefring, E. 2002 Near well reservoir monitoring through ensemble Kalman filter. Paper SPE 75235 presented at the SPE/DOE Improved Oil Recovery Symposium, Tulsa, 13–17 April.

Okano, H., Pickup, G., Christie, M.A., Subbey, S. and Monfared, H. 2005 Quantification of uncertainty in relative permeability for coarse-scale reservoir simulation. Paper SPE 94140 presented at the SPE Europec/EAGE Annual Conference, Madrid, 13–16 June.

Sasaki, Y. 1970 Numerical variational analyses with weak constraint and application to surface analyses of severe storm gust. *Mon. Weather Rev.* **98**, 899–910.

Skjervheim, J.A., Evensen, G., Aanonsen, S.I., Ruud, B.O. and Johansen, T.A. 2007 Incorporating 4D seismic data in reservoir simulation models using ensemble Kalman filter. *SPE journal*, **12**(3), 282–292. SPE-95789-PA.

Chapter 13

Optimal Experimental Design for the Large-Scale Nonlinear Ill-Posed Problem of Impedance Imaging

L. Horesh,[1] E. Haber[2] and L. Tenorio[3]
[1] *IBM T.J. Watson Research Center, USA*
[2] *University of British Columbia, Canada*
[3] *Colorado School of Mines, USA*

13.1 Introduction

Many theoretical and practical problems in science involve acquisition of data via indirect observations of a model or phenomena. Naturally, the observed data are determined by the physical properties of the model sought, as well as by the physical laws that govern the problem, however, they also depend on the experimental configuration. Unlike the model and the physical laws, the latter can be controlled by the experimenter. It is this possibility of controlling the experimental setup that leads to optimal experimental design (OED).

Large-Scale Inverse Problems and Quantification of Uncertainty Edited by L. Biegler, G. Biros, O. Ghattas, M. Heinkenschloss, D. Keyes, B. Mallick, Y. Marzouk, L. Tenorio, B. van Bloemen Waanders and K. Willcox

Optimal experimental design of well-posed inverse problems is a well established field (e.g., Fedorov (1972); Pukelsheim (2006) and references therein), yet, despite the practical necessity, experimental design of ill-posed inverse problems and in particular ill-posed nonlinear problems has remained largely unexplored. We discuss some of the intrinsic differences between well- and ill-posed experimental designs, and propose an efficient and generic framework for optimal experimental design of nonlinear ill-posed problems in the context of impedance tomography.

We use discrete data modeled as

$$d = F(m; y) + \varepsilon, \tag{13.1}$$

where d is a vector of observed data, m is a vector of the model to be recovered, F is an observation operator (aka the forward operator), y is a vector (or matrix) of experimental design parameters, and ε is a random noise vector. The design parameters y determine the experimental configuration. In the context of tomography these may define, for example, source and receiver positions, source frequencies or amplitudes. In other contexts such parameters may be ambient temperature, pressure, or any other controllable experimental setting. We assume that the problem is ill-posed in the usual sense (e.g., Hansen (1998)). In particular, there is no unique stable solution for the problem of finding m such that $F(m; y) = d$. Regularization techniques are used for finding stable and meaningful solution of such ill-posed problems. We consider a generalized Tikhonov regularization approach where the solution $\widehat{m}$ is recovered by an optimization problem of the form

$$\widehat{m} = \underset{m}{\operatorname{argmin}} \left\{ \frac{1}{2} \left\| F(m; y) - d \right\|^2 + R(m) \right\}, \tag{13.2}$$

with R being a regularization functional utilized for the incorporation of a-priori information.

Most efforts in the field of inverse problems have been primarily devoted to the development of either numerical methods (Biros and Ghattas 2005; Haber and Ascher 2001; Vogel 2001) for the solution of the optimization problem (13.2), or methods for prescribing regularization functionals and selecting their regularization parameters (Claerbout and Muir 1973; Tikhonov and Arsenin 1977; Golub *et al.* 1979; Hanke and Hansen 1993; Hansen 1998; Rudin *et al.* 1992;). Little work has addressed the selection strategy of the experimental settings y even though a proper design inherently offers improved information extraction and bias control. Figure 13.1 demonstrates the importance of experimental design in impedance tomography. The images display three recovered models; each corresponds to a particular experimental design setting, but are conducted over the very same true model. If these models were acquired for presurgical assessment of brain tumor removal, for instance, the neurosurgeon would be faced with the serious problem of deciding which region to operate.

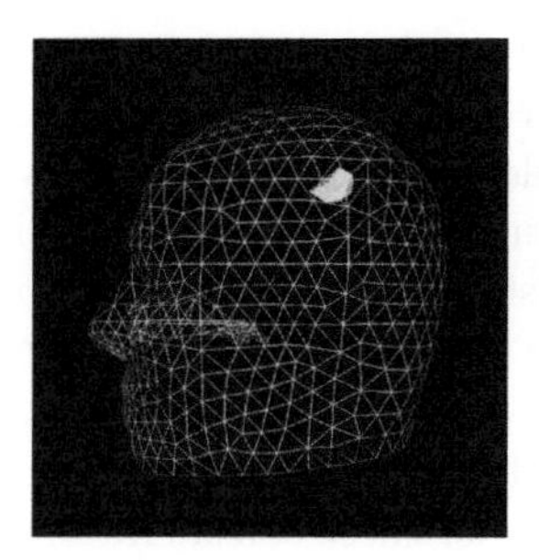
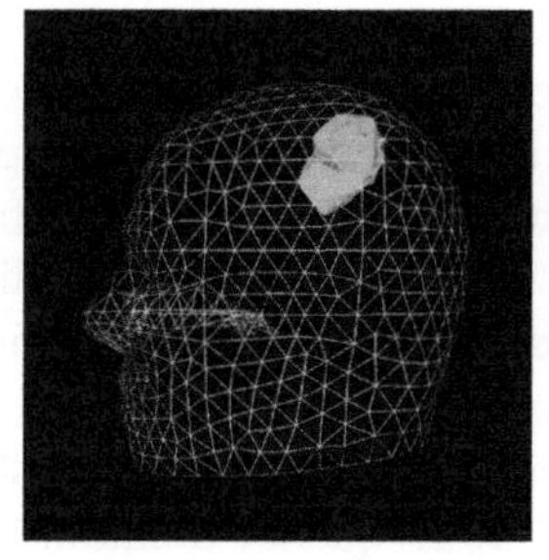
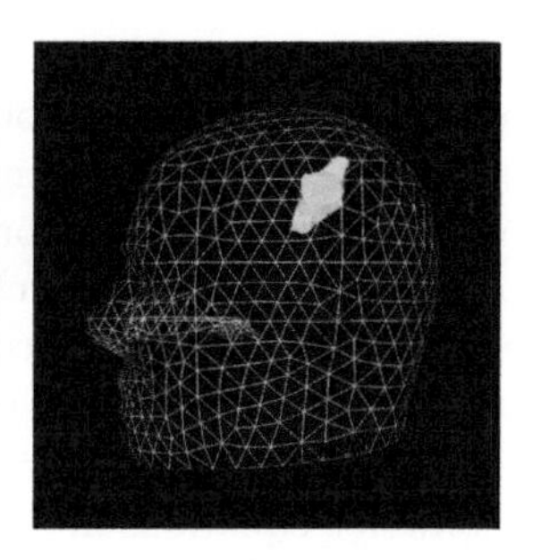

Figure 13.1 Impedance Tomography image recovery of the human head. The above models were recovered using three different data sets corresponding to three different experimental designs. Despite the great difference in localization and intensity of the localized brain tumor, the data sets were in fact acquired from the same patient.

While optimal experimental design for well-posed (over-determined) problems is well established (Pukelsheim 2006), the authors are aware of very few publications (e.g., Bardow (2008); Curtis (1999); Haber *et al.* (2008)) that address ill-posed (under-determined) problems and of even fewer that consider nonlinear problems. Still, many practical problems in biomedical imaging, geophysics, engineering as well as other physical sciences are intrinsically ill-posed and nonlinear.

In this chapter we propose a formulation for optimal experimental design of ill-posed, nonlinear inverse problems and demonstrate its application to impedance tomography. The rest of the paper is organized as follows: In Section 13.2 we introduce the basic impedance tomography problem. In Section 13.3 we briefly review different optimal experimental design formulations for linear ill-posed problems and describe the shortcomings of these approaches in overcoming the unique difficulties that arise in nonlinear ill-posed problems. In Section 13.4 we provide the rationale for our proposed formulation that is based on statistical considerations. In Section 13.5 we describe the mathematical formulation of the suggested approach and explain the optimization framework developed for its solution. Section 13.6 provides numerical results for a realistic impedance tomography model problem. In Section 13.7 we discuss the numerical results as well as various technical matters related to the implementation of such formulation. We close with some comments regarding future research.

13.2 Impedance Tomography

The inverse impedance tomography problem aims at the recovery of an impedance distribution in a domain given a set of boundary potential measurements for a known configuration of sources and receivers (Holder

2005). Acquisition is based on injection of low frequency currents from multiple sources while boundary potentials are recorded by a set of receivers. Typically, the sources and receivers are both deployed over the boundary of the domain. In the context of the estimate defined by (13.2), the model m stands for the conductivity (or admittivity) distribution in the domain and the data vector d corresponds to the electric potentials measured on the boundary.

The observation (forward) model under consideration is derived from Maxwell's equations in the low frequency domain. Under these considerations these entail the partial differential equation

$$\begin{aligned} \nabla \cdot (m \boldsymbol{\nabla} u) = 0 \qquad & u \in \Omega \\ m \boldsymbol{\nabla} u = Q \qquad & u \in \partial\Omega, \end{aligned} \tag{13.3}$$

where the model m is the conductivity, u the electric potential and Q are the sources.

Standard finite volume or finite element discretization (Ascher 2008; Vauhkonen *et al.* 1999; Vavasis 1996) leads to the construction of a discrete linear system. This system is solved for u given the conductivity model m:

$$A(m)u = D^\top M(m) D = P_Q Q, \tag{13.4}$$

where D is a sparse matrix holding the gradients of the basis functions (shape gradients) and $M(m)$ is a diagonal matrix that holds the conductivity values multiplied by the support (volumes) of the elements.

The forward model can be written as

$$F(m, V, Q) = V^\top P_V^\top A(m)^{-1} P_Q Q, \tag{13.5}$$

where the matrices P_Q and P_V map the source and measurement distributions onto the model grid.

In the context of the specific given forward model, another important physical restriction is current conservation. We assume that there is no dissipation of energy through the medium, and therefore the total amount of current driven in is equal to the total amount of current drained out. This constraint implies that $e^\top Q = 0$, where e is a vector of ones.

13.3 Optimal Experimental Design: Background

We briefly review the basic framework of OED in the context of discrete linear well-posed inverse problems where $F(m; y) = K(y)m$.

13.3.1 Optimal Experimental Design for Well-Posed Linear Problems

The data are modeled as

$$d = K(y)\, m + \varepsilon, \tag{13.6}$$

where $K(y)$ is an $\ell \times k$ matrix representation of the forward operator that acts on the model m and depends on a vector of experimental parameters y. The noise vector ε is assumed to be zero mean with independent entries of known variance σ^2. The objective of experimental design is the selection of y that leads to an optimal estimate of m.

In the well-posed case, the matrix $K(y)^\top K(y)$ is nonsingular and for fixed y the least squares (LS) estimate $\widehat{m} = (K^\top K)^{-1} K^\top d$ is unbiased with covariance matrix $\sigma^2\, C(y)^{-1}$, where $C(y) = K(y)^\top K(y)$. One can then choose y so as to obtain a good LS estimate of m. Since $\widehat{m}$ is unbiased, it is common to assess its performance using different characteristics of its covariance matrix. For example, an A-optimal type experiment design prescribes a choice of y that minimizes the trace of $C(y)^{-1}$. If instead of the trace, the determinant or the ℓ_2-norm of $C(y)^{-1}$ is used, then the design is known as D- or E-optimal, respectively (see Pukelsheim (2006) and references therein). Unsurprisingly, each optimality criterion corresponds to a different statistical interpretation: the A-optimal criterion minimizes the average variance, the D-optimality relates to the volume of an uncertainty ellipsoid (confidence region) (Beck and Arnold 1977), and the E-optimality criterion corresponds to a minimax approach. Other than the aforementioned optimality criteria, almost a complete alphabet of criteria can be found in the literature.

13.3.2 Optimal Experimental Design for Linear Ill-Posed Problems

Designs that are solely based on the covariance matrix are unsuitable for ill-posed problems where estimators of m are most likely biased. In fact, the bias may be the dominant component of the error.

Assume now that $K(y)^\top K(y)$ is singular or ill-conditioned. A regularized estimate of m can be obtained using penalized LS (Tikhonov regularization) with a smoothing penalty matrix L

$$\widehat{m} = \arg\min \frac{1}{2} \left(K(y)m - d \right)^\top \left(K(y)m - d \right) + \frac{\alpha}{2} \| \, Lm \, \|^2,$$

where $\alpha > 0$ is a fixed regularization parameter that controls the balance between the data misfit and the smoothness penalty. Assuming that $K(y)^\top K(y) + \alpha\, L^\top L$ is nonsingular, the estimator is

$$\widehat{m} = \left(K(y)^\top K(y) + \alpha\, L^\top L \right)^{-1} K(y)^\top d, \tag{13.7}$$

whose bias can be written as

$$\text{Bias}(\widehat{m}) = \mathrm{E}\,\widehat{m} - m = -\alpha\,\left(K(y)^\top K(y) + \alpha\,L^\top L\right)^{-1} L^\top L\,m. \tag{13.8}$$

Since the bias is independent of the noise level, it cannot be reduced by averaging repeated observations. The noise level affects the variability of $\widehat{m}$ around its mean $\mathrm{E}\,\widehat{m}$. Thus, this variability and the bias ought to be taken into account when choosing an estimator of m.

Define $\mathcal{B}(y,m) := \|\,\text{Bias}(\widehat{m})\,\|^2$ and $\mathcal{V}(y) = \mathrm{E}\,\|\,\widehat{m} - \mathrm{E}\,\widehat{m}\,\|^2/\sigma^2$. The sum of these two error terms provides an overall measure of the expected performance of $\widehat{m}$. This is essentially the mean squared error (MSE) of $\widehat{m}$. More precisely, the MSE of $\widehat{m}$ is defined as $\mathrm{E}\,\|\,\widehat{m} - m\,\|^2$, which can also be written as

$$\begin{aligned}\text{MSE}(\widehat{m}) &= \mathrm{E}\,\|\,\widehat{m} - \mathrm{E}\,\widehat{m} + \mathrm{E}\,\widehat{m} - m\,\|^2 = \|\,\mathrm{E}\,\widehat{m} - m\,\|^2 + \mathrm{E}\,\|\,\widehat{m} - \mathrm{E}\,\widehat{m}\,\|^2 \\ &= \|\,\text{Bias}(\widehat{m})\,\|^2 + \mathrm{E}\,\|\,\widehat{m} - \mathrm{E}\,\widehat{m}\,\|^2 = \alpha^2\,\mathcal{B}(y,m) + \sigma^2\,\mathcal{V}(y). \end{aligned} \tag{13.9}$$

The overall idea is then to define optimization problems so as to control a measure of the performance of $\widehat{m}$ that takes into account its bias and stochastic variability. A natural choice would be the MSE, nevertheless, this measure depends on the unknown m itself. However, in many practical applications it is possible to obtain examples of plausible models. For example, there are geostatistical methods to generate realizations of a given media from a single image (Journel and Kyriakidis 2004; Sarma *et al.* 2007). Also, in many applications, and in particular in medical imaging, it is possible to obtain a set of likely example models $\mathcal{M} = \{m_1, \ldots, m_s\}$. In medical imaging such models can be proposed based on prognosis studies. These studies provide useful statistics of common pathologies and their classes, and therefore can be used for the construction of models associated with likely pathologies. Basically, this set of example models enables one to implicitly define regions of interest and similarly regions of lack of interest; either of which provides valuable guidance for the experimental setup. For some applications, the designer may even feel comfortable assigning a prior distributions to the models. This would be a natural approach in the context of experimental design as it is performed prior to data collection.

Let $\mathcal{M} = \{m_1, \ldots, m_s\}$ be examples of plausible models which will be henceforth called training models. As with Bayesian optimal designs (Chaloner and Verdinelli 1995), these models are assumed to be iid samples from an unknown multivariate distribution π. We use the following sample average of the MSE

$$\widehat{\mathcal{R}}(\,y\,) = \alpha^2\,\widehat{\mathrm{E}}_\pi\,\mathcal{B}(y,m) + \sigma^2\,\mathcal{V}(y). \tag{13.10}$$

where

$$\widehat{\mathrm{E}}_\pi\,\mathcal{B}(y,m) = \frac{1}{s}\sum_{i=1}^{s}\mathcal{B}(y,m_i). \tag{13.11}$$

$\widehat{\mathcal{R}}(y)$ is an unbiased estimator of the Bayes risk $\mathrm{E}_\pi \, \mathrm{MSE}(\widehat{m})$. This type of empirical approach is commonly used in machine learning, where estimators are trained using iid samples.

13.4 Optimal Experimental Design for Nonlinear Ill-Posed Problems

Nonlinear problems are inherently more difficult. For example, there is no closed form solution of the Tikhonov estimate $\widehat{m}$ and its MSE cannot be easily separated into bias and variance components $\mathcal{B}$ and $\mathcal{V}$; linearity allowed us to find closed expressions for the expected value over the noise distribution. This time we estimate such expected value in two stages. First, we generate a noise sample ε_j that is added to $F(m_i; y)$; second, we solve the inverse problem and obtain an estimate $\widehat{m}_{i,j}$, which is then compared to m_i. We repeat this procedure for multiple noise realizations and for each training model. This sequence defines an empirical estimate of $\widehat{\mathcal{R}}(y)$

$$\widehat{\mathcal{R}}(y; \mathcal{M}, \widehat{m}) = \frac{1}{2ns} \sum_{i=1}^{s} \sum_{j=1}^{n} \| \widehat{m}_{i,j}(y) - m_i \|^2. \tag{13.12}$$

Since the design problem may still be under-determined, the design parameter y might oscillate wildly and may generate designs that overfit the training set. It is therefore important to introduce additional constraints or preferences related to the design parameters.

Numerous desirable properties may be considered for favoring one design over another. Depending on the application, some reasonable and popular preferences may be: to shorten acquisition time, to reduce the number of sources/receivers or to achieve higher resolution and promote spatial selectiveness. All these preferences pursue, of course, the final objective of finding the best model estimate given the available resources.

For most geophysical problems as well as medical applications, control over the number of sources and/or receivers is desirable. Let Q and V be, respectively, the vectors of sources and receivers and set $y = \{Q, V\}$. A sparsity preference for the number of active sources and receivers can be formulated by the addition of an ℓ_1-norm penalty to the empirical risk

$$\widehat{\mathcal{R}}_1(y; \mathcal{M}, \widehat{m}) = \frac{1}{2ns} \sum_{i=1}^{s} \sum_{j=1}^{n} \| \widehat{m}_{i,j}(y) - m_i \|^2 + \beta \, \|y\|_1 \, . \tag{13.13}$$

The idea of sparsifying by ℓ_1 penalty in optimal-control is analyzed in Stadler (2006). The design problem dictates the minimization of this regularized empirical risk while retaining the conventional inverse problem (13.2) and forward model feasible. This requirement can be formulated as a bi-level optimization problem (Alexandrov and Dennis 1994; Dempe 2000; Bard 2006; Colson *et al.* 2007; Dempe and Gadhi 2007) that reads as follows

$$\min_{y} \frac{1}{2ns} \sum_{i=1}^{s} \sum_{j=1}^{n} \| \widehat{m}_{i,j}(y) - m_i \|^2 + \beta \|y\|_1 \tag{13.14}$$

$$\text{s.t. } \widehat{m}_{i,j} = \operatorname*{argmin}_{m} \ \frac{1}{2} \|F(m;y) - d_{i,j}(m_i;y)\|^2 + R(m), \tag{13.15}$$

where $d_{i,j}(m_i;y) = F(m_i;y) + \varepsilon_j$. With a slight abuse of notation, we shall henceforth denote $\mathcal{R}_1$ by $\mathcal{R}$, and write

$$\psi(m;m_i,y) = \frac{1}{2} \|F(m;y) - d_{i,j}(m_i;y)\|^2 + R(m).$$

Before we proceed with discussing an appropriate optimization framework for the solution of the above problem, we would like to make an important distinction. Beyond the fundamental difference between linear and nonlinear inversion that distinguishes this work from our previous work (Haber *et al.* 2008), another important difference is related to the design mechanism itself. In the previous work, the aim was finding the best subset out of a given set of excitation and measurement vectors, while here their number is predefined, while their content is controlled by inducing sparsity. The current mechanism typically requires larger number of parameters, which, on the one hand offers greater solution flexibility and optimality but on the other hand, introduces higher computational demands.

In principle, it is also possible to employ Bayesian OED (Chaloner and Verdinelli 1995) for handling nonlinear OED problems of this sort. However, given the complexity of the nonlinear problem, one will most likely end up sampling models from the prior and follow the approach we propose here.

13.5 Optimization Framework

13.5.1 General Scheme

General bi-level optimization problems are difficult because the inner optimization problem may have multiple solutions and/or may be non-convex. Here, we make the assumption that the inner level is convex (or at least convex over the set of y's of interest) (Boyd and Vandenberghe 2004). This leads to simpler algorithms (Alexandrov and Dennis 1994). Thus, we replace the inner optimization problem by the necessary conditions for a minimum solving

$$\min_{y,m=(m_{i,j})} \quad \mathcal{R}(y;\mathcal{M}) = \frac{1}{2ns} \sum_{i=1}^{s} \sum_{j=1}^{n} \|m_{i,j} - m_i\|^2 + \beta \|y\|_1 \tag{13.16}$$

$$\text{s.t.} \quad c_{i,j}(m_{i,j},y) := \frac{\partial \psi_{i,j}(m;m_i,y)}{\partial m_{i,j}} = 0. \tag{13.17}$$

The necessary conditions for a minimum are (Ito and Kunisch 2008)

$$\mathcal{R}_{m_{i,j}} + (c_{i,j})_{m_{i,j}}^{\top} \lambda_{i,j} = 0 \tag{13.18a}$$

$$\mathcal{R}_y + \sum_{i,j}^{s,n} (c_{i,j})_y^{\top} \lambda_{i,j} = 0 \tag{13.18b}$$

$$c_{i,j}(m_{i,j}, y) = 0, \tag{13.18c}$$

where $\lambda_{i,j}$ are Lagrange multipliers. Although it is possible to develop an efficient method to solve the system for $m_{i,j}$ and y simultaneously, we have chosen an unconstrained optimization approach. This approach is advantageous for two reasons: first, the infrastructure of the forward simulator can be reused to solve the adjoint equation and therefore major implementation issues associated with solution of the optimality conditions as one system are avoided. Second, since the storage of all $m_{i,j}$'s and $\lambda_{i,j}$'s is not required, the complete optimization procedure can be executed by solving each inverse problem separately. More precisely, we first solve the $s \times n$ decoupled nonlinear systems (13.18c) for $m_{i,j}$ given y, then we solve the other $s \times n$ decoupled linear systems (13.18a) for $\lambda_{i,j}$. For each $m_{i,j}$ and $\lambda_{i,j}$ computed, the reduced gradient (13.18b) is updated. Thus, even when the number of training models and noise realizations is excessively large, such a design process can be conducted over modest computational architectures.

In order to avoid the non-differentiability nature of the ℓ_1-norm at zero, we use the Iterated Re-weighted Least Squares (IRLS) approximation (O'Leary 1990; Street *et al.* 1988). This approach has been successfully used for ℓ_1 inversion in many practical scenarios (Sacchi and Ulrych 1995; Vogel 2001; Whittall and Oldenburg 1992). In the IRLS approximation, the ℓ_1-norm is replaced by a smoothed version $\|x\|_{1,\varepsilon}$ of the absolute value function defined as:

$$|t|_\varepsilon := \sqrt{t^2 + \varepsilon} \quad \text{and} \quad \|x\|_{1,\varepsilon} := \sum_i |x_i|_\varepsilon.$$

As discussed in Haber *et al.* (2000), obtaining the sensitivities of $m_{i,j}$ with respect to y is straightforward; they can be written as

$$J = -\sum_{i,j}^{s,n} (c_{i,j})_{m_{i,j}}^{-1} (c_{i,j})_y.$$

The sensitivities are then used with IRLS (O'Leary 1990) to obtain an approximation of the Hessian

$$H = J^{\top} J + \beta \operatorname{diag}\left(\min(|y|^{-1}, \varepsilon^{-1})\right),$$

which in turn is used to define the update

$$y \leftarrow y - \tau H^{-1} g, \tag{13.19}$$

where τ is a line search step size and $H^{-1}g$ is computed using the conjugate gradient method. Note that computation of the update step can be performed without the explicit construction of either the matrix $(c_{i,j})_{m_{i,j}}$ or its inverse. These matrices are only accessed implicitly via matrix vector products.

13.5.2 Application to Impedance Tomography

We now discuss a specific application of the design strategy to impedance tomography. As previously discussed, we consider an experimental design problem of placing and activating sources and receivers over a predefined region of permissible locations. Such region could be the entire boundary $\partial\Omega$ of a domain Ω as often occurs in medical imaging or, alternatively, any sub-domain $\Omega_r \subseteq \Omega$.

We consider the Tikhonov regularized estimate (13.2) with $F(m; y)$ representing the forward operator (13.5) and $R(m) = \alpha \|Lm\|^2$, with L a discrete derivative operator that penalizes local impedance inhomogeneities.

At this point we comment on the choice of the regularization parameters α and β. The parameter α is chosen prior to the inversion process; we choose it so as to balance the expected misfit with the norm of $R(m)$. For iid Gaussian noise the expected value of $\|d - F(m; y)\|^2$ is $\ell\sigma^2$. Choosing $\alpha \approx \ell\sigma^2/R(m)$ yields reasonable model reconstructions. The choice of β is user-dependant. The designer may tune this parameter according to some sparsity preferences. The higher the sparsity of the active sources and receivers, the larger the value of β. Conversely better model recovery can be achieved with lower levels of β at the expense of lower sparsity. Further discussion regarding the choice of β for OED of linear problems can be found in Haber *et al.* (2008).

For our impedance tomography design problem, the optimality criterion prescribes the construction of an experimental design that minimizes the number of active sources and receivers, as well as minimizing the description error between the recovered and given training models. This configuration must comply with a feasible forward model and an inverse solution, and should be consistent with the acquisition noise level. Such design can be obtained with a sparsity requirement imposed over the source and measurement vectors in the form of ℓ_1-norm penalty. Thus, we set $y := \{V, Q\}$ and accordingly denote the forward operator by $F(m; V, Q)$. We assume that a collection of feasible, representing models $\mathcal{M} = \{m_1, \ldots, m_s\}$ is at our disposal, for which we can compute a measurement set using the forward model $F(m; V, Q)$.

The design problem given in (13.14)–(13.15) can be formulated as follows

$$\min_{V,Q} \ \mathcal{R}(V, Q; \mathcal{M}, \widehat{m}) \tag{13.20a}$$

$$\text{s.t.} \ \widehat{m} = \underset{m}{\operatorname{argmin}} \ \psi(m; V, Q), \tag{13.20b}$$

where the objective function (i.e. the regularized empirical risk $\mathcal{R}$) and the linearized constraints c, as in (13.16) and (13.17) respectively, are given by

$$\mathcal{R}(V,Q;\mathcal{M},m) = \frac{1}{2}\sum_{i,j}^{s,n} \|m_{i,j} - m_i\|^2 + \beta_1 \|V\|_1 + \beta_2 \|Q\|_1$$
$$c_{i,j}(m_{i,j};V,Q) = S_{i,j}^\top V^\top P_V^\top \left((A(m_{i,j})^{-1} - A(m_i)^{-1})P_Q Q - \varepsilon_j\right) + \alpha L^\top L m_{i,j},$$

where $S_{i,j}$ is the sensitivity matrix, that is the Fréchet derivative of the forward operator. This operator represents the sensitivity of the acquired data to small changes in the model $S_{i,j} := \partial d_{i,j}/\partial m$. For this formulation, $S_{i,j}$ can be derived by implicit differentiation using the relation (13.5) (Haber *et al.* 2000).

In order to solve (13.16)–(13.17), we shall now evaluate the remaining components of (13.18a)–(13.18b). Since the empirical risk is convex on $m_{i,j}$, we have $\mathcal{R}_{m_{i,j}} = m_{i,j} - m_i$. The derivatives of the risk with respect to the design parameters P and Q are obtained using the IRLS relation given in (13.5.1)

$$\mathcal{R}_V = \beta_1 \mathrm{diag}\left(\frac{1}{|V|_{1,\varepsilon}}\right) V, \quad \mathcal{R}_Q = \beta_2 \mathrm{diag}\left(\frac{1}{|Q|_{1,\varepsilon}}\right) Q.$$

The derivative of the linearized constraint $c_{i,j}$ with respect to the model $m_{i,j}$ is in fact the conventional Hessian of the inverse problem (13.2)

$$(c_{i,j})_{m_{i,j}} = S_{i,j}^\top S_{i,j} + \alpha L^\top L + M_{i,j},$$

where $M_{i,j}$ stands for the second order derivatives that can be computed implicitly (Haber *et al.* 2000). The derivatives $(c_{i,j})_V$ and $(c_{i,j})_Q$ can be calculated from the structures of $S_{i,j}$ and $M_{i,j}$. The reduced space gradients $g(V)$ and $g(Q)$ are obtained using the relations

$$g(V) = \mathcal{R}_V + \sum_{i,j}^{s,n} (c_{i,j})_V^\top (c_{i,j})_{m_{i,j}}^{-\top} \mathcal{R}_{m_{i,j}}$$
$$g(Q) = \mathcal{R}_Q + \sum_{i,j}^{s,n} (c_{i,j})_Q^\top (c_{i,j})_{m_{i,j}}^{-\top} \mathcal{R}_{m_{i,j}}.$$

Next, the reduced Hessians H_V and H_Q can be derived as in (13.5.1) and, similarly, the updates for V and Q at each iteration of the reduced space IRLS as in (13.19).

Remark. We have used a conventional reparametrization of the model space in order to comply with the broad dynamic range requirement and therefore set $m := \log(\gamma)$, where γ is the conductivity.

13.6 Numerical Results

We now show results of applying the optimal experimental design framework described in Section 13.5.2 to a realistic design problem of placing and activating sources and receivers for impedance imaging of a human head.

The setup involves two stages. In the first stage the source and receiver vectors are optimized to comply with the sparsity requirement using a set of given training models. In the second stage, several unseen (test) models are used to assess the performance of the obtained design.

For training purposes, three example models of impedance perturbations in the mid, left and right parietal brain lobe (marked in red on the head diagrams in Figure 13.4) are considered. The design problem consists of 16 receiver vectors V and 16 source vectors Q that were initialized at random.

Figure 13.2 shows the behavior of the empirical risk and MSE as a function of iteration number. The regularized empirical risk is reduced by an order of magnitude in the first six iterations, while the MSE itself was reduced by a factor of about two. On the 6th iteration the MSE starts to increase. At this point the tradeoff between sparsity and reconstruction accuracy begins to be evident. We stopped the optimization process when the relative change in the risk dropped below 1%.

Figure 13.3 shows the ensemble of source Q and measurement vectors V for the initial and optimized designs. The figure shows that the optimized vectors are clearly sparser than those obtained with the initial design. Another way to visualize the difference in the designs is shown in Figure 13.4, where 4 out of the 16 source vectors are plotted as colored circles at the active source locations. Again, it can be clearly observed that the optimized design requires

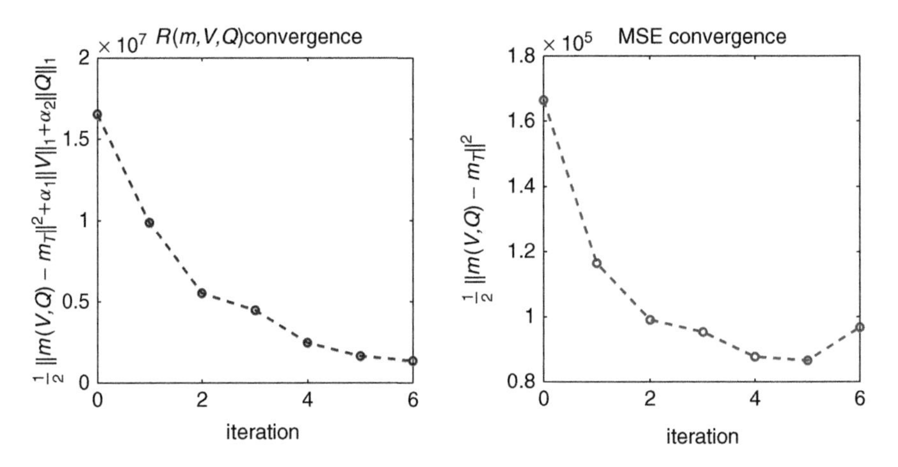

Figure 13.2 Experimental design of 16 receiver and 16 source vectors for three models. Left: Empirical risk convergence; Right: MSE convergence.

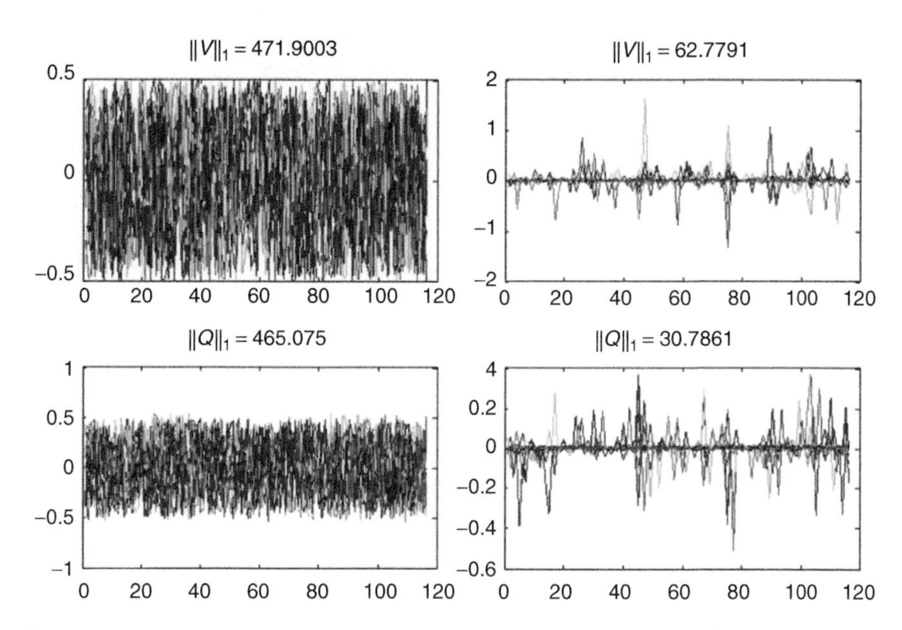

Figure 13.3 Sparsity pattern of 16 receiver vectors V (above) and 16 source vectors Q (below) for three training models. Left: Initial experimental design. Right: Optimized design after six iterations.

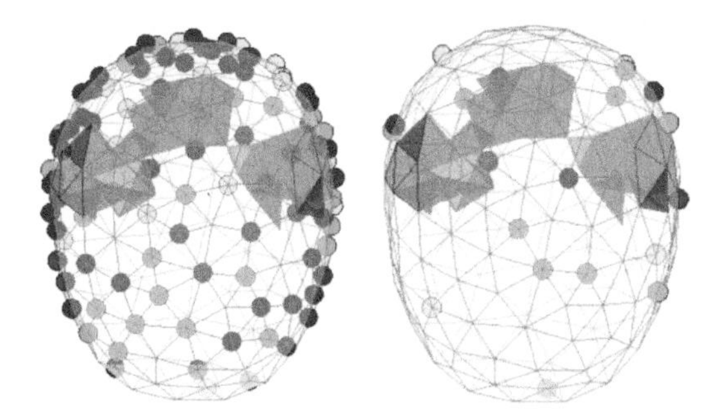

Figure 13.4 Human-head-shaped training models from a transversal projection. The red objects inside the head represent impedance perturbations as provided by the training example models. The colored dots located on vertices of the head mesh represent dominant locations for positioning sources (Only 4 out of the 16 vectors (colors) were presented here to avoid over-cluttering). Left: Initial design. Right: Optimized design.

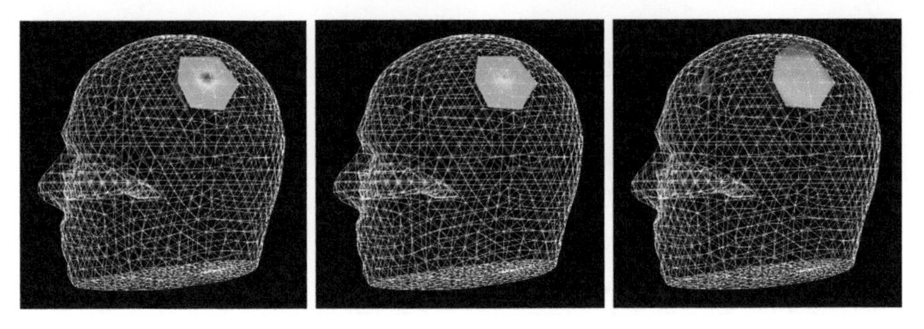

Figure 13.5 Testing set performance assessment for 16 receiver vectors V and 16 source vectors Q configuration. Left: True test model; Middle: Optimally designed recovered test model; Right: test model recovered using the initial design.

the deployment of a smaller number of active sources, and thereby complies with our original intent.

While the first three figures illustrate the performance of the design procedure at the learning stage, the performance of the obtained design for an unseen (testing) model is shown in Figure 13.5. It presents a comparison between a 3D impedance tomography model that was recovered using the initial and optimal designs. The impedance intensities of the perturbations recovered with the optimized design are distinctly better than those obtained with the initial design. The quantitative results for models obtained with the optimized design indicate that the recovered models were almost identical to the true model. Moreover, while models recovered using the initial design were cluttered with artifacts, models recovered with the optimized design show higher qualitative structural similarity with the true models.

13.7 Discussion and Conclusions

We have presented a statistical formulation for optimal experimental design of nonlinear ill-posed problems. Such problems frequently arise in medical imaging and geophysics, but also in many other fields such as chemistry and aerospace engineering. For each of the different applications, the construction of an effective design procedure requires careful thought. Nevertheless, the rationale presented in this study is generic and therefore transferable to a broad range of important applications.

The numerical results have shown that the proposed optimal experimental design can provide substantial improvement in image quality, on both the quantitative and qualitative levels. But there are several outstanding issues that still need to be addressed. The first of which is the choice of regularized empirical risk measure. This choice is crucial for obtaining an effective design. We have used an ℓ_2-norm discrepancy between training and recovered models for the empirical risk measure, and an ℓ_1-norm penalty on y to promote sparsity. These choices may seem natural for the impedance design problem considered here but in general this choice will depend on the specific application.

Obviously, the choice of regularized estimate of the inverse problem is also important. We have chosen Tikhonov regularization with an ℓ_2 data misfit and smoothness penalty to promote smooth model estimates. Such smoothness may not be appropriate in some applications. In the numerical examples above, all model examples comprised sharp parameter changes, which may have impaired, to some extent, the overall performance of the design procedure.

One defining feature of the proposed formalism is that the choice of inversion method is embedded in the design problem. Thus, different inversion frameworks will inherently result in different designs. The fact that the design reflects the inversion settings may be regarded as a hindrance, as one may expect the design to be solely related to the physical properties of the problem. Yet, in reality the choice of an inversion framework is mandatory. Moreover, since each inversion framework introduces distinctive a-priori information and individual degrees of belief one has attributed to the solution, the incorporation of such knowledge (if justified) into the design is essential.

Another fundamental issue is the source of training models. We have deliberately avoided a discussion about the retrieval of proper training examples as it is very much problem dependent issue. It was assumed that these were provided by a trusted source and as such were representative of the type of model to be recovered.

In addition to the source of models, their number is an important variable that determines how well the empirical risk approximates the true risk. A careful study of this question will require empirical process theory as in Vapnik (1995) and Rakhlin (2006).

Since several nonlinear inversion procedures are required at each design iteration, the design process is time consuming. Apart from the explicit construction of the forward model discretization matrix, the implementation of a bi-level optimization framework in this study relied on implicit calculation of matrices and parallel processing of the inversions of independent example models. This implementation enabled the processing of large-scale models despite memory limitations. Nevertheless, the overall design process is still computationally intensive and time consuming. For several applications, such as geophysical surveys, offline processing time of several hours or even a few days prior to a survey is certainly acceptable but, for some medical applications, development of faster design procedures may be essential.

Another issue that needs to be addressed is sensitivity analysis. A better insight into the interplay between the design requirement and the sensitivity gain/loss can be obtained by observing changes in the sensitivity map that evolves throughout the design process. Such analysis can illustrate and explain how sensitivity and independence of measurable information originating in regions of interest (i.e. where perturbations are expected) form higher local sensitivity compared with irrelevant areas.

The practical aim of this work is mainly to help improve current designs and, therefore, whenever the designer is able to provide a good initial design, the formulation provided here will yield an improved experimental design. In many practical settings improvement over a design at hand may suffice and, therefore, an exhaustive solution of the design problem may not be needed.

Acknowledgements

The authors wish to thank M. Benzi, R. Horesh, J. Nagy and A. Veneziani for their valuable advice and criticism of this work. In addition we wish to express our gratitude to our funding bodies: DOE (DE-FG02-05ER25696-A002), NSF (DMS 0724759, 0724717 and 0724715, CCF 0427094 and 0728877) and NIH (HL085417-01A2).

References

Alexandrov, N. and Dennis, J.E. 1994 Algorithms for bilevel optimization. In *AIAA/USAF/NASA/ISSMO Symposium on Multidisciplinary Analysis and Optimization*, pp. 810–816.

Ascher, U.M. 2008 *Numerical Methods for Evolutionary Differential Equations*. SIAM.

Bard, J.F. 2006 *Practical Bilevel Optimization: Algorithms and Applications (Nonconvex Optimization and its Applications)*. Springer.

Bardow, A. 2008 Optimal experimental design for ill-posed problems, the meter approach. *Computers and Chemical Engineering* **32**, 115–124.

Beck, J.V. and Arnold, K.J. 1977 *Parameter Estimation in Engineering and Science*. John Wiley & Sons, Ltd.

Biros, B. and Ghattas, O. 2005 Parallel Lagrange-Newton-Krylov-Schur methods for PDE-constrained optimization Parts I & II. *SIAM Journal on Scientific Computing* **27**, 687–713 & 714–739.

Boyd, S. and Vandenberghe, L. 2004 *Convex Optimization*. Cambridge University Press.

Chaloner, K. and Verdinelli, I. 1995 Bayesian experimental design: A review. *Statistical Science* **10**, 273–304.

Claerbout, J. and Muir, F. 1973 Robust modelling with erratic data. *Geophysics* **38**, 826–844.

Colson, B., Marcotte, P. and Savard, G. 2007 An overview of bilevel optimization. *Annals of Operations Research* **153**, 235–256.

Curtis, A. 1999 Optimal experimental design: a cross borehole tomographic example. *Geophysics J. Int.* **136**, 205–215.

Dempe, S. 2000 A bundle algorithm applied to bilevel programming problems with non-unique lower level solutions. *Comput. Optim. Appl.* **15**, 145–166.

Dempe, S. and Gadhi, N. 2007 Necessary optimality conditions for bilevel set optimization problems. *J. of Global Optimization* **39**, 529–542.

Fedorov, V.V. 1972 *Theory of Optimal Experiments*. Academic Press.

Golub, G.H., Heath, M. and Wahba, G. 1979 Generalized cross-validation as a method for choosing a good ridge parameter. *Technometrics* **21**, 215–223.

Haber, E. and Ascher, U. 2001 Preconditioned all-at-one methods for large, sparse parameter estimation problems. *Inverse Problems* **17**, 1847–1864.

Haber, E., Ascher, U.M. and Oldenburg, D.W. 2000 On optimization techniques for solving nonlinear inverse problems. *Inverse Problems* **16**, 1263–1280.

Haber, E., Horesh, L. and Tenorio, L. 2008 Numerical methods for experimental design of large-scale linear ill-posed inverse problems. *Inverse Problems* **24(5)**, 055012.

Hanke, M. and Hansen, P.C. 1993 Regularization methods for large-scale problems. *Surveys on Mathematics for Industry* **3**, 253–315.

Hansen, P.C. 1998 *Rank-Deficient and Discrete Ill-Posed Problems*. SIAM.

Holder, D.S. 2005 *Electrical Impedance Tomography - Methods, History, and Applications*. Institute of Physics.

Ito, K. and Kunisch, K. 2008 *Lagrange Multiplier Approach to Variational Problems and Applications*. SIAM.

Journel, A.G. and Kyriakidis, P. 2004 *Evaluation of Mineral Reserves: A Simulation Approach*. Applied Geostatistics Series. Oxford University Press.

O'Leary, D.P. 1990 Robust regression computation using iteratively reweighted least squares. *SIAM J. Matrix Anal. Appl.* **11**, 466–480.

Pukelsheim, F. 2006 *Optimal Design of Experiments*. SIAM.

Rakhlin, A. 2006 *Applications of Empirical Processes in Learning Theory: Algorithmic Stability and Generalization Bounds*. Phd Thesis, Dept. of Brain and Cognitive Sciences, Massachusetts Institute of Technology.

Rudin, L.I., Osher, S. and Fatemi, E. 1992 Nonlinear total variation based noise removal algorithms. *Physica D* **60**, 259–268.

Sacchi, M.D. and Ulrych, T.J. 1995 Improving resolution of Radon operators using a model re-weighted least squares procedure. *Journal of Seismic Exploration* **4**, 315–328.

Sarma, P., Durlofsky, L.J., Aziz, K. and Chen, W. 2007 *A New Approach to Automatic History Matching Using Kernel PCA. SPE Reservoir Simulation Symposium* Houston, Texas.

Stadler, G. 2006 Elliptic optimal control problems with ℓ_1 – control cost and applications for the placement of control devices. *Computational Optimization and Applications* 1573–2894.

Street, J.O., Carroll, R.J. and Ruppert, D. 1988 A note on computing robust regression estimates via iteratively reweighted least squares. *The American Statistician* **42**, 152–154.

Tikhonov, A.N., Arsenin, V.Y. 1977 *Solutions of Ill-Posed Problems*. Wiley.

Vapnik, V. 1995 *The Nature of Statistical Learning Theory*. Springer.

Vauhkonen, P.J., Vauhkonen, M., Savolainen, T. and Kaipio, J.P. 1999 Three-dimensional electrical impedance tomography based on the complete electrode model. *IEEE Trans. Biomedical Engineering* **46**, 1150–1160.

Vavasis, S.A. 1996 Stable finite elements for problems with wild coefficients. *SIAM J. Numer. Anal.* **33**, 890–916.

Vogel, C. 2001 *Computational Methods for Inverse Problems*. SIAM.

Whittall, K.P. and Oldenburg, D.W. 1992 *Inversion of Magnetotelluric Data for a One Dimensional Conductivity*. SEG Monograph V.5.

Chapter 14

Solving Stochastic Inverse Problems: A Sparse Grid Collocation Approach

N. Zabaras

Cornell University, USA

14.1 Introduction

In recent years there has been significant progress in quantifying and modeling the effect of input uncertainties in the response of PDEs using non-statistical methods. The presence of uncertainties is incorporated by transforming the PDEs representing the system into a set of stochastic PDEs (SPDEs). The spectral representation of the stochastic space resulted in the development of the Generalized Polynomial Chaos Expansion (GPCE) methods (Ghanem and Spanos 1991; Xiu and Karniadakis 2003). To solve large-scale problems involving high-dimensional stochastic spaces (in a scalable way) and to allow non-smooth variations of the solution in the random space, there have been recent efforts to couple the fast convergence of the Galerkin methods with the decoupled nature of Monte-Carlo sampling (Velamur Asokan and Zabaras 2005; Babuska *et al.* 2005a,b). The Smolyak algorithm has been used recently to build sparse grid interpolants in high-dimensional space (Nobile *et al.* 2008; Xiu and Hesthaven 2005; Xiu 2007). Using this method,

---

Large-Scale Inverse Problems and Quantification of Uncertainty Edited by L. Biegler, G. Biros, O. Ghattas, M. Heinkenschloss, D. Keyes, B. Mallick, Y. Marzouk, L. Tenorio, B. van Bloemen Waanders and K. Willcox

interpolation schemes for the stochastic solution can be constructed with orders of magnitude reduction in the number of sampled points. The sparse grid collocation strategy only utilizes solutions of deterministic problems to construct the stochastic solution. This allows the use of deterministic legacy codes in a stochastic setting. Furthermore, reduction in computational effort can result from adaptively constructing the representation based on the local behavior of the stochastic solution (Ma and Zabaras 2009). The mathematical framework to solve SPDEs is a thriving, mature field but with several open issues to be resolved (Ma and Zabaras 2009; Wan and Karniadakis 2005).

These developments in direct stochastic analysis raise the possibility of solving the corresponding stochastic inverse problem, specifically, *the development of a non-statistical framework that only utilizes deterministic simulators for the inverse analysis and/or design of complex systems in the presence of multiple sources of uncertainties.*

We are interested in two classes of stochastic inverse problems in one mathematical setting. The first class of problems is *stochastic inverse/estimation problems.* Consider a complex system that is inherently stochastic due to uncertainties in boundary conditions or property distribution. The problem is to reconstruct these *stochastic* boundary conditions, source terms, and/or property variations given statistical sensor data (in the form of moments or PDFs) of the dependent variables. Posing the problem in terms of available statistical sensor data encompasses most traditional scenarios: (i) when the sensor data is from a single experiment, where the stochasticity creeps in due to measurement errors/uncertainties, and (ii) when the sensor data is from statistics of multiple experiments, where the stochasticity creeps in due to uncertainties in both measurement and parameters (operating conditions, property variability).

The second class of problems is *stochastic design problems.* The motivation for posing these problems is to design operating conditions which ensure that a system exhibits a specific desired response even when the property variation in the system is uncertain. The goal is to construct the optimal stochastic input such that a system response is achieved in a probabilistic sense in the presence of other sources of uncertainty. Based on recent work in constructing realistic input models of topological and material/microstructure variability using limited and gappy data (Ganapathysubramanian and Zabaras 2008; 2007), the framework of interest has wide range applicability. Figure 14.1 shows two applications where the underlying geometric, topological and material uncertainties have to be accounted for during the design of the optimal operating conditions.

Deterministic inverse techniques (based on exact matching or least-squares optimization) lead to point estimates of unknowns without rigorously considering system uncertainties and without providing quantification of the uncertainty in the inverse or design solution (Sampath and Zabaras 2001). It is clearly necessary to incorporate the effects of uncertainty in property

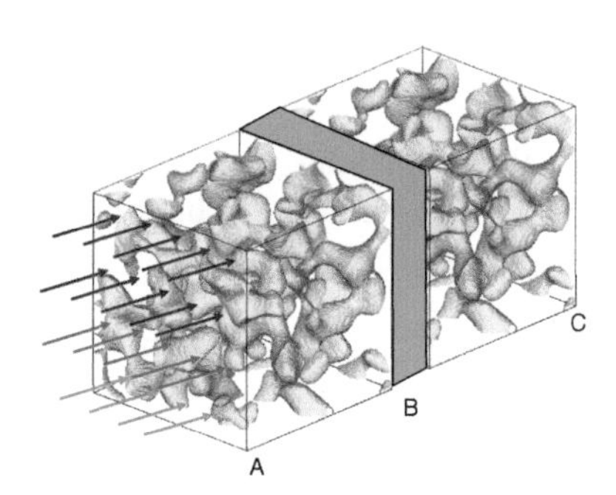

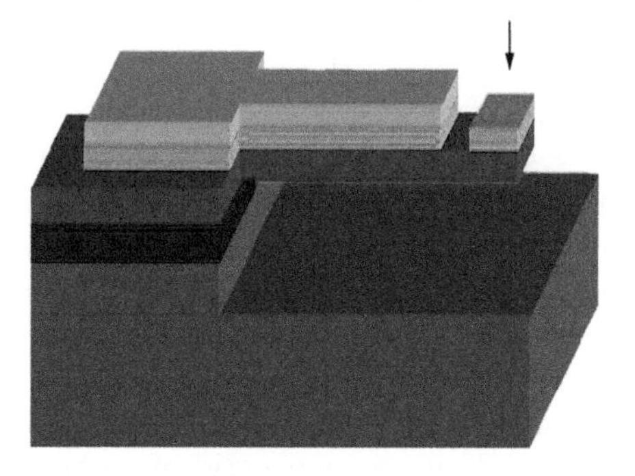

Figure 14.1 Two proof-of-concept design applications with topological, material and operating uncertainties. Left: A micro-scale heat-sink. The exact material distribution in the device is unknown (random heterogeneous material). However, we want to maintain a specific temperature profile in region 'B' where also the heat flux is zero. The stochastic design problem is to find the optimal heat flux at 'A' such that this condition is satisfied in the presence of material uncertainty. Right: The operating conditions (potential difference) across the MEMS cantilever-switch has to be designed in the presence of geometric and material uncertainties.

variation and operating conditions for the design of components and devices that perform satisfactorily in a variety of conditions. Beyond the obvious technological significance of such a framework, there are several mathematical challenges involved. New mathematical metrics and strategies have to be developed to pose and solve stochastic inverse and design problems. The high stochastic dimensionality of the direct stochastic problems necessitates the development of computationally tractable yet accurate optimization strategies with low storage overhead. One also needs to investigate the possibility of stochastic model reduction to alleviate these issues.

The framework discussed here is a non-statistical framework for stochastic design and estimation. This is in contrast to the Bayesian inference approach (Kaipio and Somersalo 2005; Somersalo and Calvetti 2007; Wang and Zabaras 2005a,b, 2006). There are a few issues with Bayesian inference that have motivated the development of alternate means of solving stochastic inverse/design problems. The higher-order statistics of the unknown are highly-dependent on the form of the prior distribution chosen. Furthermore, the inputs to this framework are a finite number of deterministic sensor measurements, i.e. this methodology provides no means of incorporating statistics and/or PDF of the measured quantities. Also, this framework has not been applied to design problems with multiple sources of uncertainty. In addition, a nonstatistical framework allows one the flexibility of working with different representation of uncertain fields and naturally incorporating the effects of correlation statistics. Finally, the probabilistic mathematical framework provides quantitative measures of convergence and the ability to selectively refine the stochastic solution.

Box 14.1 summarizes the main needed tasks for the design and control of systems in the presence of uncertainties. We will highlight a stochastic optimization framework where the uncertainty representation is based on a sparse grid collocation approach. Utilizing a sparse grid collocation strategy guarantees scalability, the ability to seamlessly incorporate multiple sources of uncertainty and more importantly relies entirely on *multiple calls to deterministic simulators* (Ma and Zabaras 2009). Issues with regularization and the mathematics of posing and solving the sensitivity equations will be considered. Using a sparse grid representation of the design variable, we propose to convert the stochastic optimization problem into a deterministic optimization problem in a higher dimensional space. This deterministic optimization problem will subsequently be solved using gradient based optimization. A method to compute the sensitivity of the dependent stochastic variables with respect to the estimated/design stochastic variables has to be formulated. This formulation arises naturally from posing the problem in the sparse grid framework and involves multiple calls to deterministic sensitivity problems. Stochastic model reduction strategies and efficient stochastic Hessian calculation strategies will accelerate the optimization framework while keeping the computational and storage overhead manageable.

Box 14.1 Tasks for the solution of stochastic inverse problems

- Develop a stochastic inverse/design framework based on recent advances in representing and solving SPDEs using nonstatistical, nonintrusive methods.
- Develop a mathematical framework that seamlessly utilizes deterministic legacy simulators (including off-the-shelf optimization algorithms) and incorporates multiple sources of uncertainty.
- Develop highly scalable algorithms for computing stochastic sensitivities and stochastic Hessian.
- Utilize stochastic model reduction strategies to accelerate the optimization framework.

14.2 Mathematical Developments

We herein introduce various mathematical and algorithmic aspects of posing and solving stochastic inverse problems. The critical questions to be addressed include: (a) representation of the stochastic estimated/design variables, (b) modeling the effect of multiple sources of uncertainties in the inverse solution and addressing the curse of dimensionality (c) the definition of appropriate stochastic metrics to compare stochastic solutions, (d) the definition of the stochastic sensitivity variables and equations and (e) the need for a mathematical framework for model reduction to accelerate the optimization problem for large-scale problems. Each of the above developments is described in some detail next.

14.2.1 The Stochastic Inverse Problem: Mathematical Problem Definition

Consider a PDE-defined system over a domain, $\mathcal{D}$. Assume that we are interested in the behavior of the system in a time interval $\mathcal{T}$. This system is affected by multiple sources of uncertainty. This is because the input conditions (initial conditions, boundary conditions, forces, material properties) cannot be known/imposed with infinite certainty. Denote the set of input conditions as $\{\boldsymbol{q}, \alpha\}$, e.g. q can represent a boundary forcing term and α a distributed material property. The dependent variable $\boldsymbol{u}$ is a function of space, time and the input parameters $\{\boldsymbol{q}, \alpha\}$. The evolution of the system is described by a set of coupled SPDEs as

$$\mathcal{B}(\boldsymbol{u}, \boldsymbol{x}, t : \{\boldsymbol{q}, \alpha\}) = 0, \tag{14.1}$$

with the appropriate boundary and initial conditions given by

$$\mathcal{L}(\boldsymbol{u}, \boldsymbol{x}, t : \{\boldsymbol{q}, \alpha\}) = 0. \tag{14.2}$$

To numerically solve Equations (14.1–14.2), it is necessary to have a finite dimensional representation of the input uncertainty Deb *et al.* (2001). Denote this finite dimensional representation as

$$\boldsymbol{q} = \mathcal{G}_1(\boldsymbol{x}, t, \boldsymbol{Y}_q), \ \boldsymbol{Y}_q \in \Gamma_q \quad (a), \qquad \alpha = \mathcal{G}_2(\boldsymbol{x}, t, \boldsymbol{Y}_\alpha), \ \boldsymbol{Y}_\alpha \in \Gamma_\alpha \quad (b), \tag{14.3}$$

where $\mathcal{G}_1 : \Gamma_q \rightarrow \Omega_q$, where Ω_q is the space of all possible realizations of $\boldsymbol{q}$. Similar definitions hold for $\mathcal{G}_2 : \Gamma_\alpha \rightarrow \Omega_\alpha$. The set of inputs belong to a tensor product space, i.e. they have spatial and temporal dependence ($\boldsymbol{q}, \alpha \in \mathcal{D} \otimes \mathcal{T}$) as well as depend on the input uncertainties ($\boldsymbol{q} \in \Gamma_q$, $\alpha \in \Gamma_\alpha$). By the Doob-Dynkin lemma, the dependent variable $\boldsymbol{u}$ also belongs to the corresponding tensor-product space, albeit one that is transformed by the differential operator, $\mathcal{B}$ ($\boldsymbol{u} \in \mathcal{D} \otimes \mathcal{T} \otimes \Gamma_\alpha \otimes \Gamma_q$).

Given a numerical representation of the input quantities (i.e. $\mathcal{G}_1$ and $\mathcal{G}_2$), we utilize sparse grid collocation strategies to solve this direct stochastic problem. The basic idea is to represent the input stochastic quantities in terms of a finite number of realizations and a set of multi-dimensional interpolating polynomials:

$$\boldsymbol{q} = \sum_{i=1}^{n_q} \boldsymbol{q}_i L_i^q(\boldsymbol{q}), \quad \alpha = \sum_{i=1}^{n_\alpha} \alpha_i L_i^\alpha(\alpha). \tag{14.4}$$

Equations (14.1–14.2) are solved for the finite set $(\{\boldsymbol{q}_i\}, \{\alpha_i\})$ of realizations of the input stochastic quantities and the stochastic solution is represented as

$$\boldsymbol{u}(.., \boldsymbol{q}, \alpha) = \sum_{i=1}^{n_q} \sum_{j=1}^{n_\alpha} \boldsymbol{u}_{i,j}(.., \boldsymbol{q}_i, \alpha_j) L_i^q(\boldsymbol{q}) L_j^\alpha(\alpha). \tag{14.5}$$

Hence given an abstract representation of the input stochastic quantities in terms of a finite number of random variables $\boldsymbol{Y}_q$ and $\boldsymbol{Y}_\alpha$, it is straightforward to solve for the stochastic dependent variable.

The inverse/design problems of interest can take several forms. In all cases, part of the boundary conditions or a source term or a material property need to be computed using some additional data provided either as sensor measurement of the dependent variable u or as desired system response. *Without loss of generality, we assume that the stochastic representation of one set of input conditions α is specified, while the representation of the other set of stochastic input conditions $\boldsymbol{q}$ has to be determined.* Mathematically, this is stated as:

$$\text{Given } \alpha = \mathcal{G}_1(\boldsymbol{x}, t, \boldsymbol{Y}_\alpha) \text{ find } \boldsymbol{q} \text{ such that } \boldsymbol{u} \text{ satisfies,}$$

$$\mathcal{B}(\boldsymbol{u}, \boldsymbol{x}, t : \{\boldsymbol{q}, \alpha\}) = 0, \tag{14.6}$$

$$\mathcal{L}(\boldsymbol{u}, \boldsymbol{x}, t : \{\boldsymbol{q}, \alpha\}) = 0, \tag{14.7}$$

$$\text{and } \boldsymbol{u} = \mathcal{H}, \tag{14.8}$$

where $\mathcal{H}$ is the desired stochastic evolution or measured stochastic system response in some locations in space and time. α can represent the known material property variability and q an unknown boundary condition.

A naive formulation would be to construct $\boldsymbol{q} = \mathcal{G}_2(\boldsymbol{Y}_q)$ given the input data and measurement. But there are inherent difficulties with this: specifically, (a) the dependent variable belongs to the tensor product space $\Gamma_u \equiv \Gamma_q \otimes \Gamma_\alpha$ and there is no simple way to decompose it back into its constituent spaces. Furthermore, it is important to emphasize that the stochastic representation, $\boldsymbol{q} = \mathcal{G}_2(\boldsymbol{Y}_q)$, of any random process is a purely abstract scheme. From a physically realizable/observable perspective, only *operations* on the stochastic representation, $\boldsymbol{q} = \mathcal{G}_2(\boldsymbol{Y}_q)$, make sense (for instance, realizations of events, moments or the PDF). These operations are *reduced* (or averaged) representations of the abstract random process. *This observability/measurability argument strongly points to the necessity of posing the inverse stochastic problem in a way that the measured or desired dependent variables are given in the form of moments or PDF.* Moreover, it is appropriate to define the expected/measured/designed quantity $\mathcal{L}(\boldsymbol{u})$ in terms of realizations, moments or PDFs. In addition, recent work (Petersdorff and Schwab 2006; Schwab and Todur 2001, 2003) suggests that the direct stochastic equations posed in the form of moments or PDF have rigorous existence and uniqueness arguments (Zabaras and Ganapathysubramanian 2008).

In this work, the solution q of the inverse problem is computed in the form of moments or PDF. For simplicity, the domain $(\boldsymbol{x}, t) \in (\mathcal{D}_q, \mathcal{T})$ of q is not shown explicitly in the following equations. That is, given $\alpha = \mathcal{G}_1(\boldsymbol{x}, t, \boldsymbol{Y}_\alpha)$

Find (a) $\langle \boldsymbol{q}^k \rangle$ or, (b) $PDF(\boldsymbol{q})$ such that $\boldsymbol{u}$ satisfies

$$\mathcal{B}(\boldsymbol{u}, \boldsymbol{x}, t : \{\boldsymbol{q}, \alpha\}) = 0, \tag{14.9}$$

and the desired behavior given by, (a) $\langle \boldsymbol{u}^k \rangle = \mathcal{H}$ or,

$$\text{(b) } PDF(\boldsymbol{u}) = \mathcal{H}. \tag{14.10}$$

14.2.2 The Stochastic Metrics and Representation of the Inverse Stochastic Solution q

As discussed in Section 14.2.1, the measurement values or the desired system response $\boldsymbol{u}$ is given at some locations within the domain in two forms, e.g. for $(\boldsymbol{x}, t) \in (\mathcal{D}_s, \mathcal{T})$: Case I: The mean and higher order moments of the system response are given, $\langle \boldsymbol{u}^k(\boldsymbol{x}, t) \rangle = \mathcal{H}_k(\boldsymbol{x}, t)$, for $k \geq 1$ and Case II: The PDF of the system response is given PDF$(\boldsymbol{u}^k(\boldsymbol{x}, t)) = \mathcal{H}(\boldsymbol{x}, t)$. Corresponding to these physically meaningful input data statistics, the inverse problem solution will construct the statistics of the designed field as follows:

Case I:

Given the known stochastic input parameters, $\alpha = \mathcal{G}_1(\boldsymbol{x}, t, \boldsymbol{Y}_\alpha)$, *find the moments of the stochastic input,* $< q^k(\boldsymbol{x}, t, \boldsymbol{Y}_q) >$, $k = 1, \ldots, p$, *such that the stochastic system defined by Eq. (14.1), results in statistics of the dependent variable corresponding to* $< \boldsymbol{u}^k(\boldsymbol{x}, t, .) > = \mathcal{H}_k(\boldsymbol{x}, t)$, $k = 1, \ldots, p$, $(\boldsymbol{x}, t) \in (\mathcal{D}_s, \mathcal{T})$.

Case II:

Given the known stochastic input parameters, $\alpha = \mathcal{G}_1(\boldsymbol{x}, t, \boldsymbol{Y}_\alpha)$, *find the PDF of the stochastic input,* $PDF[q(\boldsymbol{x}, t, .)]$, *such that the stochastic system defined by Eq. (14.1), results in statistics of the dependent variable corresponding to* $PDF[\mathbf{u}(\boldsymbol{x}, t, .)] = \mathcal{H}(\boldsymbol{x}, t)$, $(\boldsymbol{x}, t) \in (\mathcal{D}_s, \mathcal{T})$.

In design applications, the given moments or PDFs should be interpreted as 'desired' variability or performance robustness. In the context of inverse problems driven by data, this variability maybe induced either by sensor noise (in the case of a single experiment) or e.g. because of the variability of the random topology (repeated experimentation with random realizations of the medium will lead to variability in measurements even without measurement noise). All of these problems can be stated in the forms proposed above.

In this chapter, we are interested in designing for higher order statistics of the input fields. However posing and solving for the statistics directly will have the additional bottleneck of having to construct appropriate closure arguments. This will make the framework application specific. We propose to construct the complete stochastic representation instead. That is, we propose to construct $\boldsymbol{q}(\boldsymbol{x}, t, \boldsymbol{Y}_q)$ such that its statistics satisfy the two cases defined above. This presents an apparent paradox because the construction of $\boldsymbol{q}$ requires an a priori knowledge of the support of $\boldsymbol{q}$, i.e. knowledge of $\boldsymbol{Y}_q \in \Gamma_q$. However, since we are only interested in the statistics of $\boldsymbol{q}$ and these statistics are essentially integrals over the support, *they are independent of* the choice of the support, Γ_q (see Figure 14.2 for a verification using some of our preliminary work in (Zabaras and Ganapathysubramanian 2008)).

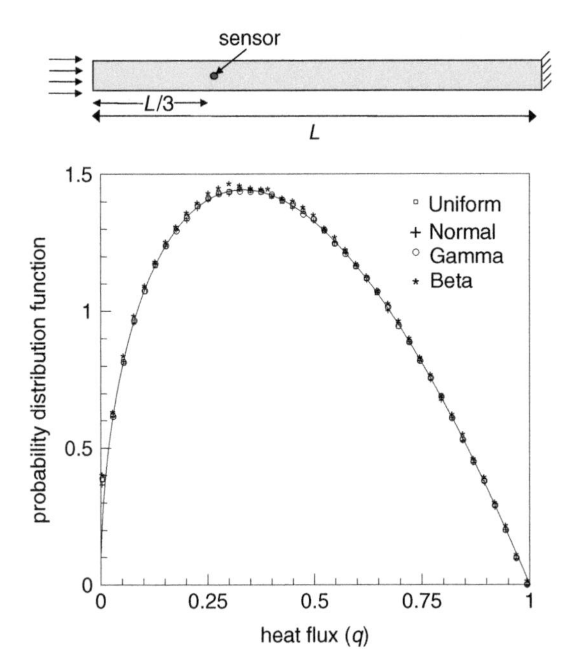

Figure 14.2 Preliminary results for a simple 1D stochastic inverse heat conduction illustrating the support independent formalism (Zabaras and Ganapathysubramanian 2008). A 1D domain of length L is considered (top figure). A specific temperature is maintained at one end. An (unknown) stochastic heat flux, q, is applied at the other end. Furthermore, the thermal conductivity is a known stochastic field. From the PDF of temperature taken at the sensor, the applied stochastic heat flux is reconstructed. The unknown heat flux is represented using four different support spaces. The PDF (and other moments) of the optimal solution is independent of the choice of the support as can be seen in the bottom figure.

Utilizing this rational, we represent the designed input stochastic field using a sparse grid representation *for an arbitrarily chosen support.* Our preliminary investigations have revealed that this representation is in fact valid and is a key step to effectively decoupling the stochastic inverse/design problem.

Without loss of generality, we can assume that $\boldsymbol{Y}_q = \{Y_q^1, \ldots, Y_q^n\}$ are uncorrelated uniform random variables defined in $\Gamma_q \equiv [0,1]^n$ as

$$q(\boldsymbol{x}, t, \boldsymbol{Y}_q) = \sum_{i=1}^{n_q} q(\boldsymbol{x}, t, \boldsymbol{Y}_q^i) L_i(\boldsymbol{Y}_q). \tag{14.11}$$

With this assumption, the problem of computing a stochastic function, $q(\boldsymbol{x}, t, \boldsymbol{Y}_q)$, is transformed into a problem of computing *a finite set of deterministic functions,* $q(\boldsymbol{x}, t, \boldsymbol{Y}_q^i), i = 1, \ldots, n_q$.

The redefined stochastic inverse/design problems can now be stated as follows:

Case I:

Given the known stochastic input parameters, $\alpha = \mathcal{G}_1(\boldsymbol{x}, t, \boldsymbol{Y}_\alpha)$, *find the* n_q *deterministic functions,* $q(\boldsymbol{x}, t, \boldsymbol{Y}_q^i)$, $i = 1, \ldots, n_q$, *such that the stochastic system defined by Equation (14.1), results in statistics of the dependent variable corresponding to* $< \boldsymbol{u}^k(\boldsymbol{x}, t, .) > = \mathcal{H}_k(\boldsymbol{x}, t)$, $k = 1, \ldots, p$, $(\boldsymbol{x}, t) \in (\mathcal{D}_s, \mathcal{T})$.

Case II:

Given the known stochastic input parameters, $\alpha = \mathcal{G}_1(\boldsymbol{x}, t, \boldsymbol{Y}_\alpha)$, *find the* n_q *deterministic functions,* $q(\boldsymbol{x}, t, \boldsymbol{Y}_q^i)$, $i = 1, \ldots, n_q$, *such that the stochastic system defined by Equation (14.1), results in statistics of the dependent variable corresponding to* $PDF[\boldsymbol{u}(\boldsymbol{x}, t, .)] = \mathcal{H}(\boldsymbol{x}, t)$, $(\boldsymbol{x}, t) \in (\mathcal{D}_s, \mathcal{T})$.

Equation (14.9) together with the known initial conditions, α and the sensor/desired conditions define an *ill-posed* problem that can be solved for the stochastic field $\boldsymbol{q}$. In the proposed work, we assume that a solution to the inverse problem exists in the sense of Tikhonov. That is, we look for a solution $\{q_i^*\}_{i=1}^{n_q}$ such that:

$$\mathcal{CF}[\{q_i^*\}_{i=1}^{n_q}] \leq \mathcal{CF}[\{q_i\}_{i=1}^{n_q}], \forall\ q(\boldsymbol{x}, t, \boldsymbol{Y}_q) = \sum_{i=1}^{n_q} q_i L_i(\boldsymbol{Y}_q). \tag{14.12}$$

Here, $\mathcal{CF}[\{q_i\}_{i=1}^{n_q}]$ is a cost functional that quantifies how well the given data are matched from the simulated solution for a guessed random function

q. For a given $\{q_i\}_{i=1}^{n_q}$, the cost functional is computed using the dependent stochastic variable, $\boldsymbol{u}(\boldsymbol{x}, t, \boldsymbol{Y}_q, \boldsymbol{Y}_\alpha : \{q_i\}_{i=1}^{n_q})$. This is obtained from the solution of the direct stochastic problem in Equation (14.9) with $\boldsymbol{q}$ as a (known) stochastic function, that is, using $\{q_i\}_{i=1}^{n_q}$ as the guessed stochastic input conditions along with the given conditions α. We will utilize quadratic cost functionals to allow use of the extensive (deterministic) quadratic optimization tools available.

For Case I, the cost functional measures how well the stochastic solution satisfies the given moments:

$$\mathcal{CF}[\{\boldsymbol{q}_i\}_{i=1}^{n_q}] = \frac{1}{2}\int_{\mathcal{D}_s}\int_{t=0}^{t_{max}}\sum_{k=1}^{p}[\langle \boldsymbol{u}(\boldsymbol{x}, t, .)^k\rangle - \mathcal{H}_k(\boldsymbol{x}, t)]^2 d\boldsymbol{x}dt, \tag{14.13}$$

where $\langle \boldsymbol{u}(\boldsymbol{x}, t, .)^k\rangle$ is the k-th moment of the dependent variable and $\mathcal{D}_s$ is the region over which sensor measurements are available. In Case II, the cost functional measures how well the stochastic solution satisfies the measured/desired PDF. Since the PDF, the CDF and the inverse CDF represent the same distribution in different ways, for the sake of computational effectiveness, we propose to define the cost functional in terms of the inverse CDF. This choice is motivated by the fact that the inverse CDF has a fixed, known support [0, 1] and this greatly simplifies defining the cost functional. Denote $\Upsilon^{-1}[f, \boldsymbol{Y}]$ as the inverse cumulative distribution of the random variable f. $\boldsymbol{Y}$ is the spanning variable, $\boldsymbol{Y} \in [0, 1]$. The inverse CDF is utilized because it is computationally easy to represent it using the collocation based polynomial interpolation representation. We define the cost functional in this case as:

$$\mathcal{CF}[\{\boldsymbol{q}_i\}_{i=1}^{n_q}] = \frac{1}{2}\int_{\mathcal{D}_s}\int_{t=0}^{t_{max}}(\Upsilon^{-1}[\boldsymbol{u}(\boldsymbol{x}, t), \boldsymbol{Y}] - \Upsilon^{-1}[\mathcal{H}(\boldsymbol{x}, t), \boldsymbol{Y}])^2 d\boldsymbol{x}dt. \tag{14.14}$$

The stochastic inverse problem has now been converted into a deterministic optimization problem in a larger dimensional space – the space where the collocated values of q at each $(\boldsymbol{x}, t) \in (\mathcal{D}_q, \mathcal{T})$ lie. Without spatial or temporal regularization (see also Section 14.2.5), we treat the nodal values of q at each finite element point $(\boldsymbol{x}_j, t_k)$ in the discretization of $(\mathcal{D}_q, \mathcal{T})$ as independent uncorrelated random variables. Their collocated values $\{\boldsymbol{q}_i(\boldsymbol{x}_j, t_k)\}_{i=1}^{n_q}$ (on the same support space Γ_q) define the unknowns of the overall optimization problem.

14.2.3 Solving the Direct Stochastic Problem: Adaptivity Sparse Grid Collocation

As part of the inverse analysis, one needs to be able to efficiently construct the solution $\boldsymbol{u}$ to the underlying SPDEs. We utilize a sparse grid collocation strategy. This framework represents the stochastic solution as a polynomial approximation. This interpolant is constructed via independent function calls to the deterministic PDE with different realizations of the stochastic input. The sparse grid collocation strategy follows naturally from a full-tensor product representation of a multi-variate function in terms of univariate interpolation formulae. Consider a univariate interpolation operator $\mathcal{U}$ to represent a one-dimensional function f in terms of realizations of f at some sampling points. That is, $\mathcal{U}^i = \sum_{x^i \in X^i} a_{x^i} \cdot f(x^i)$ with the set of support nodes $X^i = \{x_1^i, \ldots, x_{m^i}^i\}, x_k \in [0,1], 1 \leq k \leq m_i$ and the polynomial basis functions a_{x^i}. A multi-dimensional function $f(x_1, \ldots, x_N)$ can be represented as product of one-dimensional functions as $\mathcal{U}^N f = (\mathcal{U}^{i_1} \otimes \ldots \otimes \mathcal{U}^{i_N})(u)$.

The tensor product representation quickly suffers the 'curse-of-dimensionality' problem – where if one utilized $\mathcal{O}(k)$ points in one dimension, one would require $\mathcal{O}(k^N)$ points in N-dimensions. *The sparse grid collocation strategy selectively chooses points in this uniform N-dimensional sampling* to significantly reduce the number of sampling points. It has been shown the number of points reduces from $\mathcal{O}(k^N)$ to $\mathcal{O}((log(k))^N)$ (Gerstner and Griebel 1998). Here *we utilize a hierarchical basis based approach towards adaptivity that resolves the issues of locality and curse-of-dimensionality.* This borrows ideas from wavelet-based representation of functions. One of the key motivations towards using hierarchical basis functions is their linear scaling with dimensionality. This naturally results in rigorous convergence estimates for the adaptive sparse grid methods developed. This includes construction of appropriate hierarchical surpluses (described later in this section) that relate the interpolant to the local variance error. Such construction offers a natural means of adaptively refining the stochastic solution. We briefly describe the proposed developments below.

Hierarchical-Basis Based Adaptive Sparse Collocation

The key to incorporating scalable adaptivity is defining incremental or hierarchical linear interpolants (Ma and Zabaras 2009). Define the incremental interpolant (Klimke 2006) as $\Delta^i = \mathcal{U}^i - \mathcal{U}^{i-1}$, where i is the level of interpolation used in the interpolation operator. Denote the number of interpolation points in multiple dimensions by the index $\mathbf{i} = (i_1, \ldots, i_N)$ with $|\mathbf{i}| = i_1 + \ldots + i_N$. The conventional Smolyak algorithm is given by $\mathcal{A}(q,N) = \sum_{|\mathbf{i}| \leq q} \left(\Delta^{i_1} \otimes \ldots \otimes \Delta^{i_N}\right)$. To compute $\mathcal{A}(q,N)$, one needs to compute the function values at the sparse grid points given by $\mathcal{H}(q,N) = \bigcup_{q-N+1 \leq |\mathbf{i}| \leq q} \left(X^{i_1} \times \ldots \times X^{i_N}\right)$ where $X^i = \{y_1^i, \ldots, y_{m_i}^i\}$ are the set of points used by $\mathcal{U}^i$.

One should select the set X^i in a nested fashion such that $X^i \subset X^{i+1}$ to obtain many recurring points with increasing q. The basic idea towards adaptivity here is to use hierarchical surplus as an error indicator and only refine the grid points whose hierarchical surplus is larger than a pre-defined threshold. The hierarchical surplus at any point is just the value of the incremental interpolant. That is

$$\Delta^i(f) = \sum_{j=1}^{m_i^\Delta} a_j^i \cdot \underbrace{(f(x_j^i) - \mathcal{U}^{i-1}(f)(x_j^i))}_{w_j^i}, \tag{14.15}$$

where w_j^i are the hierarchical surpluses. We can apply the above equation to obtain the sparse grid interpolation formula for the multivariate case in a hierarchical form as $\mathcal{A}_{q,N}(f) = \mathcal{A}_{q-1,N}(f) + \Delta\mathcal{A}_{q,N}(f)$ and

$$\mathcal{A}_{q-1,N}(f) = \sum_{|\mathbf{i}|\leq q-1} (\Delta^{i_1} \otimes \ldots \otimes \Delta^{i_N}), \tag{14.16}$$

$$\Delta\mathcal{A}_{q,N}(f) = \sum_{|\mathbf{i}|=q} \sum_{\mathbf{j}} \left(a_{j_1}^{i_1} \otimes \ldots \otimes a_{j_N}^{i_N}\right) \cdot w_{\mathbf{j}}^{\mathbf{i}}, \tag{14.17}$$

$$w_{\mathbf{j}}^{\mathbf{i}} = f(x_{j_1}^{i_1}, \ldots, x_{j_N}^{i_N}) - \mathcal{A}_{q-1,N}(f)(x_{j_1}^{i_1}, \ldots, x_{j_N}^{i_N}). \tag{14.18}$$

Defining the interpolation strategy in terms of the hierarchical surpluses provides a rigorous framework for adaptively interpolating multivariate functions. For continuous functions, the hierarchical surpluses tend to zero as the interpolation level tends to infinity. Furthermore, for non-smooth functions, details about the singularities are indicated by the magnitude of the hierarchical surplus (Ma and Zabaras 2009).

Choice of Nested Points

Our preliminary work on these aspects (Ma and Zabaras 2009) has revealed that the choice of the univariate sampling points is absolutely critical in determining the scalability of the framework. With an aim to make this framework scale efficiently to high-dimensions, one can utilize three types of nested grids: the Clenshaw-Curtis-type grid, the Chebyshev type grid and the maximum norm grid. For these grids, the number of points required as the dimensionality increases scales relatively slowly.

Data Structure for Dimensional Scalability

As the dimensionality of the problem increases, the memory requirements for storing and constructing the interpolant increases substantially. One needs to utilize a tree-based data structure to efficiently store and retrieve the hierarchical surpluses. Given some user defined tolerance, ε, the generalized sparse grid strategy is utilized to construct interpolants. The interpolant utilizes points *until the hierarchical surpluses associated with all most recently added points becomes less than the tolerance.* The choice of the new points to be sampled is governed by the tree structure (the daughter nodes of a node whose surplus $> \varepsilon$ are sampled in the next stage of the adaptive procedure).

Error Control through Variance Estimation

The hierarchical surpluses can further be utilized to provide estimates on the convergence of the stochastic solution. By developing explicit relations between the hierarchical surpluses to extract information about the local variance of the solution, one can adaptively sample only those regions whose variance is larger than a prescribed tolerance. Our preliminary investigations into the adaptive framework (Ma and Zabaras 2009) have shown the feasibility of this approach. Using adaptive hierarchical interpolants, we were able to solve SPDEs driven by random inputs that lie in a 100-dimensional space – this being an order of magnitude increase over conventional stochastic solution strategies.

14.2.4 Stochastic Sensitivity Equations and Gradient-Based Optimization Framework

We utilize a gradient based optimization strategy to design the optimal stochastic input $\{\boldsymbol{q}_i\}$. The first step is to compute the gradient of the cost functional with respect to the design variables. The directional derivative, $\mathrm{D}_{\Delta q_v}\mathcal{CF}[\{q_i\}_{i=1}^{n_q}]$ of the cost functionals have to be computed:

Gradient of the cost functional – Case I:

$$\mathrm{D}_{\Delta q_v}\mathcal{CF}[\{q_i\}_{i=1}^{n_q}]=\int_{\mathcal{D}_s}\int_{t=0}^{t_{max}}\sum_{k=1}^{p}[\langle \boldsymbol{u}(\boldsymbol{x},t,.)^k\rangle-\mathcal{H}_k(\boldsymbol{x},t)]\mathrm{D}_{\Delta q_v}(\boldsymbol{u}(\boldsymbol{x},t,.))d\boldsymbol{x}dt.$$

Gradient of the cost functional – Case II:

$$\mathrm{D}_{\Delta q_v}\mathcal{CF}[\{q_i\}_{i=1}^{n_q}] = \int_{\mathcal{D}_s}\int_{t=0}^{t_{max}}(\Upsilon^{-1}[\boldsymbol{u}(\boldsymbol{x},t),\,\boldsymbol{Y}]-\Upsilon^{-1}[\mathcal{H}(\boldsymbol{x},t),\,\boldsymbol{Y}])\mathrm{D}_{\Delta q_v}(\boldsymbol{u}(\boldsymbol{x},t,.))d\boldsymbol{x}dt. \quad (14.19)$$

Recall that the stochastic variable $\boldsymbol{u}$ is represented in terms of the stochastic interpolants as

$$\boldsymbol{u}=\sum_{i=1}^{n_q}\sum_{j=1}^{n_\alpha}\boldsymbol{u}(\boldsymbol{x},t,\,\boldsymbol{Y}_q^i,\,\boldsymbol{Y}_\alpha^j)L_q^i(\boldsymbol{Y}_q)L_\alpha^j(\boldsymbol{Y}_\alpha). \quad (14.20)$$

The calculation of the stochastic gradient requires the calculation of the directional derivative of $\boldsymbol{u}$ w.r.t. q_v, i.e. of $\mathrm{D}_{\Delta q_v}\boldsymbol{u}(\boldsymbol{x},t,\,\boldsymbol{Y}_q^r,\,\boldsymbol{Y}_\alpha^e)$. Denote the directional derivative of the stochastic variable with respect to each design variable as $\hat{\boldsymbol{u}}(\boldsymbol{x}, t, \boldsymbol{Y}_q, \boldsymbol{Y}_\alpha : \{q_i\}_{i=1}^{n_q}, \Delta q_v) \equiv \mathrm{D}_{\Delta q_v} \boldsymbol{u}(\boldsymbol{x}, t, \boldsymbol{Y}_q, \boldsymbol{Y}_\alpha : \{q_i\}_{i=1}^{n_q})$. This defines the *sensitivity field of the dependent variable* $\boldsymbol{u}$ *as the linear in* Δq_v *part of* $\boldsymbol{u}(\boldsymbol{x}, t, \boldsymbol{Y}_q, \boldsymbol{Y}_\alpha : \{q_i\}_{i=1, i\neq v}^{n_q}, q_v + \Delta q_v)$, *where* Δq_v *is a perturbation to one of the unknown (collocated values of* q*) variables.*

$$\boldsymbol{u}(\boldsymbol{x}, t, \boldsymbol{Y}_q, \boldsymbol{Y}_\alpha : \{q_i\}_{i=1, i\neq v}^{n_q}, q_v + \Delta q_v) =$$
$$\boldsymbol{u}(\boldsymbol{x}, t, \boldsymbol{Y}_q, \boldsymbol{Y}_\alpha : \{q_i\}_{i=1}^{n_q}) + \hat{\boldsymbol{u}}(\boldsymbol{x}, t, \boldsymbol{Y}_q, \boldsymbol{Y}_\alpha : \{q_i\}_{i=1}^{n_q}, \Delta q_v) + h.o.t. \quad (14.21)$$

The stochastic sensitivity equations are simply obtained by taking the directional derivative of the equations that define the parametric direct problem used to compute $\boldsymbol{u}$ for each input q, i.e. the direct equations and the boundary conditions are linearized w.r.t. the design variables $\{q_i\}$. We denote directional derivatives of operators and fields with ˆ. These directional derivatives are taken with respect to each of the n_q collocation points used to represent q. The sensitivity equations take the following general form:

Continuum stochastic sensitivity equations:

$$\hat{\mathcal{B}}(\hat{\boldsymbol{u}}, \boldsymbol{u}, \boldsymbol{x}, t, \boldsymbol{Y}_q, \boldsymbol{Y}_\alpha, \{q_i\}_{i=1}^{n_q}, \Delta q_v) = 0, \quad (14.22)$$
$$\hat{\mathcal{L}}(\hat{\boldsymbol{u}}, \boldsymbol{u}, \boldsymbol{x}, t, \boldsymbol{Y}_q, \boldsymbol{Y}_\alpha, \{q_i\}_{i=1}^{n_q}, \Delta q_v) = 0. \quad (14.23)$$

The sensitivity of the stochastic variable computed at the $\boldsymbol{Y}_q^v$ *collocation point depends only on perturbations to the* q_v*-th design variable.* Thus the sensitivity equations for perturbation Δq_v to each design variable $q_v(x_j, t_k)$ are SPDEs defined in an M dimensional support space where $M = n_q \times n_\alpha$. These equations are solved with the sparse grid collocation method highlighted earlier for the direct analysis. Thus the complete stochastic sensitivity field w.r.t. q_v can be re-constructed using *only deterministic sensitivity problems* as

$$\hat{\boldsymbol{u}}(\boldsymbol{x}, t, \boldsymbol{Y}_q, \boldsymbol{Y}_\alpha : \{q_k\}_{k=1}^{n_q}, \Delta q_v) =$$
$$\sum_{r=1}^{n_q} \sum_{e=1}^{n_\alpha} \hat{\boldsymbol{u}}(\boldsymbol{x}, t, \boldsymbol{Y}_q^r, \boldsymbol{Y}_\alpha^e : \{q_k\}_{k=1}^{n_q}, \Delta q_v) L_r(\boldsymbol{Y}_q) L_e(\boldsymbol{Y}_\alpha). \quad (14.24)$$

Simplified sensitivity calculations for linear problems are given in (Zabaras and Ganapathysubramanian 2008). Let us now denote the gradient of the cost functional with respect to the design variables $\{q_i\}_{i=1}^{n_q}$ as $\boldsymbol{d} = \{d_i\}^T$. The gradient of the cost functional can be written in terms of the directional derivative (the directional derivative is just the gradient computed in a specific direction). Since we have a scheme to compute the directional derivative by solving the continuum stochastic sensitivity equations, the gradient of the cost-function is then simply given as $\boldsymbol{d} = \{d_v\}^T = \{\mathrm{D}_{\Delta q_v} \mathcal{CF}[\{q_i\}_{i=1}^{n_q}] / \Delta q_v\}^T$. Preliminary implementation of these ideas within a steepest descent approach for a stochastic diffusion problem are reported in our work in (Zabaras and Ganapathysubramanian 2008).

There are two aspects of these developments that play a crucial part towards making the framework scalable. The first involves developing mathematical strategies to make the solution of the direct stochastic problem more efficient – through the incorporation of adaptivity. This part of the proposed work is discussed in a previous section. The second involves developing mathematical frameworks that make the stochastic optimization problem efficient – through fast computation of the stochastic Hessian and accelerated convergence.

Efficient Optimization: Incorporating Stochastic Gradient Information

There is significant work that needs to be considered in the optimization to reduce the computational overheard including using high fidelity optimization schemes utilizing the Hessian information. One could investigate the use of stochastic quasi-Newtonian schemes for computing the Hessian, specifically BFGS like schemes (Bashir *et al.* 2007; Dennis and More 1997). The key advantage is that the computational complexity is $\mathcal{O}(n_q M)$ while the convergence rate is improved from linear (e.g. for the steepest descent method) to quadratic (for quasi-Newton schemes). The stochastic Hessian can be computed as a set of decoupled deterministic Hessians at the M collocated points. For the stochastic optimization problem, at each search iteration k, the Hessian at each collocation point can be updated independently of the calculations at other collocation points. This ensures that these computations are easily scalable.

Efficient Optimization: Hierarchical Optimization Strategies

The stochastic design field $\boldsymbol{q}$ is typically represented as a set of n_q deterministic fields. The computational complexity of the optimization scales with increasing n_q as does the accuracy of representation. Our preliminary numerical investigations in (Zabaras and Ganapathysubramanian 2008) revealed the following trends (a) The optimization problem with coarser representation (i.e. smaller n_q) of the design variable converges faster initially, but becomes very slowly converging later and (2) The optimization problem with coarser representation is numerically faster to solve. These two observations naturally lead to the possibly of formulating a *hierarchical stochastic optimization* problem, where a coarser problem is solved to quickly compute a coarse solution that is in turn used as an initial guess to solve a finer problem. This idea (similar to the accelerated convergence using multi-grid methods) seems to offer great promise in terms of rapid solutions to inverse problems through solution of a hierarchy of coarser optimization problems. One can utilize this strategy

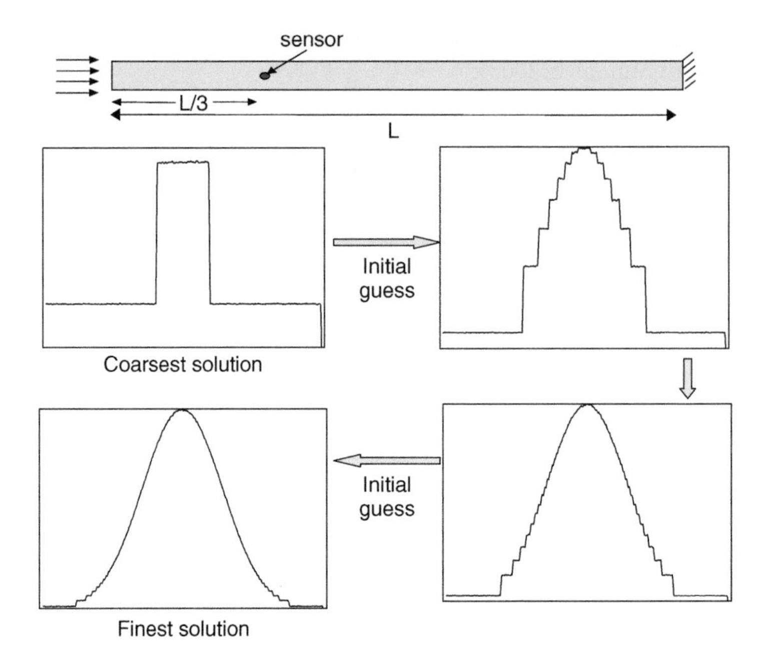

Figure 14.3 A simple 1D stochastic inverse heat conduction illustrating the hierarchical optimization strategy (Zabaras and Ganapathysubramanian 2008). A 1D domain of length L is considered (top figure). A specific temperature is maintained on the right end. An (unknown) stochastic heat flux, q, is applied at the left end. Furthermore, the thermal conductivity is a known stochastic field. From the PDF of temperature taken at the sensor, the applied stochastic heat flux is reconstructed. The optimization is accelerated by first solving for a coarse representation of the stochastic variable and utilizing this solution as an initial guess to a refined optimization problem. This resulted in two orders of magnitude reduction in iterations.

to accelerate the optimization framework. A key issue that has to be satisfactorily resolved is when to stop the current optimization and use the solution as the initial guess for a refined optimization problem. This can be decided based on a trade-off between accuracy of representation and computational effort (Figure 14.3).

Analytical Gradient Calculations: The Continuum Stochastic Adjoint Equations

For some classes of problems the gradient of the cost functional can be analytically computed using appropriate adjoint variables (Sampath and Zabaras 2001). In our recent work (Velamur Asokan and Zabaras 2004), we have extended this to formulating the stochastic adjoint equations to analytically compute the stochastic gradient. The continuum stochastic adjoint equations provide a direct means of computing the gradient while the continuum stochastic sensitivity equations provide the step size. One can utilize sparse grid collocation to solve the stochastic adjoint equations to explicitly compute the stochastic gradient. This strategy does not require the explicit computation of the stochastic Hessian. While our earlier implementation of these techniques for a stochastic inverse heat conduction problem using a GPCE approach revealed slow convergence in higher-order GPCE terms, one can revisit these techniques using the adaptive hierarchical sparse grid collocation discussed in Section 14.2.3.

14.2.5 Incorporating Correlation Statistics and Investigating Regularization

In classical inverse problems (Alifanov 1994), the ill-posedness requires some form of regularization or smoothing assumptions for computing the solution. But in the context of stochastic inverse problems, we are searching for a solution *in a distribution sense* in a much larger space (tensor product space of spatio-temporal and stochastic variations). This expanded search space essentially guarantees a stochastic solution (Kaipio and Somersalo 2005). Nevertheless, regularization arguments play a very significant role in reducing the computational effort by imposing some smoothness criteria on the designed stochastic solution. One needs to investigate mathematical approaches to stochastic regularization as a purely computational strategy. A potential approach is two-pronged: (a) utilize spatial (temporal) regularization via parametrization of the spatial (temporal) representation of the unknown stochastic fields, and (b) utilize stochastic regularization via parametrization of the stochastic field.

Spatial (Temporal) Parametrization of the Unknown Stochastic Fields

We here extend deterministic parametric representation of fields in a straightforward manner for the spatial representation of stochastic fields. Consider a Bézier representation of $\boldsymbol{q}(\boldsymbol{x}, \boldsymbol{Y})$- a stochastic field in two spatial dimensions. A Bézier curve of order (m_{x1}, m_{x2}) is defined by a set of $(m_{x1}+1) \times (m_{x2}+1)$ control points $P_{i,j}$ as

$$q(x1, x2, \boldsymbol{Y}) = \sum_{i=0}^{m_{x1}} \sum_{j=0}^{m_{x2}} P_{i,j}(\boldsymbol{Y}) B_i^{m_{x1}}(x1) B_j^{m_{x2}}(x2), \tag{14.25}$$

$$\text{where,} \quad B_i^{m_x}(x) = \frac{m_x!}{i!(m_x - i)!} x^i (1-x)^{m_x - i}. \tag{14.26}$$

The design variables are now the stochastic control values $P_{i,j}(\boldsymbol{Y})$. These variables can be subsequently represented in terms of n_q collocated values (as in Equation 14.11) in the form:

$$P_{i,j}(\boldsymbol{Y}) = \sum_{r=1}^{n_q} P_{i,j}^r(\boldsymbol{Y}_r) L_r(\boldsymbol{Y}). \tag{14.27}$$

The sensitivity equations are also defined in terms of these n_q deterministic parameters. Since the choice of the parametrization controls the spatial smoothness of the solution, it is necessary to construct convergence estimates as the spatial parametrization is varied. Some preliminary work on convergence of this approximation is given in (Zabaras and Ganapathysubramanian 2008). Similar arguments can be utilized to parameterize the temporal variation of the stochastic field.

Stochastic Regularization of the Unknown Stochastic Fields

An alternative strategy is to consider some stochastic regularization. This can be imposed by enforcing some correlation structure on the unknown stochastic field. The correlation kernel $C(\boldsymbol{x})$ determines the stochastic field and imposes some smoothness on its spatial variability. One can convert the problem of designing the stochastic field $\boldsymbol{q}$ to the problem of designing its correlation structure $C(\boldsymbol{x})$. The unknown correlation function can be represented for example in terms of its spectral expansion

$$C(r) = \sum_{i=1}^{K} a_i \sin(ix/L), \tag{14.28}$$

where L is the characteristic length of the system and $\{a_i\}$ is the set of K Fourier modes of the correlation function (and we have assumed that the field is isotropic, without loss of generality). By representing the correlation this way, the stochastic optimization problem is converted to computing the K modal coefficients $\{a_i\}$. Given any correlation kernel, the Karhunen-Loève expansion can be utilized to explicitly represent the stochastic field. The choice of the number of modes K used in the representation of the unknown correlation provides a measure of the regularization. Defining and constructing the gradient of the cost functional is important along with developing a strategy for estimating the number of modes required based on the statistics of the measurements.

Designing/estimating the correlation statistics also has physical significance. Spatial correlation statistics provides some notion of how easy or difficult it is to physically realize the designed process. A very large or small correlation signifies expensive manufacturing, fabrication or operating conditions. This introduces the possibility of computing the designed fields with some constraints to obtain physically accessible fields. The above discussed strategies naturally take into account these issues.

14.2.6 Stochastic Low-Dimensional Modeling

In the stochastic inverse/design problem, we need to solve the direct problem and the sensitivity problem for all of the collocation points at each iteration. For large stochastic dimensionality, each iteration in the optimization procedure will take substantial amount of computational time. The solution procedure can be accelerated significantly if a reduced order model of the system can be constructed (Holmes *et al.* 1996). *This reduced representation can either be created off-line or adaptively modified on-the-fly.*

There has been recent work on extending the classical proper orthogonal decomposition (POD) method to stochastic evolution equations (Venturi *et al.* 2008). A compact expansion of the stochastic dependent variable has been proposed into stochastic spatial modes. We consider a random field $\boldsymbol{u}(\boldsymbol{x}, t; \boldsymbol{Y})$ in a space-time domain and we look for biorthogonal representations as

$$\boldsymbol{u}(\boldsymbol{x}, t; \boldsymbol{Y}) = \sum_{i=1}^{K} b_i^{(h)}(t)\, \boldsymbol{a}_i^{(h)}(\boldsymbol{x}; \boldsymbol{Y}). \tag{14.29}$$

The spatial-stochastic modes $\boldsymbol{a}_i^{(h)}(\boldsymbol{x}; \boldsymbol{Y})$ are constructed from snapshots of the stochastic field by solving an eigenvalue problem resulting from minimizing the residual of the projection of the stochastic field onto the basis functions. The residual is calculated based on some norm $\{,.,\}_h$. This stochastic counterpart of the POD framework will produce low-order ODEs (in terms of the temporal coefficients $b_i^{(h)}(t)$) that are independent of the stochastic dimensionality of the system. It is important that care be taken in choosing the appropriate norm over which to construct the spatial stochastic basis (inner products based on the mean, variance and standard deviation have been suggested (Venturi *et al.* 2008)). One can construct these spatial-stochastic modes $\boldsymbol{a}_i$ using sparse grid collocation. That is, the stochastic modes are represented as

$$\boldsymbol{a}_k^{(h)}(\boldsymbol{x}; \mathbf{Y}) = \sum_{i=1}^{M} \hat{\boldsymbol{a}}_{ki}^{(h)}(\boldsymbol{x}, \mathbf{Y}_i) L_i(\mathbf{Y}). \tag{14.30}$$

An arbitrary stochastic field is represented in terms of these stochastic modes as

$$\boldsymbol{u}(\boldsymbol{x}, t; \boldsymbol{Y}) = \sum_{i=1}^{K} b_i^{(h)}(t) \sum_{j=1}^{M} \boldsymbol{a}_{ij}^{(h)}(\boldsymbol{x}; \mathbf{Y}_j) L_j(\mathbf{Y}). \tag{14.31}$$

The unknowns are the deterministic coefficients $\{b_i\}$. Inserting this representation, Equation (14.31) into Equations (14.1)–(14.2) and using the orthogonality properties of the modes results in a set of ODEs for the modal coefficients, $\{b_i\}$. The cost functional, its gradient and the stochastic optimization problem can now be posed in terms of this set of ODEs instead of the set of SPDEs. Once can utilize this strategy for the control/optimization of SPDEs.

14.3 Numerical Examples

Let us consider a two-dimensional domain $\mathcal{D} \equiv [-0.5, 0.5] \times [-0.5, 0.5]$ with the temperature at the right boundary specified as $\theta = 0.5$ (Figure 14.4). At $t = 0$, the temperature in the domain is $\theta = 0$. The spatial variation of the thermal diffusivity follows an exponential correlation with correlation length $b = 10$ and mean value of $\alpha = 10$. We consider $s = 41$ equally spaced sensors inside the domain at a distance $d = 0.1$ from the left boundary with each sensor collecting data over the time interval [0, 0.5]. This data is given in terms of PDF of the temperature at each of these sensor location at 50 equally spaced time intervals in time range [0, 0.5]. The inverse problem of interest is posed as follows: *Identify the PDF of the heat flux on the left boundary in the time range* [0, 0.5] *such that the experimental measurements are reconstructed.*

Let us expand the thermal diffusivity using a Karhunen-Loève expansion:

$$\alpha(x, y, \omega_\alpha) = \alpha_{mean}(x, y) + \sum_{i=1}^{m} \sqrt{\lambda_i} f_i(x, y) \xi_i, \tag{14.32}$$

where λ_i and f_i are the eigenvalues and eigenvectors of the correlation kernel. Figure 14.5 plots the first few eigenvalues of the correlation matrix. The first three eigenvalues represent about 96 % of the variation. Correspondingly, the thermal diffusivity is represented using three random variables $\xi_i, i = 1, 2, 3$. Figure 14.6 shows the eigenmodes corresponding to these random variables.

Computing the 'Experimental' PDF

A direct problem is solved using an assumed stochastic variation for the heat flux applied on the left boundary. The PDFs of the temperature variation at the sensor locations are computed. This serves as the 'experimental' data that is used to run the inverse problem. The spatial variation of the heat

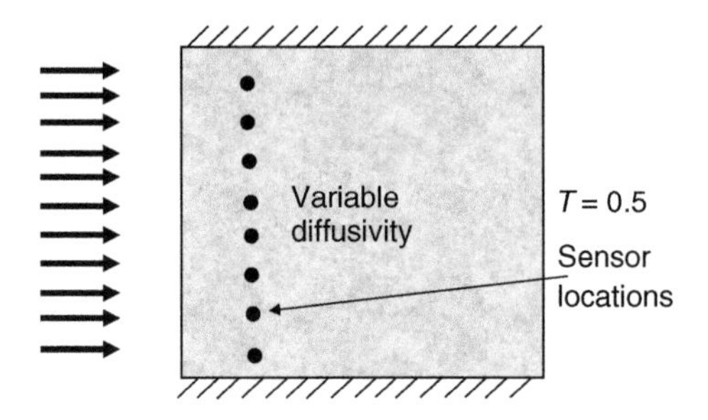

Figure 14.4 Schematic of the problem of interest: The temperature is given at the sensor locations shown by the dark circles. The spatial variation of the thermal diffusivity is defined by a known correlation kernel. We are interested to compute the unknown stochastic heat flux on the left boundary.

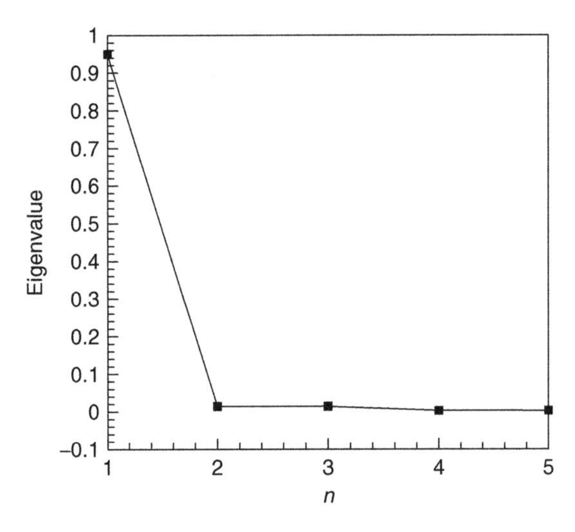

Figure 14.5 The eigenvalues of the correlation kernel that represents the thermal diffusivity variation.

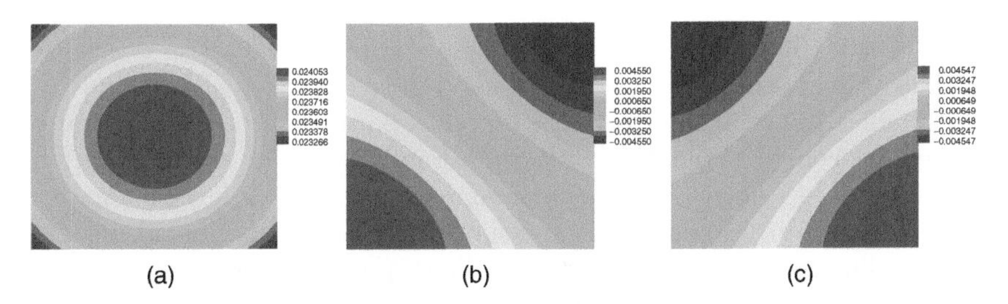

Figure 14.6 The first three modes of the Karhunen-Loève expansion of the thermal diffusivity.

flux is assumed to have an exponential correlation $C_1(y_1, y_1) = exp(-|y_1 - y_2|/b_1)$, with $b_1 = 0.5$. This stochastic heat flux is expanded using a Karhunen-Loève expansion and the eigenvalues are plotted in Figure 14.7. The first three eigenvalues represent about $\sim 94\,\%$ of the variation. The applied heat flux is thus represented using three random variables.

The heat flux applied has a mean value of 20 and is given a time-dependent damping term $exp(-\beta t)$ with $\beta = 2.0$. The input heat flux is thus of the form:

$$q_{exp}(y, t, \omega_q) = e^{-\beta t}\Big[20.0 + 5.0\sum_{i=1}^{3}\sqrt{\lambda_i}g_i(y)\xi_i\Big]. \tag{14.33}$$

The direct problem to obtain the 'experimental' statistics is solved using a 80×80 quad element discretization of the spatial domain using

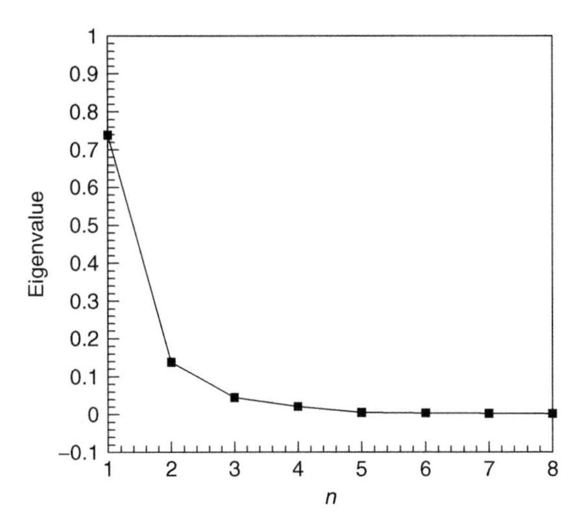

Figure 14.7 The eigenvalues of the correlation kernel that represents the 'experimental' heat flux.

$\Delta t = 0.0025$. In this problem, the temperature variation resides in a six-dimensional stochastic space (3D uncertainty in the thermal diffusivity and 3D uncertainty in the stochastic flux). A sparse grid collocation strategy is used to compute the temperature variability at the sensor locations. The thermal diffusivity variation is represented using a level 6 depth of interpolation (1073 points in 3D space). Similarly, 1073 points are used for the heat flux variation. The stochastic direct problem involves the solution of $1073 \times 1073 = 1.151 \times 10^6$ direct deterministic problems.

The PDF of the applied 'experimental' heat flux at different times at $(x, y) = (-0.5, 0.0)$ is given in Figure 14.8. Notice that the initially diffuse PDF peaks and shifts towards a value of zero with increasing time because of the damping effect. This is clearly seen in Figure 14.9 that plots the time variation of the mean flux applied.

The PDF of the resultant 'experimental' temperature at location $(x, y) = (-0.4, 0.0)$ is given in Figure 14.10. The initially diffuse PDF slowly peaks and shifts towards zero with increasing time because of the damping effect. The effect of the uncertain thermal diffusivity is also seen in the bimodal structure of the resultant PDF.

The Optimization Problem: Computational Details and Results

A 40×40 quad element discretization of the domain is utilized to solve the inverse problem. The time domain is discretized into $n_t = 50$ equal time steps. The total number of nodal points on the left vertical boundary is $n_y = 41$ and the heat flux at each of these nodal points is assumed to be an independent

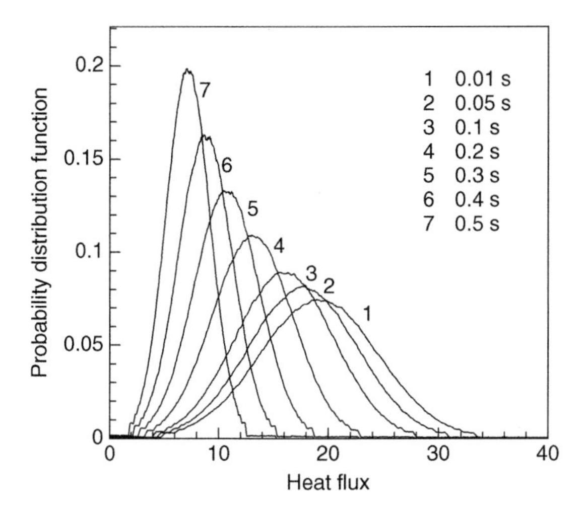

Figure 14.8 The 'experimental' PDF of the heat flux at the location $(x, y) = (-0.5, 0.0)$ at different times.

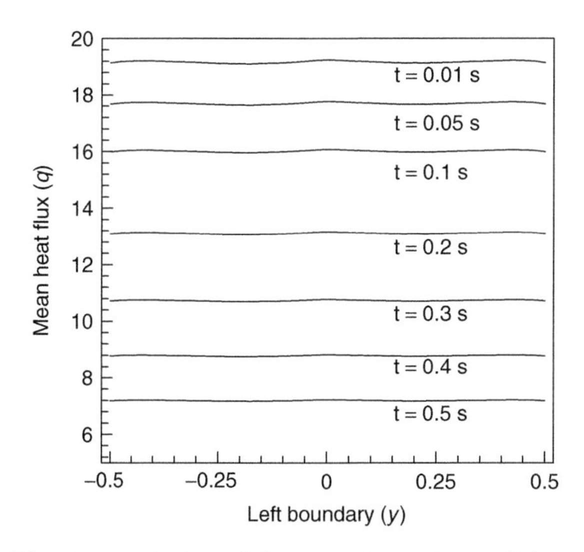

Figure 14.9 The time variation of the mean 'experimental' heat flux applied.

random variable. The variation in the heat flux at each nodal point at each time instant has to be estimated. The total number of random variables that have to be estimated is equal to $n_y \times n_t = 2050$. The thermal diffusivity variation is represented using a level 5 depth of interpolation corresponding to using $n_\alpha = 441$ realizations of the thermal diffusivity field. Each of the random

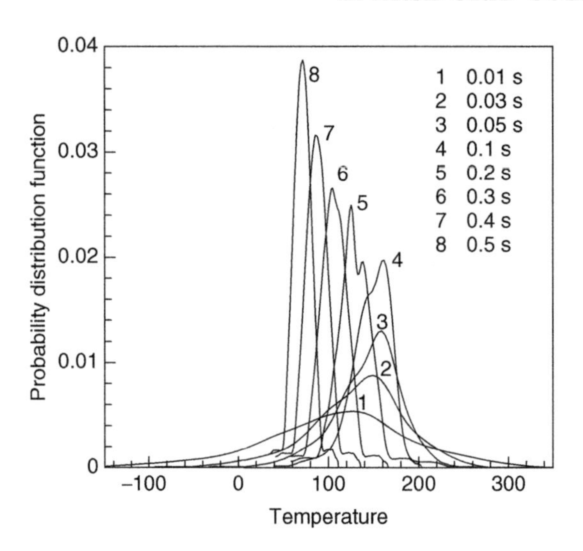

Figure 14.10 The 'experimental' PDF of temperature at the sensor location $(x, y) = (-0.4, 0.0)$ at different times.

fluxes is represented as

$$q(x, t, \xi) = \sum_{i=1}^{n_q} q(x, t, \xi_i) L_i(\xi). \tag{14.34}$$

Without loss of generality, we assume that the heat flux can be represented using one uniform random variable. That is, $\xi = U[0, 1]$. A level 6 depth of interpolation is used to represent each random variable. This corresponds to $n_q = 65$. The total number of design variables is consequently $n_q \times n_y \times n_t = 133250$.

The optimization problem requires the estimation of the sensitivity of the temperature at the sensor locations to perturbations to each of the design variables. A set of decoupled deterministic sensitivity problems were run to construct the stochastic temperature sensitivity. The number of such deterministic sensitivity problems run is $n_\alpha \times n_y \times n_t = 0.9 \times 10^6$. Each iteration of the optimization problem requires the solution of the stochastic forward problem. The stochastic forward problem is solved as a set of decoupled direct deterministic problems. The total number of such direct deterministic problems was $N_{run} = n_\alpha \times n_q = 28665$. Each deterministic problem requires the solution on a 40×40 quad grid for n_t time steps. The number of degrees of freedom (DOF) in each deterministic problem is $N_{det} = (41)^2 \times n_t = 84050$. The total number of DOF in *each* direct stochastic solve of the optimization algorithm is $N_{stochastic} = N_{det} \times N_{run} = 2.41 \times 10^9$. Thus, more than a *billion* DOF are solved at each iteration of the stochastic optimization problem.

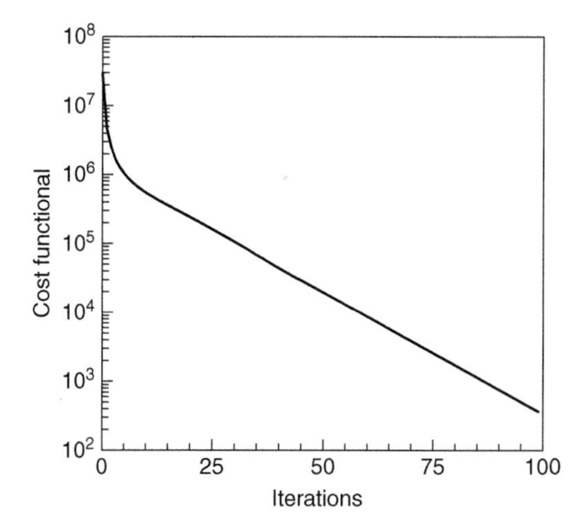

Figure 14.11 Reduction in the cost functional.

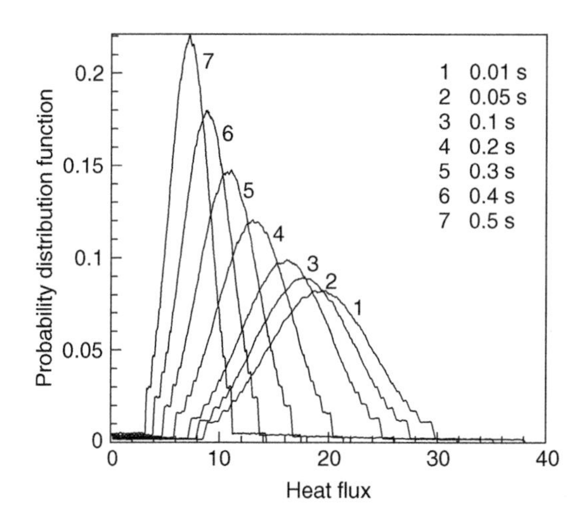

Figure 14.12 The reconstructed PDF of heat flux at the boundary location $(x, y) = (-0.5, 0.0)$ using a depth of interpolation 6).

The reduction in the cost functional with the number of iterations of the stochastic optimization is shown in Figure 14.11. The optimization problem was run using our in-house Linux super computing cluster. Forty nodes, corresponding to 160 processors were utilized for the current problem. Each optimization iteration of the problem took 74 minutes to complete.

Figure 14.12 plots the time evolution of the PDF of the heat flux at one location on the boundary. The stochastic heat fluxes are reconstructed very well though there is some pixelation near the tails of the PDF.

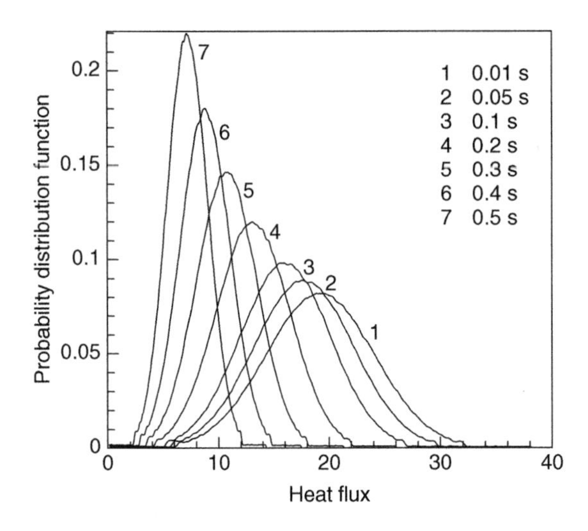

Figure 14.13 The reconstructed PDF of heat flux at the boundary location $(x, y) = (-0.5, 0.0)$ at different times (with depth of interpolation 8).

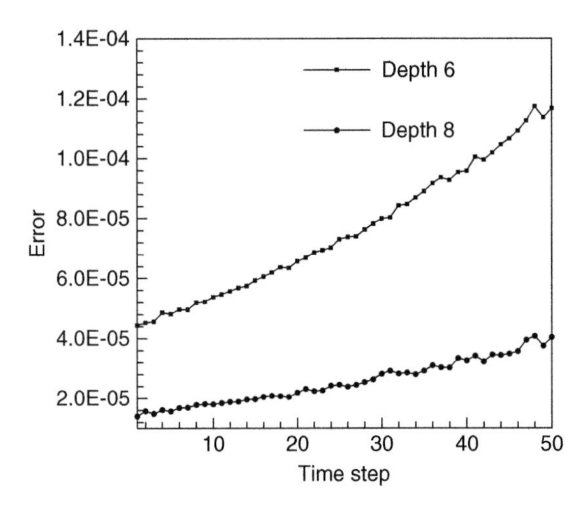

Figure 14.14 Error in the heat flux between the actual and reconstructed solutions.

To see if this pixellation disappears when a finer representation of the stochastic flux is used, another optimization was run using a depth of interpolation 8 of the stochastic heat flux. This corresponds to $n_q = 257$ design variables to represent each unknown. We utilize the hierarchical stochastic optimization method to solve this larger optimization problem. That is, the solution of the previous optimization problem is used as the initial guess of this finer stochastic optimization problem. Figure 14.13 plots the time evolution of the PDF of the heat flux at the same location on the boundary. The pixellation near the tails of the PDF in Figure 14.12 is smoothed out by using a higher-depth of interpolation of the stochastic space.

The difference between the exact and the reconstructed stochastic heat flux is defined in terms of the error, $e = \sum_{i=1}^{n_y}(\Upsilon^{-1}[q_{ex}(\boldsymbol{x}_i), u] - \Upsilon^{-1}[q_{rc}(\boldsymbol{x}_i), u])^2$, where $\Upsilon^{-1}(., u)$ is the corresponding inverse CDF of the heat flux at each nodal point on the left boundary and q_{ex} denotes the actual heat flux while q_{rc} denotes the reconstructed solution. Figure 14.14 plots the error for the two optimization problems.

14.4 Summary

A scalable methodology was introduced that provides the ability to perform design and estimation in the presence of multiple sources of uncertainties while relying purely on the use of deterministic simulators. This is accomplished through constructing the solution of the stochastic direct problem using sparse grid interpolation strategies. A physically motivated framework based on arguments of measurability/observability of random processes is used to pose the stochastic inverse problem. The stochastic sensitivity solution is constructed via repeated calls to deterministic sensitivity/direct problems. The stochastic optimization problem is converted to a deterministic optimization problem in a higher-dimensional space. This naturally allows us to utilize mature deterministic optimization algorithms to solve the stochastic optimization. We also briefly discussed a hierarchical stochastic optimization algorithm which seems to provide significant computational gains. We are currently investigating extensions of the proposed stochastic optimization framework (e.g. coupling with more advanced optimization tools, using model reduction for the stochastic direct and sensitivity problems) along with applications to multiscale estimation.

References

Alifanov, O.M. 1994 *Inverse Heat Transfer Problems*, Springer-Verlag, Berlin.

Babuska, I., Tempone, R. and Zouraris, G.E. 2005a Solving elliptic boundary value problems with uncertain coefficients by the finite element method: the stochastic formulation. *Comput. Methods Appl. Mech. Engrg.* **194** 1251–1294.

Babuska, I., Nobile, F. and Tempone, R. 2005b A stochastic collocation method for elliptic partial differential equations with random input data. *ICES Report* 05–47.

Bashir, O., Ghattas, O., Hill, J., van Bloemen Waanders, B. and Willcox, K. 2007 *Hessian-Based Model Reduction for Large-Scale Data Assimilation Problems, Lecture Notes in Computer Science*, Springer Berlin/Heidelberg.

Deb, M.K., Babuska, I.K. and Oden, J.T. 2001 Solution of stochastic partial differential equations using the Galerkin finite element techniques. *Comput. Methods Appl. Mech. Engrg.* **190** 6359–6372.

Dennis, J.E. and More, J.J. 1997 Quasi Newton methods, motivation and theory. *SIAM Review* **19/1** 46 89.

Ganapathysubramanian, B. and Zabaras, N. 2007 Modelling diffusion in random heterogeneous media: Data-driven models, stochastic collocation and the variational multi-scale method. *J. Comp. Physics* **226** 326–353.

Ganapathysubramanian, B. and Zabaras, N. 2008 A non-linear dimension reduction methodology for generating data-driven stochastic input models. *J. Comp. Physics* **227** 6612–6637.

Gerstner, T. and Griebel, M. 1998 Numerical integration using sparse grids. *Numerical Algorithms* **18** 209–232.

Ghanem, R.G. and Spanos, P.D. 1991 *Stochastic Finite Elements: A Spectral Approach* Dover publications.

Holmes, P., Lumley, J.L. and Berkooz, G. 1996 *Turbulence, Coherent Structures, Dynamical Systems and Symmetry*. Cambridge University Press.

Kaipio, J. and Somersalo, E. 2005 *Statistical and Computational Inverse Problems*, Springer, New York.

Klimke, A. 2006 *Uncertainty modeling using fuzzy arithmetic and sparse grids*, Ph.D. Thesis, Universitt Stuttgart, Shaker Verlag, Aachen.

Ma, X. and Zabaras, N. 2009 An adaptive hierarchical sparse grid collocation algorithm for the solution of stochastic differential equations. *J. Comp. Physics* **228**, 3084–3113.

Nobile, F., Tempone, R. and Webster, C. 2008 A sparse grid stochastic collocation method for elliptic partial differential equations with random input data. *SIAM J. Numer. Anal.* **46** 2309–2345.

Petersdorff, T. and Schwab, Ch. 2006 Sparse finite element methods for operator equations with stochastic data. *Applications of Mathematics* **51** 145–180.

Sampath, R. and Zabaras, N. 2001 A functional optimization approach to an inverse magneto-convection problem. *Comput. Methods Appl. Mech. Engrg.* **190** 2063–2097.

Schwab, Ch. and Todur, R.A. 2001 Sparse finite elements for elliptic problems with stochastic loading. *Numerische Mathematik* **95**, 707–734.

Schwab, Ch. and Todur, R.A. 2003 Sparse finite elements for stochastic elliptic problems–higher moments. *Computing* **71** 43–63.

Somersalo, E. and Calvetti, D. 2007 *Subjective Computing: Introduction to Bayesian Scientific Computing*, Springer Verlag.

Velamur Asokan, B. and Zabaras, N. 2004 Stochastic inverse heat conduction using a spectral approach. *Int. J. Numer. Meth. Eng.* **60**, 1569–1593.

Velamur Asokan, B. and Zabaras, N. 2005 Using stochastic analysis to capture unstable equilibrium in natural convection. *J. Comp. Physics* **209** 134–153.

Venturi, D., Wan, X. and Karniadakis, G.E. 2008 Stochastic low-dimensional modeling of random laminar wake past a circular cylinder. *J. Fluid Mechanics* **606**, 339–367.

Wan, X. and Karniadakis, G.E. 2005 An adaptive multi-element generalized polynomial chaos method for stochastic differential equations. *J. Comp. Physics* **209**, 617–642.

Wang, J. and Zabaras, N. 2005a Hierarchical Bayesian models for inverse problems in heat conduction. *Inverse Problems* **21**, 183–206.

Wang, J. and Zabaras, N. 2005b Using Bayesian statistics in the estimation of heat source in radiation. *Int. J. Heat Mass Transfer* **48**, 15–29.

Wang, J. and Zabaras, N. 2006 A Markov random field model to contamination source identification in porous media flow. *Int. J. Heat Mass Transfer* **49**, 939–950.

Xiu, D. 2007 Efficient collocation approach for parametric uncertainty analysis. *Comm. Computational Physics* **2**, 293–309.

Xiu, D. and Hesthaven, J.S. 2005 High-order collocation methods for the differential equation with random inputs. *SIAM J. Sci. Comput.* **27**, 1118–1139.

Xiu, D. and Karniadakis, G.E. 2003 Modeling uncertainty in flow simulations via generalized polynomial chaos. *J. Comp. Physics* **187**, 137–167.

Zabaras, N. and Ganapathysubramanian, B. 2008 A scalable framework for the solution of stochastic inverse problems using a sparse grid collocation approach. *J. Comp. Physics* **227**, 4697–4735.

Chapter 15

Uncertainty Analysis for Seismic Inverse Problems: Two Practical Examples

F. Delbos, C. Duffet and D. Sinoquet

Institut Francais du petrole, Rueil-Malmaison, France

15.1 Introduction

In the oil industry, many applications lead to large size inverse problems: the unknown parameters describing the subsoil cannot be determined directly by measurements (or only locally), thus parameter estimation problems are usually formulated as inverse problems. Forward simulators compute synthetic measurable data from those parameters. In this chapter, we focus on seismic inverse problems. Among them we can cite:

- the wave velocity determination by traveltime inversion which is an in verse
- the stratigraphic inversion;
- history matching and 4D seismic inversion.

Large-Scale Inverse Problems and Quantification of Uncertainty Edited by L. Biegler, G. Biros, O. Ghattas, M. Heinkenschloss, D. Keyes, B. Mallick, Y. Marzouk, L. Tenorio, B. van Bloemen Waanders and K. Willcox

The Wave Velocity Determination by Traveltime Inversion

This inverse problem belongs to the seismic imaging workflow which consists of processing seismic data in order to obtain geologically interpretable seismic sections (migrated sections in time or depth domain). In exploration phases, geologists can recognize the main structures from these images and locate the interesting parts of the subsoil. This processing consists first of the determination of a velocity macro-model from the traveltimes of the reflected seismic waves and then in a migration step to transform the raw seismic data in migrated time or depth domain (see Ehinger and Lailly 1995).

The Stratigraphic Inversion for Reservoir Delineation and Characterization

This aims at characterizing a static model of the reservoir with quantitative properties (permeabilities, porosities ...) and qualitative properties (geological facies, faults). The stratigraphic inversion determines the elastic properties of the subsoil from time migrated seismic sections, well data and geological prior information (Brac *et al.* 1988; Tonellot *et al.* 2001). These properties are then interpreted in terms of quantitative and qualitative petro-physical properties via geostatistical approaches (Nivlet *et al.* 2006).

History Matching and 4D Seismic Inversion for Dynamic Reservoir Characterization (Berthet *et al.* 2007)

During the production steps of a field, the available data are inverted to determine the dynamic properties of the reservoir in order to make decisions regarding the production scheme and to predict the oil production of the field in the future. The integrated data are the production data acquired at production/injection wells (bottom-hole pressure, gas-oil ratio, oil rate) and 4D seismic data (variations of seismic impedances from 3D seismic data acquired at different calendar times). A fluid flow simulator and a petro-elastic model are used in order to compute the synthetic data from the reservoir parameters (porosities, permeabilities, faults, productivity index of the wells ...).

For these three practical inverse problems which are part of a general workflow with some multiple interpretation steps, it is crucial to estimate the uncertainties on the solution. The noise in data, the modeling errors in forward problems, the under-determination inherent to inverse problems lead to the non-uniqueness of the solution, the obtained solution being only one among others. Thus, in order to avoid misleading in the interpretation of the solutions and to propagate the uncertainties in the different steps of the workflow, a careful posterior uncertainty analysis should be done.

Classically, numerical methods for estimating posterior uncertainties can be divided into two large classes, both based on Bayesian approach: the first one consists of sampling the posterior probability distribution with no assumption on the prior probability density function (prior PDF) (Contreras *et al.* 2005; Escobar *et al.* 2006; Haas and Dubrule 1994; Malinverno and Leaney 2000; Mosegaard and Tarantola 1995 among others). In the second class of methods, under the hypothesis of Gaussian prior PDF, a linearized uncertainty analysis is performed. In our applications, where the size of the model space and of the data space are high-dimensional, these stochastic methods are often rejected for practical issues of CPU cost. Indeed, to be effective, they often need a large number of simulations as for instance in Monte Carlo techniques.

In this chapter, we describe the methodologies developed for uncertainty analysis of the first two seismic inverse problems we mentioned above with a Bayesian formalism. Applications on realistic datasets are presented for both seismic inverse problems.

15.2 Traveltime Inversion for Velocity Determination

Geophysical methods for imaging complex geological subsurface in petroleum exploration requires the determination of an accurate propagation velocity model. Seismic reflection tomography (see Bishop *et al.* 1985) turns out to be an efficient method for that: this method allows us to determine a seismic velocity distribution from traveltimes data associated with seismic waves reflecting on geological surfaces. This inverse problem is formulated as a least-squares minimization problem which consists of the minimization of the mismatch between the observed traveltimes and the traveltimes computed by the forward problem (solved by a ray tracing method).

Classically, the uncertainties on the solution model are estimated thanks to the analysis of the posterior covariance matrix obtained in the linearized framework (see Gouveia and Scales 1997; Tarantola 2005 among others). The computation of this matrix is generally expensive for 3D problems; among others, Berryman (2000) proposed a method to compute approximations of the covariance and resolution matrices from the different iterates of conjugate gradient algorithm used to solve the linearized problem. But the physical interpretation of the terms of this large size matrix may be cumbersome when we want to go further than the classical but restrictive analysis of the diagonal terms. Two different methods based on the linearized inverse problem completed by a nonlinear approach are proposed and applied on a real dataset.

15.2.1 Characteristics and Formulation

Let us first present the traveltime inversion method for velocity determination. The chosen model representation is a *blocky* velocity model where the velocity distribution is described by slowly varying layer velocities delimited

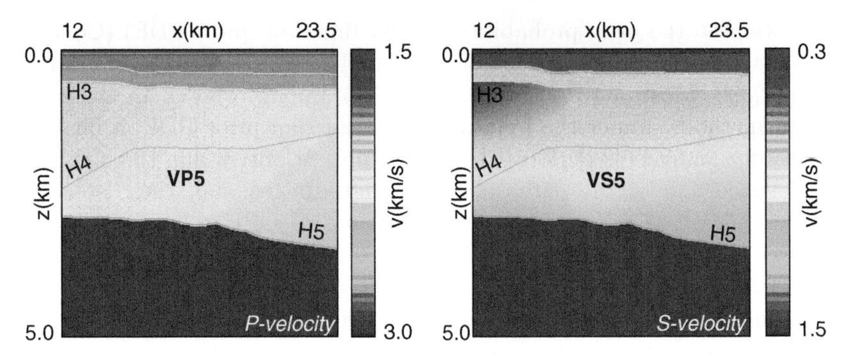

Figure 15.1 Solution velocity model obtained by tomography. Left: P velocity. Right: S velocity. The RMS value of the traveltime misfits is $6.2ms$. The anisotropy parameters values are: $\eta = 6.29\,\%$ and $\delta = -4.43\,\%$.

by interfaces (see Figure 15.1 for an example). The model is thus composed of two kinds of parameters: parameters describing the velocity variations within the layers and parameters describing the geometry of the interfaces delimiting the velocity blocks.[1] Moreover, the velocity anisotropy is modeled by two parameters η and δ (see Stopin 2001 for more details).

The forward problem consists of computing traveltimes of reflected seismic waves, given a model, an acquisition survey (locations of the sources and the receivers) and signatures (generally reflectors where the waves reflect). It is solved by a ray tracing method which is a high frequency approximation of the wave equation (see Cervený 1987 among others). We denote by $T(m)$ the forward modeling operator, that gives, for a specified reflector, the traveltimes associated with all source-receiver pairs.

Reflection traveltime tomography is the corresponding inverse problem: its purpose is to adjust m such that $T(m)$ best matches a vector of traveltimes T^{obs} picked on seismic data. A natural formulation of this problem is to minimize the nonlinear least squares objective function which arises under the assumption of Gaussian PDF for data and prior model (Tarantola 2005, p. 63–64).

$$f(m) := \| T(m) - T^{obs} \|^2_{C_d^{-1}} + \| m - m_{\text{prior}} \|^2_{C_m{}^{-1}}, \tag{15.1}$$

where $\| \, . \, \|_A$ denotes the norm defined by A: $\| x \|_A = x^T A x$, measurements of T follow a Gaussian PDF centered at T^{obs} with a covariance matrix C_d ($T \sim \mathcal{N}(T^{obs}, C_d)$), the unknown model m is a sample of a Gaussian PDF centered at m_{prior}, the prior model (coming, for instance, from geological

[1] In our approach (Jurado *et al.* 2000), the subsurface model m is composed of 2D or 3D B-spline functions describing velocity variations in a layer ($v(x, y) + k.z$ or $v(x, y, z)$) and 2D B-spline functions describing the interfaces ($Z(x, y)$, $Y(z, x)$ or $X(y, z)$).

knowledge and/or additional information resulting from well measurements) and prior covariance matrix C_m (in practice, the prior information term of the cost function is a weighted H^2–norm of the differences between current model and prior model), $m \sim \mathcal{N}(m^{\text{prior}}, C_m)$.

15.2.2 Optimization Method

This large size (up to $\sim$ 100 000–500 000 traveltime data and $\sim$ 5000–10 000 unknowns) nonlinear least-squares problem is solved classically by a Gauss-Newton method based on successive linearizations of the forward operator $T(m)$ which needs the computation of its Jacobian matrix. The computation of the derivatives of T with respect to the model is cheap thanks to the Fermat principle[2] (see Bishop *et al.* 1985). The resulting approximation of the Hessian matrix of the cost function is designed by $H(m) = J(m)^T C_d^{-1} J(m) + {C_m}^{-1}$.

Then, the quadratic approximation

$$F_k(\delta m) := ||J_k \delta m - \delta T_k^{obs}||^2_{C_d^{-1}} + ||m_k - m_{\text{prior}} + \delta m||^2_{{C_m}^{-1}} \qquad (15.2)$$

of the objective function f (see equation 15.1) is minimized at each Gauss-Newton iteration k by a preconditioned conjugate gradient (m_k is the current model, δm is the unknown model perturbation, $J_k = \frac{\partial T}{\partial m}(m_k)$ is the Jacobian matrix evaluated at m_k, and $\delta T_k^{obs} = T^{obs} - T^{cal}(m_k)$). Furthermore, some additional constraints on the model may be introduced via equality or inequality constraints to discriminate the eventual local minima (as illustrated in subsection 15.2.4). A sequential quadratic programming (SQP) approach based on an augmented Lagrangian method is used to solve this nonlinear constrained optimization problem (for more details on the optimization algorithm see Delbos *et al.* 2006).

15.2.3 Uncertainty Analysis

A classical approach to quantify uncertainties consists of the analysis of the Hessian matrix (or its inverse: the posterior covariance matrix) associated with the linearized problem (15.2) around the solution m_{post}. This approach is only valid in the vicinity of the solution model, the size of the vicinity depending on the nonlinearity of the forward map (Tarantola 2005).

The bi-linear form associated with the Hessian matrix measures the influence of a model perturbation δm on the quadratic cost function defined around

[2]The derivatives of travel times with respect to velocity parameters are calculated by integrating a perturbation of the velocity along the rays calculated in the background model. The derivatives with respect to interface parameters are the integration of the background velocity along the shortened (negative contribution) or extended (positive contribution) part of the ray resulting from the interface displacement (we consider only the vertical displacement of impact point).

the solution m_{post}:

$$F(\delta m) - F(0) = \frac{1}{2}\delta m^T (J_{m_{\text{post}}}^T C_d^{-1} J_{m_{\text{post}}} + C_m{}^{-1})\delta m \tag{15.3}$$

with $J_{m_{\text{post}}} = J(m_{\text{post}})$ and $F(0) = f(m_{\text{post}})$.

The posterior covariance matrix is thus defined by the inverse of $H(m_{\text{post}})$ (see Tarantola 2005)

$$C_{\text{post}} = (J_{m_{\text{post}}}^T C_d^{-1} J_{m_{\text{post}}} + C_m{}^{-1})^{-1}. \tag{15.4}$$

The credible region of models can be characterized by the contour lines

$$(m - m_{\text{post}})^T C_{\text{post}}^{-1} (m - m_{\text{post}}) = constant, \tag{15.5}$$

which are ellipsoïds of center m_{post} and correspond also to contour lines of the posterior Gaussian PDF:

$$\exp\left(-\frac{1}{2}\delta m^T C_{\text{post}}^{-1} \delta m\right). \tag{15.6}$$

The diagonal terms of C_{post} are the uncertainties on the parameters describing the model and the off-diagonal terms are the correlations between these uncertainties. For instance, the probability that the true model parameter p_i verifies

$$-2(C_{\text{post}})_{i,i} \leq p_i - p_{\text{post}} \leq 2(C_{\text{post}})_{i,i}, \tag{15.7}$$

independently of the values of the other model parameters, is about 95 %. To take into account the correlations between the parameters, we should study the 95 % contour line of the PDF. The axes of the ellipsoïds (15.5) are defined by the eigenvectors of C_{post}, the square root of the eigenvalues giving the uncertainties on the associated eigenvector.

Remark 15.2.1 In practice, the diagonal terms of C_{post} provide the uncertainties on the B-spline parameters, unknowns of the discretized inverse problem: they are not physical quantities. We would rather compute uncertainties on physical quantities, for instance, the evaluation of the B-spline functions in the physical domain.

Remark 15.2.2 Following Delprat-Jannaud and Lailly (1992), the eigenvector decomposition of the posterior covariance matrix gives access to the worst/best determined model directions and the associated uncertainties. The best (resp. worst) determined model direction corresponds to the smallest (resp. highest) eigenvalue of the posterior covariance matrix, i.e. highest (resp. smallest) perturbation of the quadratic cost function. The uncertainties associated with these eigenvectors, namely the square root of the eigenvalues, are meaningless, the eigenvectors being composed of mixed velocity and interface parameters, combinations which are usually difficult to link to physical quantities.

From those remarks, we propose two methods to quantify geological uncertainties on the solution model which avoid the expensive computation and the cumbersome analysis of the generally large posterior covariance matrix C_{post}:

- The first proposed method to quantify uncertainties is the simulation of admissible models from the posterior PDF. The simulations directly provide interpretable results, i.e. physical models. The method (see for instance Duffet and Sinoquet 2006; Parker 1994) consists of random simulations of model perturbations following the posterior PDF. A Cholesky factorization of the Hessian matrix is necessary which limits the feasibility of this method for large datasets.
- The second proposed method allows us to deal with large size models at a reasonable cost and provides uncertainties on chosen physical quantities. The approach consists of building macro-parameters (MP) with a geophysical interest. These macro-parameters are linear combinations of the inverted parameters such as the mean of the velocity variations in a zone, the slope of an interface, the average thickness of a layer, etc. Grenié (2001) has introduced the notion of macro-parameter (his main motivation being to avoid numerical problems in the inversion of the complete Hessian). We propose here a generalization of his work (general definition of macro-parameters) which allows the computation of uncertainties for large size 3D problems.

We define a macro-parameter as:

$$P = Bp, \tag{15.8}$$

where p is the n model parameters vector (B-spline parameters in our case), P is the n_{MP} macro-parameters vector and B is the condensation matrix. We compute the posterior covariance matrix $\tilde{C}_{\text{post}}$ in the macro-parameter space:

$$\tilde{C}_{\text{post}} = BC_{\text{post}}B^T. \tag{15.9}$$

Note that $\tilde{C}_{\text{post}}$ (a $n_{MP} \times n_{MP}$ matrix) is small compared to C_{post} (a $n \times n$ matrix) since $n_{MP} \ll n$. To obtain this matrix, we do not need to compute the whole inverse of the Hessian, $C_{\text{post}} = H^{-1}$, in the parameter space. Indeed, to obtain $\tilde{C}_{\text{post}} = \tilde{H}^{-1} = BH^{-1}B^T$ in the MP space, we just need $H^{-1}B_j^T$, where B_j^T are the different columns of B^T. We thus solve n_{MP} linear systems, $\tilde{H}_j^{-1} = BH^{-1}B_j^T$, that are similar to the linearized problem we solve at each Gauss-Newton iteration. Thus, the computational cost for one MP is comparable to one iteration of the inversion process.

15.2.4 Application

We consider a 2D real data set already studied by Broto *et al.* (2003) with PP and PS[3] data. 45338 traveltime data were interpreted and an uncertainty of $5\,ms$ (resp. $8\,ms$) are associated with PP data (resp. PS data). A layer-stripping approach (separate inversion of each velocity layer from the shallowest layer to the deepest one) provides the velocity model of Figure 15.1.

The model is described by four interfaces, corresponding to the interpreted events ($H1$, $H3$, $H4$ et $H5$) which define three layers with only lateral velocity variations for the two first upper layers ($v(x)$, there is no vertical variations) and the deepest layer stretching from $H3$ to $H5$ with a 2D velocity $V\ P5(x, z)$ and $V\ S5(x, z)$ (Figure 15.1). V P5 and V S5 define respectively the vertical velocity propagation of the P-waves and the velocity propagation of the S-waves. This model is composed of 4588 parameters, 592 for the interfaces and 3936 for the velocities. In the deepest layer, we assume a Vertical Transverse Isotropic velocity field characterized by V P5 and V S5 and the two parameters η and δ measuring respectively the velocity an-ellipticity and the near vertical velocity propagation of the P-waves (see for details Thomsen 1986 and Stopin 2001). The values of η and δ in the model correspond to a strongly anisotropic medium. The traveltime misfits for the solution model obtained by traveltime inversion are consistent with the data uncertainties (RMS $= 6.2\,$ms).

As already shown by Stopin (2001), it turns out that the anisotropy parameter δ is strongly unidentifiable from seismic data. The value of δ parameter was obtained in Broto *et al.* (2003) by a trial and error approach in order to match approximately the depth of $H5$ horizon given at well location.

We propose to carry on a posterior uncertainty analysis in order to quantify the uncertainties on this solution model: we focus on the anisotropic layer delimited by $H3$ and $H5$ (velocities $V\ P5$ and $V\ S5$).

Application of the Simulation Method

Figure 15.2 shows the 100 simulated admissible models obtained from the solution model of Figure 15.1: variations of velocities $V\ P5$ along x and z directions and histograms of the anisotropy parameters η and δ. From these simulations, we observe that the highest uncertainties on the lateral velocity variations are located at the boundaries of the model, areas that are not well illuminated by the rays. Indeed, for a slice at constant $z = 2.7\,$km, we observe uncertainties around $690\,$m/s for $V\ P5$ at the boundaries of the model and uncertainties around $450\,$m/s in the illuminated parts of the model. For the anisotropy parameters η and δ, we notice uncertainties around $0.3\,\%$ on η and $2\,\%$ on δ.

[3] PS-wave results from the conversion at the reflector of a down-going P-wave (compressional wave) into an up-going S-wave (shear wave). In opposition to the PS-wave, the pure P mode is often called the PP-wave.

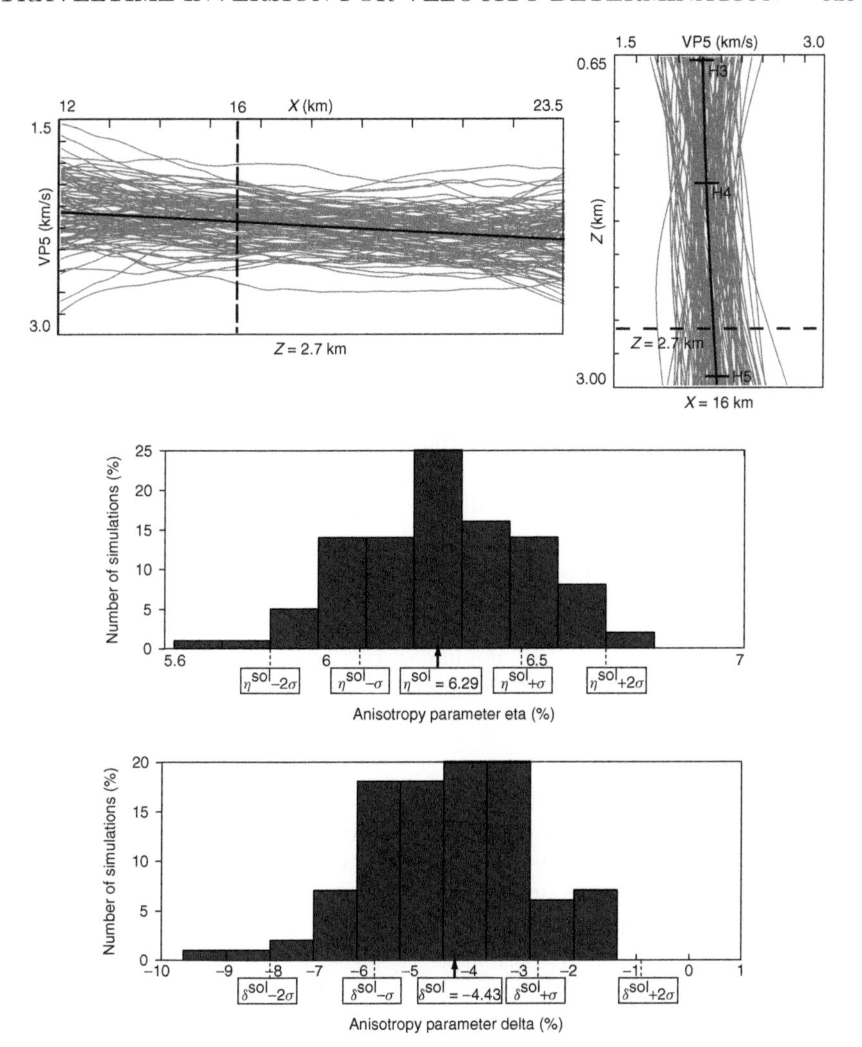

Figure 15.2 100 simulated velocity admissible models : P-velocity variations along x axis at $z = 2.7$ km (top-left), along z axis at $x = 16$ km (top-right), anisotropy parameters η and δ (bottom). σ corresponds to the standard deviation on the anisotropy parameter η (resp. δ) given by the macro-parameter method.

This method is quite attractive for the straightforward interpretation of the results despite its cost (cost of the Choleski decomposition): it provides admissible physical models leading to small perturbations of the quadratic cost functions (15.2).

Application of the MP Method

We choose simple MP, the mean of the velocity variations and the mean of the interface depth in the illuminated part of the layer (region covered by rays). The results are listed in Table 15.1. The uncertainty on anisotropy parameter δ is high: $\pm 74\,\%$ (if we consider twice relative standard deviation) and we observe high correlations between η and δ (0.93) and also correlations between anisotropy parameters and $H5$ depth (-0.33 and -0.36). We also notice the bad determination of the velocities: $\pm 22\,\%$ (relative value). All these results are consistent with the results obtained by the first method. This method with its general formalism allows us to highlight easily strong coupling between the chosen MP. It provides interesting information on the correlations of macro-parameter uncertainties at a lower computational cost than simulations of the admissible model parameters using the complete posterior covariance matrix. We could also perform Gaussian simulations of the macro-parameters using the reduced posterior covariance matrix.

Limitations of the Linearized Approach

As already mentioned, the two proposed methods rely on the quadratic approximation (15.2) of the nonlinear cost function. Traveltime misfits have been computed for 20 simulated models chosen among the 100 models obtained by simulation method. There is only one model which is

Table 15.1 Standard deviations in bold (square roots of the diagonal terms of the posterior covariance matrix) associated with the MP 'mean in the illuminated part of the model' and the normalized correlations between the uncertainties on the different defined MP. This uncertainty analysis is performed for the solution model of Figure 15.1. The normalized values of the standard deviations ($100\sigma/|value^{sol}|$) are also indicated in italic.

	VP5	***VS5***	η	δ	$H4$	***H5***
VP5	**475.1 m/s** *(22 %)*	0.002	−0.03	−0.02	0.005	0.01
VS5	0.002	**168.9 m/s** *(22 %)*	−0.04	−0.03	0.005	0.01
η	−0.03	−0.04	**0.22 %** *(4 %)*	0.93	−0.16	−0.33
δ	−0.02	−0.03	0.93	**1.6 %** *(37 %)*	-0.17	−0.36
$H4$	0.005	0.005	−0.16	−0.17	**77.1 m** *(5 %)*	0.06
H5	0.01	0.01	−0.33	−0.36	0.06	**80.3 m** *(3 %)*

not acceptable (11.6 ms). The others are admissible models with a RMS traveltime misfits bounded by 7 ms. The simulation method has then provided model perturbations that correspond to small perturbations of the quadratic cost function but also to small perturbations of the nonlinear cost function (RMS of traveltime misfits remain small). It shows for this example, the good agreement between the quadratic cost function and the nonlinear one around the solution model. Nevertheless, we should keep in mind that this approach is only valid in the vicinity of the solution model (linearized framework) and complex cases may require a nonlinear approach. Indeed, the simulated model with a RMS traveltime misfits of 11.6 ms indicates that some of the simulated models may produce unacceptable traveltimes misfits ($RMS \approx 11$ ms and MAX = 25 ms for PP data and = 41 ms for PS data). It shows the limitation of the quadratic approximation of the nonlinear cost function and thus the limitation of the linearized methods of uncertainty quantification. Moreover Delbos *et al.* (2006) have shown that a tomographic inversion with constraints on the location of the interfaces at well locations provided a very different model (model 2 described in Table 15.2) from the solution model of Figure 15.1. The two models verify the traveltime data with the expected accuracy. This second model does not belong to the range of possible models detected by the two methods we proposed: indeed, δ is equal to 15.94 % whereas the expected range of values was between −8 % and −1 % (see for instance Figure 15.2). It shows again the limitations of the linearized approach.

A Pragmatic Nonlinear Uncertainty Analysis

To perform a nonlinear analysis we have chosen an experimental approach which consists of performing several constrained inversions allowing us to test different geological scenarios to try to delimit the space of admissible

Table 15.2 Three different solution models obtained by traveltime inversion with different constraints.

	Model 1	*Model 2*	*Model 3*
RMS of traveltime misfits	6.0 ms	6.0 ms	6.37 ms
mean of the P-velocity	2.849 km/s	3.623 km/s	2.895 km/s
mean of the S-velocity	1.229 km/s	1.621 km/s	1.213 km/s
η	6.29 %	8.27 %	6.2 %
δ	−4.4 %	15.94 %	2 %
mean of the depth of the reflector H4	1.838 km	1.755 km	1.659 km
mean of the depth of the reflector H5	3.167 km	3.021 km	2.882 km

solutions. For instance, we could test different hypothesis on the values of the anisotropy parameters for which a strong uncertainty has been detected by the linearized approach. If we introduce a constraint on the anisotropy parameter δ, such as $\delta > 0$, we find the model 3 (Table 15.2): this result expands the range of admissible models (variability of δ detected by MP approach was $-4.43 \pm 3.2\,\%$). In the same way, we can test the variability of the anisotropy parameter η. We notice (see Table 15.2) that the value of the anisotropy parameter η of the model 2 is 8.27 %, but all the simulated values of this parameter are around 6 % and the variability of η provided by MP method is $\pm 0.44\,\%$. Nevertheless, tomographic inversion under the constraint $\eta \leq 4\,\%$ or $\eta \geq 8.5\,\%$ does not provide a model that satisfies both these constraints and the data with the expected accuracy.

For this middle size inverse problem, we have shown that different methods of uncertainty analysis may be applied in practice: the simulation method which computes the range of admissible models (under the linearization assumptions), the macro-parameter approach which allows to focus the uncertainty analysis on geological quantities and finally, a nonlinear approach based on constrained optimization which tests geological scenarii and complete the linearized approaches.

15.3 Prestack Stratigraphic Inversion

Prestack stratigraphic inversion aims at determining the subsurface elastic parameters (P-impedance, S-impedance and density) from prestack seismic data. It belongs to one step of the seismic reservoir characterization problem which consists of building relations between seismic data and reservoir properties and allows a better characterization of reservoirs. This characterization depends on quantitative properties (permeability, porosity, net to gross etc. ...) as well as qualitative properties (geological facies). The general workflow of reservoir characterization at IFP is divided into two steps:

- well to seismic calibration (see Lucet *et al.* 2000), prior model building and prestack stratigraphic inversion,
- qualitative and / or quantitative interpretation.

The first step is devoted to compute the elastic parameters of a subsurface model while eliminating laterally uncorrelated noise. This step integrates different types of data: seismic data, well logs and structural information. The second step consists of interpreting the elastic parameters previously obtained at the first step in terms of lithologies or reservoir properties using qualitative or quantitative geostatistical approaches. These two steps allow to define a seismic constraint for the construction of static reservoir model. In addition they are also used in the construction of the dynamic reservoir model (history matching constrained by 4D seismic, see Nivlet *et al.* 2006).

In this section we propose an application of a deterministic uncertainty analysis in prestack stratigraphic inversion. The deterministic approaches are usually based on a Gaussian assumption: all the PDF(s) (prior and posterior) are assumed Gaussian. This strong hypothesis leads to an explicit form of posterior PDF (see Section 15.2.3 and Tarantola 2005). The numerical computation of this PDF for stratigraphic inversion has been the subject of numerous studies. Among these studies, we note in particular the frequency method proposed by Buland *et al.* (2003). This method, based on a fast Fourier transform algorithm, is in practice very effective in propagating the uncertainties from the inversion up to interpretation. However, it is submitted to stringent assumptions: linearized inversion and strict stationarity of the prior PDF (which implies a single stratigraphic unit). Other studies focus on the relaxation of the Gaussian assumption in the deterministic approach. For example, Malinverno and Briggs (2004) suggest a Bayesian hierarchical approach in which samples are made by MCMC in the space of hyper-parameters. To our knowledge, this work has not been applied to the inversion of 3D seismic data.

Our work at IFP belongs to the class of deterministic methods with Gaussian assumptions. A linearized posterior uncertainty analysis is possible in the vicinity of the solution by the analysis of the posterior covariance matrix as for traveltime inversion (see Section 15.2). The exact computation of this matrix consists of inverting the Hessian of the objective function at the solution. For 3D problems this task is generally very expensive or even not realizable, the dimension of the model space being several millions. In case of white noise covariance function in the model space, the posterior covariance matrix is block diagonal. It is then possible to sequentially invert block diagonal matrices of reasonable dimension (see Delbos *et al.* 2008). In other cases, when the prior covariance function in the model space is exponential, we present an alternative approach: the use of preconditioning methods in order to compute an approximation of the diagonal of the posterior covariance matrix. Studying the structure of the Hessian, we argue which of these methods are best fitted for stratigraphic inversion. Limitations and extensions of the chosen method are discussed. And the proposed method is illustrated on a realistic PP data set.

15.3.1 Characteristics and Formulation

Let us first recall the problem of interest and introduce some notations. We consider discrete models that describe the P-wave impedance Z_P, the S-wave impedance Z_S and density ρ in the subsurface. The subsurface is discretized on a regular grid composed of more than ten millions of points in real 3D applications: $n_x * n_y * n_t \sim 1.E + 7$, where $n_x \sim 100$ and $n_y \sim 100$ are the number of seismic traces in the horizontal directions and $n_t \sim 1000$ is the number time samples. The model vector m is defined as $m^T = (Z_P^T, Z_S^T, \rho^T)$.

The observed seismic data are composed of n_θ cubes (n_θ being around 3 to 8) sampled on the same regular grid as the model and d_θ^{obs} is the seismic data cube corresponding to the reflection angle θ.

The input for the Bayesian prestack stratigraphic inversion are:

- the forward model which is a 1D convolution operator defined by

$$d_\theta^{\text{synt}}(m) := r_\theta(m) * w_\theta, \tag{15.10}$$

  where d_θ^{synt} is the synthetic data, r_θ is the Aki-Richards reflectivity series corresponding to the current model m and to the angle θ and w_θ is the wavelet assumed to be known at angle θ (see Aki and Richards 2002 for more details about the forward problem).

- the conditional distribution of the data d_θ^{obs} given the model m is Gaussian $\mathcal{N}(d_\theta^{\text{synt}}(m), C_d)$. The matrix C_d is diagonal (kernel of a white noise covariance operator) which means that the errors on the observed seismic data d_θ^{obs} are assumed uncorrelated. This matrix is defined by the hyper-parameters σ_d which are the standard deviations of the errors on the observed seismic data.

- the prior distribution for m is Gaussian $\mathcal{N}(m_{\text{prior}}, C_m)$. The matrix C_m and the vector m_{prior} describe a Gaussian prior PDF in the model space. The matrix C_m, as defined by Tonellot *et al.* (1999) is the kernel of the covariance operator in the model space which is exponential along the correlation lines and a white noise in the orthogonal direction (see Example 5.5 of Tarantola 2005 for an example of an exponential covariance function). C_m is defined by the following hyper-parameters: the prior model standard deviations σ_m and the correlation lengths λ (in x and y directions).

15.3.2 Optimization Method

The geological term, Equation (15.13), of the stratigraphic objective function is crucial to reduce the under-determination of the stratigraphic inverse problem: adding geological information enables first to have a well posed problem and second to reduce the range of solution models that fit the seismic data. The maximum a posteriori model (MAP), noted m_{post}, is the solution model of the nonlinear least-squares optimization problem

$$\min_{m \in \mathbb{R}^n} f(m) := f_S(m) + f_G(m), \tag{15.11}$$

where n is the number of unknowns. The objective function $f : \mathbb{R}^n \to \mathbb{R}$ of this optimization problem is composed of two terms: a seismic and a geologic one. The seismic term

$$f_S(m) := \sum_\theta \| d_\theta^{\text{synt}}(m) - d_\theta^{\text{obs}} \|^2_{C_d^{-1}} \tag{15.12}$$

measures the total mismatch between the synthetic seismic data d_θ^{synt} associated with the current model m with the angle θ and the observed seismic data. The geological term

$$f_G(m) := \| m - m_{\text{prior}} \|^2_{C_m{}^{-1}} \tag{15.13}$$

measures the mismatch between the a priori and predicted model parameters, where m_{prior} is the elastic prior model.

The nonlinear least squares objective function f is minimized using a standard nonlinear conjugate gradient (NCG) algorithm globalized by a line search method. More details about the prestack stratigraphic inversion algorithm are given by Tonellot *et al.* (2001). In practice, only around 10–20 NCG iterations are sufficient to provide a good estimate of the MAP model m_{post}. Note that a BFGS or a l-BFGS algorithm can also be applied to solve problem (15.11). Since the problem is weakly nonlinear, the iterates of both methods are very close. Then, the NCG algorithm has the advantage that it requires the less storage in memory.

15.3.3 Uncertainty Analysis

The stratigraphic inversion being a very large inverse problem, except for methods subject to hard extra assumptions on the prior information (see Buland *et al.* 2003) complete and accurate estimation of posterior uncertainties of this problem is currently unrealistic. Moreover, even if one were able to estimate these uncertainties, there still remains the question of properly using such a vast amount of information. This is why we restrict ourselves to the estimation of the diagonal elements of the covariance matrix C_{post}. By a linearization of the operator r_θ around the MAP model m_{post} and Gaussian assumptions (see Section 15.2.3 for more details about this approach), an explicit form of the posterior PDF can be obtained. This distribution is defined by the MAP model m_{post} and by the posterior covariance matrix

$$C_{\text{post}} = \left[\left(\sum_\theta J^T W_\theta C_d^{-1} W_\theta^T J \right) + C_m{}^{-1} \right]^{-1} = [H_S + H_G]^{-1} = H^{-1}, \tag{15.14}$$

where J is the Jacobian matrix of r_θ around m_{post} and W_θ is the convolution matrix of the wavelet at angle θ. It is straightforward to show that the first term is a Gauss-Newton approximation of the Hessian H_S of f_S around m_{post}. The second term, which is the inverse of the covariance prior matrix on the model is the Hessian H_G of f_G. Then computing C_{post} requires two steps: the computation of the matrices H_S and ${C_m}^{-1}$ and then the computation of the diagonal elements of C_{post}.

For the first step we can use an analytical formula. Delbos *et al.* (2008) gives an explicit expression of the elements of the matrix H_S and Tonellot (2000) shows that the inverse of C_m is given by an explicit analytical formula including first derivatives of m with respect to spatial coordinates. The inverse of C_m can then be computed using a finite element approximation.

The key contribution of this application is to the second step: how to compute an approximation of the diagonal elements of C_{post} which is the inverse of the large matrix H. The novel idea we present here is to use the well-known preconditioning methods of linear algebra. These methods are classically used to speed up the convergence of iterative methods (for example the conjugate gradient algorithm) when solving large linear system. It consists of transforming the linear system $Ax = b$ to the equivalent linear system

$$P^{-1/2}AP^{-1/2}P^{1/2}x = P^{-1/2}b, \tag{15.15}$$

where P is the symmetric positive definite preconditioning matrix. The objective of preconditioning is then to choose P as close as possible to A ($P \approx A$) so that the condition number of $P^{-1}A$ is close to 1 or the eigenvalues of $P^{-1}A$ are more clustered. The novel idea we propose in this section is to hijack this classical use of preconditioning methods in order to get an approximation of the diagonal elements of the posterior covariance matrix C_{post}. In our case, it consists of choosing a preconditioner P for H such that P^{-1} can be fast computed and such that the diagonal elements of C_{post} can be approximated by the diagonal elements of P^{-1}:

$$\text{DIAG}(P^{-1}) \approx \text{DIAG}(C_{\text{post}}).$$

Among all the preconditioning methods we particularly focused our attention on two different methods that seem better suited to our goal:

- sparse approximate inverse methods (SPAI),
- polynomial preconditioner methods.

Details about the first approach, the SPAI methods, can be found in the work of Chow (1999). It consists of transforming the inverse matrix problem $HM = I$ into the optimization problem

$$\min_M \| HM - I \|_F^2 := \sum_{j=1}^{n} \| (HM - I)\delta_j \|_2^2,$$

where $\| \, . \, \|_F$ is the Frobenius norm and δ_j is a vector with zeros in all components except in the jth position where the value is one. It is then equivalent to solve n optimization problems

$$\min_{M_j} \| HM_j - \delta_j \|_2^2, \quad j \in 1, ..., n, \tag{15.16}$$

where for each problem the jth column M_j of M is the unknown. These optimization problems are generally solved using a QR decomposition of H (see Grote *et al.* 1997). In order to facilitate the solution of the optimization problems, a sparsity pattern is imposed on M, so as to reduce the number of degrees of freedom. The choice of this sparse structure is crucial to get a good approximation of M and to limit CPU computational cost. Indeed, the larger the number of degrees of freedom, the better the approximation is but the higher the computational cost of the optimization problems. The difficulty is therefore to find a compromise between these two conflicting objectives. Alleaume and Sinoquet (2005) shows that SPAI methods can be used to obtain a good approximation of the diagonal elements of C_{post} on 2D models. But their main drawback is that they are not tractable for 3D models. In fact, the matrix H is then too large to be stored in memory which is required by the method.

The second approach consists of using polynomial preconditioner methods. Meurant (1999) describes a wide variety of polynomial preconditioners and their practical implementation. Among this wide variety we have adopted the Truncated Neumann series method. Given a matrix A with spectral radius $\rho(A) < 1$, this method consists of approximating the inverse of a matrix $(I + A)$ by

$$(I + A)^{-1} \approx \sum_{k=0}^{l} (-A)^k, \tag{15.17}$$

where l is the order of the Neumann polynomial. The simplest idea is to generate a few terms of the Neumann series (in practice only degrees $l = 1$ and $l = 3$ are used). The even degrees (for example $l = 2$ or $l = 4$) are not used as it was shown by Dubois *et al.* (1979) that a Neumann polynomial of odd degree k is more efficient than the Neumann polynomial of even degree $k + 1$. For 3D prestack inversion, due to the properties of the covariance matrices C_d and C_m, the matrix H to be inverted has the following $N \times N$ block structure:

$$H = \begin{bmatrix} D_1 & E_1 & 0 & F_1 & 0 & 0 & 0 \\ E_1^T & D_2 & E_2 & 0 & F_2 & 0 & 0 \\ 0 & \ddots & \ddots & \ddots & 0 & \ddots & 0 \\ F_1^T & 0 & \ddots & \ddots & \ddots & 0 & F_{N-n_x} \\ 0 & F_2^T & 0 & \ddots & \ddots & \ddots & 0 \\ 0 & 0 & \ddots & 0 & \ddots & \ddots & E_{N-1} \\ 0 & 0 & 0 & F_{N-n_x}^T & 0 & E_{N-1}^T & D_N \end{bmatrix}, \tag{15.18}$$

where $N = n_x * n_y$ is the number of seismic traces per seismic cube and D_i, E_i and F_i are block matrices of dimension $(3 * n_t) \times (3 * n_t)$. This matrix structure arises quite often in partial differential equations of 3D problems. Note that in case of 2D prestack inversion (inversion of angle dependent seismic sections) the structure is identical except for the block matrices F_i which are null (the matrix H being then a tridiagonal block matrix). Using block structure (15.18), the matrix H can be rewritten as

$$H = D + E + F = D\left[I + D^{-1}(E + F)\right],$$

where D, E and F are the matrices built from structure (15.18) keeping respectively blocks $(D_i)_{i=1,N}$, $(E_i)_{i=1,N}$ and $(F_i)_{i=1,N}$ (and their transposed). Using Equation (15.17) and assuming that $\rho(D^{-1}(E + F)) < 1$, we can then approximate C_{post} (the inverse of H) by the truncated Neumann polynomial at the order l

$$C_{\text{post}} \approx P_l^{-1} := D^{-1} \sum_{k=0}^{l} \left[-D^{-1}(E + F)\right]^k. \qquad (15.19)$$

Note that, even for $l = 1$, this last polynomial preconditioner approximation is difficult to compute (H being large). But, remember that we are only looking for the diagonal elements of C_{post}:

$$\text{DIAG}(C_{\text{post}}) \approx \text{DIAG}\left(D^{-1}\right) + \sum_{k=1}^{l} \text{DIAG}\left(D^{-1}\left[-D^{-1}(E + F)\right]^{2k}\right).$$

Furthermore, a nice property (not demonstrated here) is that the diagonal blocks of $(D^{-1}(E + F))^k$ are null for each odd k. Hence, the first and third order approximation of the diagonal elements of C_{post} are

$$\text{DIAG}(P_1^{-1}) = \text{DIAG}\left(D^{-1}\right) \quad \text{and} \qquad (15.20)$$

$$\text{DIAG}(P_3^{-1}) = \text{DIAG}\left(D^{-1}\right) + \text{DIAG}\left(D^{-1}(D^{-1}(E + F))^2\right). \qquad (15.21)$$

The main advantage of both polynomial preconditioner approximations is that the computation of the diagonal elements belonging to the ith diagonal block involves a limited number of blocks of H:

$$\text{DIAG}(P_1^{-1})_i = \text{DIAG}\left(D_i^{-1}\right) \quad \text{and} \qquad (15.22)$$

$$\begin{aligned} \text{DIAG}(P_3^{-1})_i = {} & \text{DIAG}\left(D_i^{-1}\right) \\ & + \text{DIAG}\left(D_i^{-2}\left[E_{i-1}^T D_{i-1}^{-1} E_{i-1} + E_i^T D_{i+1}^{-1} E_i \right.\right. \qquad (15.23) \\ & + \left.\left. F_{i-n_x}^T D_{i-n_x}^{-1} F_{i-nx} + F_{i+n_x-1}^T D_{i+n_x}^{-1} F_{i+nx-1}\right]\right). \end{aligned}$$

Using these last two formulae, we have implemented an efficient algorithm to compute the first and third order polynomial approximations of the diagonal elements of C_{post}. Note that this algorithm is intrinsically parallelizable which enables us to save CPU time and allows an uncertainty analysis to be possible for huge 3D stratigraphic inversion problems.

15.3.4 Application

In order to compute the exact standard deviations on the MAP model m_{post} and to compare them with approximated standard deviations, a small size 2D synthetic application is studied. The exact model, see Figure 15.3, is a limited 2D cross-line of a 3D realistic model representative of a turbiditic complex channel in a deep offshore environment (see Bourgeois *et al.* 2005). The inversion window is 360 m long and 200 m thick. This area is discretized on regular a grid of 19 traces (one each $\Delta_x = 20\,m$) and of 100 time samples (one each $\Delta_t = 2\,ms$) which leads to $n = 5700$ unknowns. Using the exact model, the P-P Aki-Richards reflectivity formula and a 1D convolution operator (see Equation (15.10)), a 2D seismic dataset is generated. Four angle-limited stacks are generated ($n_\theta = 4$) : 05°–15°, 15°–25°, 25 and 35°–45° degrees. The wavelet used for each angle is a 5–10–60–80 Hz band-pass filter. The prior model, displayed in Figure 15.3 is the exact model filtered with a 20-40 Hz low-pass filter. The covariance matrices C_d and C_m are filled up with the following hyper-parameters: standard deviations σ_d on the seismic data error of 5 % (with respect to the signal average energy) and standard deviations σ_m on the prior model error of 10 % (with respect to the average values of the prior elastic parameters). Five correlation lengths λ_x defining the exponential covariance matrix C_m have been chosen between Δ_x and $10\Delta_x$. Then, for each λ_x, a MAP model m_{post} is computed through the inversion algorithm (see Section 15.3.2). For each posterior model m_{post}, we have plotted, in Figure 15.4, its 100 standard deviations σ_{post} ($(\sigma_{\text{post}}{}^2)_i$ being the ith diagonal elements of C_{post}) associated to the 9th trace and to the

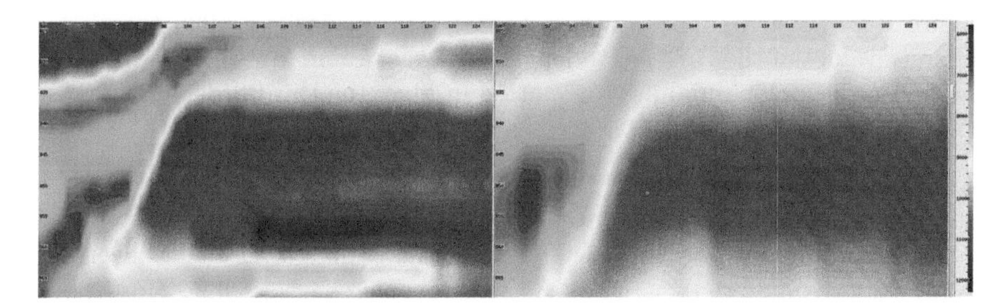

Figure 15.3 Exact P-impedance model (left); prior P-impedance model (right).

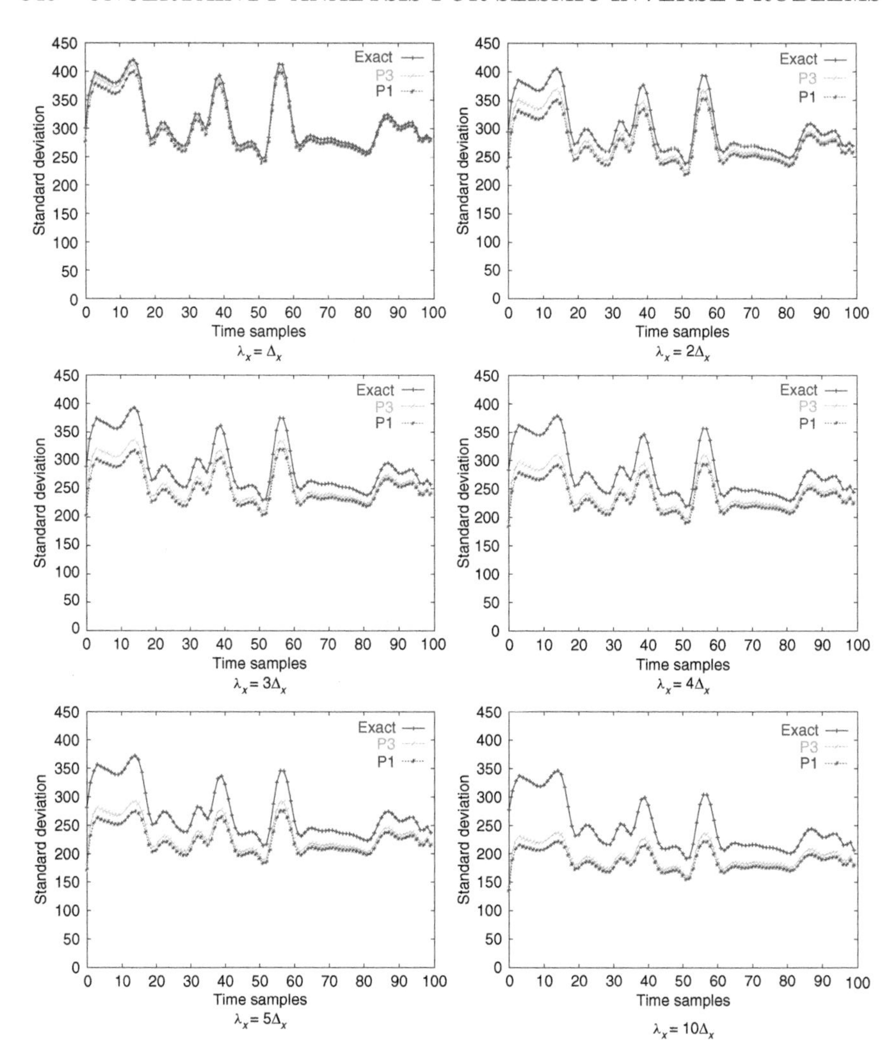

Figure 15.4 P-impedance posterior standard deviations (9th trace) for different values of the correlation length λ_x: exact values are displayed in red, the first order polynomial approximation in blue, the third order polynomial approximation in green.

P-impedance parameter. Posterior standard deviations have been computed using three different methods: the exact method (matrix H being inverted) and the approximated polynomial methods (described in previous subsection) at the order 1 and 3. In these figures, we can observe that polynomial approximations recover the global trend of exact standard deviations on the MAP model. Moreover, Table 15.3 shows that polynomial preconditioner

Table 15.3 Relative mean error of posterior standard deviations for the P_1 and P_3 polynomial approximations versus λ_x.

λ_x/Δ_x	1	2	3	4	5	10
Mean error on P_1 (%)	3.2	9.0	12.7	15.2	17.2	23.2
Mean error on P_3 (%)	1.6	6.2	9.6	11.9	13.9	19.8

methods provide reliable approximation of the standard deviations when the correlation lengths defining the prior exponential covariance matrix C_m are small (less than 4 inter-trace length). Finally, we can also notice that the third order polynomial approximation gives slightly better approximations (the mean error being smaller) than the one at the first order.

15.4 Conclusions

In practice, uncertainty analysis methods for large size seismic inverse problems assume often Gaussian PDF and a quasi-linearity of the forward operator: they rely on the study of the inverse of the Gauss-Newton Hessian approximation of the nonlinear objective function in the vicinity of the solution. Despite these simplifying assumptions, the size of the studied inverse problems makes this study cumbersome even nontractable. In this chapter, we have illustrated how, in practice, this difficulty has been overcome for two seismic inverse problems by taking into account the specificities of the applications: structure of the Hessian matrix, the definition of relevant physical macro-parameters and the characteristics of the forward problem (CPU time, constraints on memory storage).

Applications on realistic datasets have shown the potential of the proposed methods: simulation of admissible models and macro-parameter approach for traveltime inversion; polynomial preconditioner approach for stratigraphic inversion. Besides the classical Bayesian approach, a pragmatic nonlinear uncertainty analysis for traveltime inversion has been proposed to complete the linearized analysis: some geological hypothesis are tested by introducing additional constraints in the optimization problem in order to distinguish the eventual local minima.

References

Aki, K. and Richards, P.G. 2002 *Quantitative Seismology*, 2nd edn. University Science Books.

Alleaume, A. and Sinoquet, D. 2005 Estimation des incertitudes dans le modèle solution de l'inversion sismique. *IFP Internal Report*, **58902**.

Berryman, J.G. 2000 Analysis of approximate inverses in Tomography II. Iterative inverses *Optimization and Engineering* **1** 437–473.

Berthet, P., Cap, J., Ding, D.Y., Lerat, O., Roggero, F. and Screiber, P.-E. 2007 Matching of production history and 4D seismic data: application to the Girassol Field, offshore Angola. *SPE Annual Technical Conference and Exhibition, Society of Petroleum Engineers*, SPE 109929.

Bishop, T.N., Bube, K.P., Cutler, R.T. *et al.* 1985 Tomographic determination of velocity and depth in laterally varying media *Geophysics* **50** 6 903–923.

Bourgeois, A., Labat, K., Euzen, T., Froidevaux, P. and Le, Bras C. 2005 How to build a geological and geophysical realistic synthetic seismic data set for benchmarking reservoir studies? *75th Annual International Meeting, SEG*, Expended Abstracts, 1378–1382.

Brac, J., Dequirez, P.Y., Hervé, F. *et al.* 1988 Inversion with a priori information: an approach to integrated stratigraphic interpretation *Annual Meeting of the Society of Exploration Geophysicists*, Anaheim, California.

Broto, K., Ehinger, A. and Kommedal, J. 2003 Anisotropic traveltime tomography for depth consistent imaging of PP and PS data. *The Leading Edge* 114–119.

Buland, A., Kolbjørnsen, O. and Omre, H. 2003 Rapid spatially coupled AVO inversion in the Fourier domain. *Geophysics*, **68**, No. 3, 824–836.

Cervenỳ, V. 1987 Raytracing algorithms in three-dimensional laterally varying layered structures. *Seismic Tomography Edited by Nolet G* 99–133.

Contreras, A., Torres-Verdin, C., Kvien, K., Fasnacht, T. and Chesters, W. 2005 AVA Stochastic inversion of pre-stack seismic data and well logs for 3D reservoir modeling. *67th Conference & Exhibition, EAGE*, F014.

Chow, E. 1999 A priori sparsity patterns for parallel sparse approximate inverse preconditioners. *SIAM Journal on Scientific Computing*, **21**, 5, 1804–1822.

Delbos, F., Gilbert, J.C., Glowinski, R. and Sinoquet, D. 2006 Constrained optimization in seismic reflection tomography: an SQP augmented Lagrangian approach. *Geophys. J. Int.*, **164**, 670–684.

Delbos, F., Sghiouer, K. and Sinoquet, D. 2008 Uncertainty analysis in prestack stratigraphic inversion: a first insight. *70th Conference & Exhibition, EAGE*, I015.

Delprat-Jannaud, F. and Lailly, P. 1992 What information on the earth model do reflection travel times hold? *Journal of Geophysical Research* **97** 19827–19844

Dubois, P.F., Geenbaum, A. and Rodrigue, G.H. 1979 Approximating the inverse of a matrix for use in iterative algorithms on vector processors. *Computing*, **22**, 257–268.

Duffet, C. and Sinoquet, D. 2006 Quantifying uncertainties on the solution model of seismic tomography. *Inverse Problems*, **22**, 525–538.

Ehinger, A. and Lailly, P. 1995 Velocity model determination by the SMART method, Part 1: Theory. *65th Ann. Internat. Mtg., Soc. Expl. Geophys., Expanded Abstracts*, 739–742.

Escobar, I., Williamson, P., Cherrett, A. *et al.* 2006 Fast geostatistical stochastic inversion in a stratigraphic grid. *76th Annual International Meeting, SEG*, Expanded Abstracts, 2067–2071.

Gouveia, W.P. and Scales, J.A. 1997 Resolution of seismic waveform inversion: Bayes versus Occam *Inverse problems* 323–349.

Grenié, D. 2001 *Tomographie de temps de trajet adapté au problème de l'imagerie 3D en exploration pétroliére: théorie, mise en oeuvre logicielle et applications*, Ph.D. thesis Univ. Paris VII.

Grote, M.J. and Huckle, T. 1997 Parallel preconditioning with sparse approximate inverses. *SIAM Journal on Scientific Computing*, **18**, 3, 838–853.

Haas, A. and Dubrule, O. 1994 Geostatistical inversion: a sequential method of stochastic reservoir modeling constrained by seismic data. *First Break*, **12**, 561–569.

Jurado, F., Sinoquet, D. and Ehinger, A. 1996 3D reflection tomography designed for complex structures. *Expanded Abstracts 66th Ann. Internat. Mtg. Soc. Expl. Geophys.* 711–714.

Lucet, N., Déquirez, P.Y. and Lailly, F. 2000 Well to seismic calibration: a multi-well analysis to extract one wavelet. *70th Annual International Meeting, SEG*, Expanded Abstracts, 1615–1618.

Malinverno, A. and Leaney, S. 2000 A Monte Carlo method to quantify uncertainty in the inversion of zero-offset VSP data *70th Ann. Mtg. Soc. Expl. Geophys.*

Malinverno, A. and Briggs, V. 2004 Expanded uncertainty quantification in inverse problems: Hierarchical Bayes and empirical Bayes. *Geophysics*, **69**, n°4, pp. 1005–1016.

Meurant, G. 1999 *Computer Solution of Large Linear System*. Elsevier.

Mosegaard, K. and Tarantola, A. 1995 Monte Carlo sampling of solutions to inverse problems. *J. Geophys. res.* **100** 12 431–447

Nivlet, P., Tonellot, T., Sexton, P., Piazza, J.L., Lefeuvre, F. and Duplantier, O. 2006 Building 4D constraint for history matching from stratigraphic pre-stack inversion: Application to the Girassol field. *68th Conference & Exhibition, EAGE, incorporating SPE Europec 2006*.

Parker, R. 1994 *Geophysical Inverse Theory*. Springer, Berlin.

Stopin, A. 2001 *Reflection tomography of compressional and shear modes for determination of anisotropic velocity models*, Ph.D. thesis of Université Louis Pasteur Strasbourg I.

Tarantola, A. 2005 *Inverse Problem Theory and Methods for Model Parameter Estimation* (Philadelphia: Society for Industrial and Applied Mathematics) p 342.

Thomsen, L. 1986 Weak elastic anisotropy *Geophysics* **51** 1954–1966.

Tonellot, T., Macé, D., Richard, V. and Cuer, M. 1999 Prestack elastic waveform inversion using a priori information. *69st Annual International Meeting, SEG*, Expanded Abstracts, 227–230.

Tonellot, T. 2000 *Introduction d'informations a priori dans l'inversion linéarisée élastique de données sismiques de surface avant sommation*, Ph.D. thesis, Université René Descartes–Paris V.

Tonellot, T., Macé, D. and Richard, V. 2001 Joint stratigraphic inversion of angle-limited stacks. *71st Annual International Meeting, SEG*, Expended Abstract, 227–230.

Chapter 16

Solution of Inverse Problems Using Discrete ODE Adjoints

A. Sandu

Virginia Polytechnic Institute and State University, Blacksburg, USA

16.1 Introduction

Data assimilation is the process by which model predictions utilize real measurements to obtain an optimal representation of the state of the physical system. Data assimilation combines imperfect information from three different sources: the model, which encapsulates our knowledge of the physical laws that govern the evolution of the system; prior information about the state of the system; and time-distributed measurements of the true state. The optimal estimation problem is posed in a statistical framework in order to rigorously account for uncertainties in the model, prior, and measurements.

In the variational approach data assimilation is formulated as a 'model-constrained' numerical optimization problem. The function to minimize is a metric related to the probability density of the uncertain model solution. (For example, one can minimize the negative logarithm of the posterior probability density to obtain a maximum likelihood estimate; or can minimize the

Large-Scale Inverse Problems and Quantification of Uncertainty Edited by L. Biegler, G. Biros, O. Ghattas, M. Heinkenschloss, D. Keyes, B. Mallick, Y. Marzouk, L. Tenorio, B. van Bloemen Waanders and K. Willcox

variance of the posterior probability density to obtain a minimum variance estimate, etc.) The optimization problem is constrained by the model equations, which link the uncertain model parameters (which need to be estimated) to the model output (used in the formulation of the cost function).

An essential computational ingredient for data assimilation with large-scale, time-dependent systems, is the construction of gradients using adjoints of ordinary differential equation models. Discrete adjoints are popular since they can be constructed automatically by reverse mode automatic differentiation. In this chapter we analyze the consistency and stability properties of discrete adjoints of two widely used classes of time stepping methods: Runge-Kutta and linear multistep.

Consider an ordinary differential equation (ODE)

$$y' = f(y)\,, \quad y\,(t_{\mathrm{ini}}) = y_{\mathrm{ini}}\,, \quad t_{\mathrm{ini}} \le t \le t_{\mathrm{end}}\,, \quad y \in \mathbb{R}^d. \tag{16.1}$$

We are interested in the following inverse problem: find the initial conditions for which the cost function below is minimized,

$$\overline{y}_{\mathrm{opt}} = \arg\min_{y_{\mathrm{ini}}} \overline{\Psi}\,(y_{\mathrm{ini}}) \quad \text{subject to (16.1);} \qquad \overline{\Psi}(y_{\mathrm{ini}}) = g\big(y(t_{\mathrm{end}})\big). \tag{16.2}$$

The general optimization problem (16.2) arises in many important applications including control, shape optimization, parameter identification, data assimilation, etc. This cost function depends on the initial conditions of (16.1). We note that the formulation (16.1)–(16.2) implies no loss of generality. Problems where the solution depends on a set of arbitrary parameters can be transformed into problems where the parameters are the initial values. Similarly, cost functions that involve an integral of the solution along the entire trajectory can be transformed into cost functions that depend on the final state only via the introduction of quadrature variables. Besides the theoretical convenience, this transformation is also useful in practice as expensive numerical quadrature formulas (for the evaluation of integral cost functions) are replaced by the numerical integration of a few additional variables.

To solve (16.1)–(16.2) via a gradient based optimization procedure one needs to compute the derivatives of the cost function $\overline{\Psi}$ with respect to the initial conditions. This can be done effectively using continuous or discrete adjoint approaches.

In the *continuous adjoint* ('differentiate-then-discretize') approach (Miehe and Sandu 2008) one derives the adjoint ODE associated with (16.1)

$$\overline{\lambda}' = -J^T\Big(t, y(t)\Big)\,\overline{\lambda}\,, \quad \overline{\lambda}\,(t_{\mathrm{end}}) = \left(\frac{\partial g}{\partial y}\big(y(t_{\mathrm{end}})\big)\right)^T, \quad t_{\mathrm{end}} \ge t \ge t_{\mathrm{ini}}\,. \tag{16.3}$$

Here $J = \partial f / \partial y$ is the Jacobian of the ODE function. The system (16.3) is solved backwards in time from t_{end} to t_{ini} to obtain the gradients of the cost function with respect to the state (Miehe and Sandu 2008). Note that the continuous adjoint equation (16.3) depends on the forward solution $y(t)$ via the argument of the Jacobian. For a computer implementation the continuous adjoint ODE (16.3) is discretized and numerical solutions $\overline{\lambda}_n \approx \overline{\lambda}(t_n)$ are obtained at $t_{\rm end} = t_N > t_{N-1} > \cdots > t_1 > t_0 = t_{\rm ini}$.

In the *discrete adjoint* ('discretize-then-differentiate') approach (Miehe and Sandu 2008) one starts with a discretization of (16.1) which gives the numerical solutions $y_n \approx y(t_n)$

$$y_0 = y_{\rm ini}\,, \quad y_n = \mathcal{M}_n\left(y_0, \cdots, y_{n-1}\right)\,, \quad n = 1, \cdots, N. \tag{16.4}$$

The numerical solution at the final time is $y_N \approx y\left(t_{\rm end}\right)$. The optimization problem (16.2) is reformulated in terms of this numerical solution,

$$y_{\rm opt} = \arg\min_{y_{\rm ini}} \Psi\left(y_{\rm ini}\right) \quad \text{subject to (16.4);} \quad \Psi\left(y_{\rm ini}\right) = g\left(y_N\right). \tag{16.5}$$

The gradient of (16.5) is computed directly from (16.4) using the transposed chain rule. This calculation and produces the discrete adjoint variables λ_N, λ_{N-1}, $\cdots$, λ_0

$$\lambda_N = \left(\frac{\partial g}{\partial y}\left(y_N\right)\right)^T, \quad \lambda_n = 0\,, \quad n = N-1, \cdots, 0\,, \tag{16.6}$$
$$\lambda_\ell = \lambda_\ell + \left(\frac{\partial \mathcal{M}_n}{\partial y_\ell}\left(y_0, \cdots, y_{n-1}\right)\right)^T \lambda_n\,, \quad \ell = n-1, \cdots, 0\,, \quad n = N, \cdots, 0.$$

Note that the discrete adjoint equation (16.6) depends on the forward numerical solution $y_0, \cdots, y_N$ via the arguments of the discrete model. The computational process (16.6) gives the sensitivities of the numerical cost function (16.5) with respect to changes in the forward numerical solution (16.4). Discrete adjoints are useful in optimization since they provide the gradients of the numerical function that is being minimized. They can be obtained by reverse mode automatic differentiation.

This chapter analyzes some of the properties of the discrete adjoint variables λ_n and their relationship with the continuous adjoint variables $\overline{\lambda}(t_n)$. This analysis is important for two reasons. First, in practical sensitivity analysis studies, one computes discrete sensitivities but interprets them as sensitivities of the continuous system. Next, in the practical solution of inverse problems one solves the discrete version (16.5) to obtain $y_{\rm opt} = \operatorname{argmin} \Psi$. This solution is intended to approximate the optimum $\overline{y}_{\rm opt} = \operatorname{argmin} \overline{\Psi}$ of the original problem (16.2). How accurate is this approximation? We have that $\nabla\overline{\Psi}\left(\overline{y}_{\rm opt}\right) = 0$, $\nabla\Psi\left(y_{\rm opt}\right) = 0$, and

$$\begin{aligned}
0 &= \nabla\overline{\Psi}\left(\overline{y}_{\rm opt}\right) - \nabla\overline{\Psi}\left(y_{\rm opt}\right) + \nabla\overline{\Psi}\left(y_{\rm opt}\right) - \nabla\Psi\left(y_{\rm opt}\right) \\
&= \nabla^2\,\overline{\Psi}\left(\overline{y}_{\rm opt}\right)\cdot\left(\overline{y}_{\rm opt} - y_{\rm opt}\right) + \nabla\overline{\Psi}\left(y_{\rm opt}\right) - \nabla\Psi\left(y_{\rm opt}\right) + \mathcal{O}\left(\left\|\overline{y}_{\rm opt} - y_{\rm opt}\right\|^2\right)
\end{aligned}$$

Assuming the high order terms are small, the following bound is obtained for the difference between the optimal solutions:

$$\|\overline{y}_{\rm opt} - y_{\rm opt}\| \leq {\rm cond}\Big(\nabla^2\,\overline{\Psi}\,(\overline{y}_{\rm opt})\Big) \cdot \max_{y_0} \Big\|\overline{\lambda}(t_0) - \lambda_0\Big\|.$$

The magnitude of the solution difference depends on the condition number of the Hessian (the conditioning of the inverse problem) and on the difference between the continuous and the discrete adjoint solutions at the initial time. A good approximation is obtained when the difference between the continuous and the discrete gradients decreases rapidly with the numerical method step size, $\|\overline{\lambda}(t_0) - \lambda_0\| = \mathcal{O}(h^p)$, where p is related to the order of the forward time discretization method (16.4).

Consistency properties of discrete Runge-Kutta adjoints have been studied by Hager (2000), who gives additional order conditions necessary in the context of control problems. This theory is further refined by Bonans and Laurent-Varin (2006). Walther (2007) has studied the effects of reverse mode automatic differentiation on explicit Runge-Kutta methods, and found that the order of the discretization is preserved by discrete adjoints up to order four. Giles (2000) has discussed Runge-Kutta adjoints in the context of steady state flows. Baguer and Romisch (1995) have constructed discrete adjoints for linear multistep methods in the context of control problems.

This chapter provides a comprehensive analysis of the properties of discrete adjoints for ODE solvers based on the work of the author. This analysis includes discrete adjoints of Runge Kutta methods (Sandu 2006) and of linear multistep methods (Sandu 2008), adjoints of adaptive step methods (Alexe and Sandu 2008), efficient implementation of discrete adjoints (Miehe and Sandu 2008; Sandu *et al.* 2003), and their use in data assimilation applications (Carmichael *et al.* 2008).

16.2 Runge-Kutta Methods

A general s-stage Runge-Kutta discretization method is defined as (Hairer *et al.* 1993)

$$\begin{aligned} y_{n+1} &= y_n + h \sum_{i=1}^{s} b_i\, k_i\,, \quad h = t_{n+1} - t_n\,, \\ k_i &= f\left(t_n + c_i h,\, Y_i\right)\,, \quad Y_i = y_n + h \sum_{j=1}^{s} a_{i,j} k_j\,, \end{aligned} \tag{16.7}$$

where the coefficients $a_{i,j}$, b_i and c_i are prescribed for the desired accuracy and stability properties. If $a_{i,j} \neq 0$ for $j \geq i$ Equations (16.7) form a nonlinear system which needs to be solved for the stage derivative values k_i.

Hager (2000) has shown that taking the discrete adjoint (16.6) of the Runge-Kutta method (16.7) leads to the following computational process

$$\begin{aligned} \lambda_n &= \lambda_{n+1} + \sum_{j=1}^{s} \theta_j \,, \\ \theta_i &= h\, J^T(t_n + c_i h, Y_i) \left(b_i \lambda_{n+1} + \sum_{j=1}^{s} a_{j,i} \theta_j \right) \,, \quad i = 1, \cdots, s. \end{aligned} \tag{16.8}$$

16.2.1 Accuracy of the Discrete Adjoint RK Method

We regard the discrete adjoint equation (16.8) as a numerical method applied to the adjoint ODE (16.3) and look to assess its accuracy. In this section we make the assumption that the systems of (non)linear algebraic equations appearing in forward implicit methods (16.7) and in their discrete adjoints (16.8) are solved exactly.

For the results of this section it is convenient to represent the Runge-Kutta method (16.7) in Henrici notation (Hairer *et al.* 1993)

$$y_{n+1} = y_n + h_n\, \Theta(y_n, h_n, t_n) \,, \tag{16.9}$$

where Θ is called the increment function of the method. The corresponding discrete Runge-Kutta adjoint (16.8) is then represented as

$$\lambda_n = \big(I_d + h_n \Theta_y(y_n, h_n, t_n)\big)^T \cdot \lambda_{n+1} \,, \tag{16.10}$$

where Θ_y is the Jacobian of the increment function. Here and in what follows we denote by I_k the $k \times k$ identity matrix (for some k) and by $\mathbf{1}_k$ a column vector with k entries, all equal to one.

We denote the sequence of discretization step sizes in the interval $[t_{\text{ini}}, t_{\text{end}}]$ and the maximum step size by

$$h_n = t_{n+1} - t_n, \quad h = \{h_1, \cdots, h_N\}, \quad |h| = \max_{1 \le n \le N} h_n. \tag{16.11}$$

The number of steps depends on the step discretization sequence, $N = N(h)$.

Let U be a neighborhood of the exact forward solution $\{(t, y(t)),\ t \in [t_{\text{ini}}, t_{\text{end}}]\}$. We will assume that, in this neighborhood, the function $f(t, y)$ and its Jacobian $J(t, y)$ are Lipschitz continuous in y. A direct consequence (Hairer *et al.* 1993) is that the increment function and its Jacobian are also Lipschitz continuous,

$$\begin{aligned} \forall h \in [0, H]\,, \quad \forall y, z \in U : \ & \big\|\Theta(y, h, t) - \Theta(z, h, t)\big\| \le L\ \|y - z\| \,, \\ & \big\|\Theta_y(y, h, t) - \Theta_y(z, h, t)\big\| \le K\ \|y - z\| \,. \end{aligned}$$

Proposition 16.1 (Local Error Analysis) *Assume the Runge-Kutta method (16.9) is an order p consistent approximation of the ODE (16.1),*

$$y(t_{n+1}) - y(t_n) - h_n\,\Theta\big(y(t_n), h_n, t_n\big) = \mathcal{O}\big(h^{p+1}\big) \quad \forall h\,:\,|h| \le H\,.$$

Then its discrete adjoint (16.10) is an order p consistent approximation of the adjoint ODE (16.3)

$$\overline{\lambda}(t_n) - \big(I_d + h_n \Theta_y\big(y(t_n), h_n, t_n\big)\big)^T \cdot \overline{\lambda}(t_{n+1}) = \mathcal{O}\big(h^{p+1}\big) \quad \forall h\,:\,|h| \le H\,. \tag{16.12}$$

Proof: The proof can be found in (Sandu 2006). The analysis of the order of discrete adjoints is based on the concept of elementary differentials from the theory of order conditions explained in Hairer *et al.* (1993). The proof works for both implicit and explicit methods and is valid for any order p. ∎

The analysis carried out in Proposition 16.1 shows that one discrete adjoint step has the same order of consistency as the forward step. The formulation of the discrete adjoint method requires the forward solution (which is precomputed and stored). The discrete adjoint method cannot use the exact forward trajectory values $y(t_n)$; in practice it is formulated based on a numerically computed forward solution y_n. Most often the same Runge-Kutta method is used for both the forward trajectory and the discrete adjoint calculations. Note from (16.8) that one needs to checkpoint (or recompute) both the full step values y_n and the intermediate stage values Y_i at each time step.

Proposition 16.2 (Global Error Analysis) *Assume that the Runge-Kutta method (16.7) is consistent of order (at least) p, and its solution converges with order p to the ODE solution (16.1)*

$$\begin{aligned} \Delta y_n = y_n - y(t_n)\,, \quad & \|\Delta y_n\| \in \mathcal{O}\left(h^p\right)\,, \\ & \forall h\,:\,|h| \le H\,, \quad n = 1, \cdots, N(h). \end{aligned}$$

Then the discrete Runge-Kutta adjoint solution (16.8) converges with order p to the adjoint ODE solution (16.3),

$$\begin{aligned} \Delta\lambda_n = \lambda_n - \overline{\lambda}(t_n)\,, \quad & \|\lambda_n - \overline{\lambda}(t_n)\| \in \mathcal{O}\left(h^p\right)\,, \\ & \forall h\,:\,|h| \le H\,, \quad n = N(h) - 1, \cdots, 0. \end{aligned}$$

Proof: The forward method is convergent therefore it is stable. Specifically, the following linear stability condition holds

$$\left\| \prod_{i=n}^{n+\ell} \big(I_d + h_i \Theta_y\big(y_i, h_i, t_i\big)\big) \right\| \le M \quad \forall n, \ell. \tag{16.13}$$

The discrete adjoint model (16.10) depends on the numerical solution y_n. Express the discrete adjoint as a function of the exact forward solution

$$\begin{aligned}\lambda_n &= \left(I_d + h_n \Theta_y(y_n, h_n, t_n)\right)^T \cdot \lambda_{n+1} \\ &= \left(I_d + h_n \Theta_y(y_n, h_n, t_n)\right)^T \cdot \Delta\lambda_{n+1} + \left(I_d + h_n \Theta_y(y_n, h_n, t_n)\right)^T \cdot \overline{\lambda}(t_{n+1}) \\ &= \left(I_d + h_n \Theta_y(y_n, h_n, t_n)\right)^T \cdot \Delta\lambda_{n+1} + \left(I_d + h_n \Theta_y(y(t_n), h_n, t_n)\right)^T \cdot \overline{\lambda}(t_{n+1}) \\ &\quad + h_n \left(\Theta_y(y_n, h_n, t_n) - \Theta_y(y(t_n), h_n, t_n)\right)^T \cdot \overline{\lambda}(t_{n+1})\end{aligned}$$

Using (16.12) we obtain

$$\Delta\lambda_n = \left(I_d + h_n \Theta_y(y_n, h_n, t_n)\right)^T \cdot \Delta\lambda_{n+1} + \varepsilon_n^{\text{dadj}} + \varepsilon_n^{\text{fwd}} \tag{16.14}$$

where

$$\begin{aligned}\varepsilon_n^{\text{dadj}} &= \left(I_d + h_n \Theta_y(y(t_n), h_n, t_n)\right)^T \cdot \overline{\lambda}(t_{n+1}) - \overline{\lambda}(t_n) \\ \varepsilon_n^{\text{fwd}} &= h_n \left(\Theta_y(y_n, h_n, t_n) - \Theta_y(y(t_n), h_n, t_n)\right)^T \cdot \overline{\lambda}(t_{n+1})\end{aligned}$$

Equation (16.14) describes the propagation of the discrete adjoint global error $\Delta\lambda_n$. The local error at step n, $\varepsilon_n = \varepsilon_n^{\text{dadj}} + \varepsilon_n^{\text{fwd}}$, is given by a term ($\varepsilon_n^{\text{fwd}}$) coming from the global error of the forward solution plus the local truncation error of the discrete adjoint formula ($\varepsilon_n^{\text{fwd}}$). The local truncation error analysis (Proposition 16.1) shows that

$$\|\varepsilon_n^{\text{dadj}}\| \le C\, h_n^{p+1} \le C\, h_n\, |h|^p$$

The Lipschitz continuity of the Jacobian of the increment function gives

$$\left\|\Theta_y(y_n, h_n, t_n) - \Theta_y(y(t_n), h_n, t_n)\right\| \le K\, \|\Delta y_n\| \in \mathcal{O}(|h|^p).$$

Under the assumption that the adjoint system is smooth we have a bounded continuous adjoint solution $\|\overline{\lambda}(t)\|$ on $t \in [t_{\text{ini}}, t_{\text{end}}]$. Therefore

$$\|\varepsilon_n^{\text{fwd}}\| \le h_n\, K \cdot \|\Delta y_n\| \cdot \max_{t_{\text{ini}} \le t \le t_{\text{end}}} \|\overline{\lambda}(t)\| \le C\, h_n\, |h|^p.$$

Consequently the total local error for the discrete adjoint is bounded by

$$\|\varepsilon_n\| \le C\, h_n\, |h|^p.$$

From the global error recurrence (16.14), using the fact that $\lambda_N = \overline{\lambda}(t_N)$, and using (16.13), we have that

$$\Delta\lambda_\ell = \sum_{m=N-1}^{\ell} \left(\prod_{n=m}^{\ell} \left(I_d + h_n \Theta_y(y_n, h_n, t_n)\right)^T \right) \cdot \varepsilon_m$$

$$\|\Delta\lambda_\ell\| \le M \sum_{m=N-1}^{\ell} \varepsilon_m \le M\, C\, |h|^p \sum_{m=N-1}^{\ell} h_m = M\, C\, (t_N - t_\ell)\, |h|^p.$$

Therefore the global discrete adjoint error satisfies the bound

$$\|\Delta\lambda_\ell\| \in \mathcal{O}\big(|h|^p\big)\,, \quad \forall \ell = 0, \cdots, N-1.$$

This proves that the discrete Runge-Kutta adjoint solution converges with order p to the continuous adjoint ODE solution *at any discretization point* t_ℓ (including the initial time $t_0 = t_{\rm ini}$ as well as any intermediate time step). ■

16.3 Adaptive Steps

When the Runge-Kutta method is used with an adaptive step size strategy a new step size h_{n+1} is computed based on both the current solution (via an error estimate)and on the current step size

$$y_{n+1} = y_n + h_n\, \Theta\,(y_n, h_n, t_n) \tag{16.15a}$$

$$h_{n+1} = \mathcal{S}\,(y_n, h_n, t_n) \tag{16.15b}$$

$$t_{n+1} = t_n + h_n\,, \qquad n = 0, \cdots, N-1\,. \tag{16.15c}$$

Here $\mathcal{S}$ denotes the step size control mechanism. The effect of the variable step control mechanism on the discrete adjoint has been studied in (Alexe and Sandu 2008). The n-th step in the discrete adjoint of (16.15a–16.15c) is

$$\lambda_n = \left(I_d + h_n \Theta_y\right)^T \lambda_{n+1} + \mathcal{S}_y^T\, \mu_{n+1} \tag{16.16a}$$

$$\mu_n = \left(\Theta + h_n \Theta_h\right)^T\, \lambda_{n+1} + \mathcal{S}_h^T\, \mu_{n+1} + \nu_{n+1} \tag{16.16b}$$

$$\nu_n = h_n \Theta_t^T \lambda_{n+1} + \mathcal{S}_t^T\, \mu_{n+1} + \nu_{n+1}\,, \qquad n = N-1, \cdots, 0\,. \tag{16.16c}$$

Here subscripts denote partial derivatives and we omit the function arguments for simplicity. The last term in (16.16a) is a side-effect of the differentiation of the time step controller mechanism (16.15b). The adjoint derivative μ of the time step h is an $\mathcal{O}(1)$ term that can destroy the convergence of the discrete adjoint solution. To counter this effect one can set the non-physical adjoint derivatives μ_{n+1} to zero before calculating (16.16a), or implement a correction of the form $\lambda_n \leftarrow \lambda_n - \mathcal{S}_y^T\, \mu_{n+1}$ after calculating (16.16a).

16.3.1 Efficient Implementation of Implicit RK Adjoints

In this section we discuss the implementation of discrete adjoints of implicit Runge-Kutta methods following Miehe and Sandu (2008). For simplicity we consider only methods with nonsingular coefficient matrix A. We only consider direct methods for solving the linear algebra systems. Iterative linear system solvers are a possibility for large systems; they fall outside the scope of the discussion in this section. For implementation purposes an implicit RK method (16.7) is written in the equivalent form (Hairer and Wanner 1991, Section IV.8)

$$Z = (hA \otimes I_d) \cdot F(Z), \quad y_{n+1} = y_n + \left(b^T A^{-1} \otimes I_d\right) \cdot Z. \tag{16.17}$$

where

$$A \otimes B = \left(a_{i,j} B\right)_{1 \le i,j \le s}, \quad Z = \begin{bmatrix} Z_1 \\ \vdots \\ Z_s \end{bmatrix}, \quad F(Z) = \begin{bmatrix} f\,(T_1,\, y_n + Z_1\,) \\ \vdots \\ f\,(T_s,\, y_n + Z_s\,) \end{bmatrix}.$$

Replacing the nonlinear system in k_i by a nonlinear system in Z_i has numerical advantages for stiff systems where f has a large Lipschitz constant. The nonlinear system (16.17) in Z can be solved by Newton iterations of the form

$$\begin{aligned} \left[I_{ds} - h\mathcal{J}\right] \Delta Z^{[m]} &= Z^{[m]} - (hA \otimes I_d) \cdot F\left(Z^{[m]}\right), \\ Z^{[m+1]} &= Z^{[m]} - \Delta Z^{[m]}. \end{aligned} \tag{16.18}$$

where

$$\mathcal{J} = \left(a_{i,j} J\,(T_j,\, y_n + Z_j)\right)_{1 \le i,j \le s} \tag{16.19}$$

The cost of the Newton iterations is dominated by the LU decomposition of the $ds \times ds$ matrix $I_{ds} - h\mathcal{J}$. A more efficient approach is provided by evaluating all the Jacobians at the beginning of the current step

$$\mathcal{J} \approx \left(a_{i,j} J\,(t_n,\, y_n)\right)_{1 \le i,j \le s} = A \otimes J\,(t_n,\, y_n\,) \tag{16.20}$$

The same LU decomposition of $I_{ds} - hA \otimes J\,(t_n,\, y_n\,)$ is used for every iteration m in (16.18). With full linear algebra the computational work is of $\mathcal{O}(d^3 s^3)$. For singly diagonally implicit RK methods (SDIRK) a single real LU decomposition of dimension d is necessary, and the computational work is of $\mathcal{O}(d^3)$. For fully implicit RK methods (FIRK) a transformation that diagonalizes A (Hairer and Wanner 1991) replaces the large system by several complex and one real LU decompositions of dimension d and reduces the computational work to $\mathcal{O}(sd^3)$.

Similar techniques are applied to efficiently implement the discrete Runge-Kutta adjoint. Using the matrix (16.19) and the notation

$$U = \begin{bmatrix} u_1 \\ \vdots \\ u_s \end{bmatrix}, \qquad G = h \begin{bmatrix} b_1\, J^T(T_1, Y_1)\, \lambda_{n+1} \\ \vdots \\ b_s\, J^T(T_s, Y_s)\, \lambda_{n+1} \end{bmatrix},$$

the discrete adjoint method (16.8) can be written compactly as

$$\lambda_n = \lambda_{n+1} + (\mathbf{1}_s \otimes I_d) \cdot U\,, \quad \left[I_{ds} - h\mathcal{J}\right]^T \cdot U = G. \tag{16.21}$$

Note that the same system matrix $I_{ds} - h\mathcal{J}$ is used in both the adjoint solution (16.21) and in the forward iterative solution (16.18). In order to avoid forming the LU decomposition of $I_{ds} - h\mathcal{J}$ we again approximate $\mathcal{J} \approx A \otimes J(t_n, y_n)$ and solve the linear system (16.21) by Newton iterations. The relevant LU decompositions can, in principle, be checkpointed during the forward run and reused in the adjoint run.

16.3.2 Iterative Solvers

Implicit Runge-Kutta methods employ an iterative procedure to solve a nonlinear system of equations at each step. The linearization and adjoint operations should, in principle, be applied to these iterations as well (16.18). It has been shown (Christianson 1994; Giering and Kaminski 1998; Giles 2001; Griewank 2003) that the resulting adjoint iterations converge to the solution of the discrete adjoint method (16.21), but at an additional computational cost. For this reason we only consider a direct solution approach to the adjoint linear system (16.21).

16.3.3 Considerations on the Formal Discrete RK Adjoints

Hager (2000) has shown that for $b_i \neq 0$ the Runge-Kutta adjoint (16.8) can be rewritten as another Runge-Kutta method applied to the continuous adjoint Equation (16.3)

$$\begin{aligned} \lambda_n &= \lambda_{n+1} + h \sum_{i=1}^{s} \overline{b}_i \, \ell_i\,, \quad \ell_i = J^T\left(t_{n+1} - \widehat{c}_i h,\, Y_{s+1-i}\right) \Lambda_i\,, \\ \Lambda_i &= \lambda_{n+1} + h \sum_{j=1}^{s} \overline{a}_{i,j} \ell_j\,, \end{aligned} \tag{16.22}$$

where

$$\overline{b}_i = b_{s+1-i}\,, \quad \widehat{c}_i = 1 - c_{s+1-i}\,, \quad \overline{a}_{i,j} = \frac{a_{s+1-j,s+1-i} \cdot b_{s+1-j}}{b_{s+1-i}}. \tag{16.23}$$

We will call (16.22) the *formal discrete adjoint method* of (16.7) (Miehe and Sandu 2008). To some extent the properties of the discrete adjoint are related to the properties of the formal adjoint method (16.22). Note that each stage solution Λ_i is evaluated at the time moment:

$$\overline{c}_i = \sum_{j=1}^{s} \overline{a}_{i,j} = \frac{1}{b_{s+1-i}} \sum_{j=1}^{s} a_{s+1-j,s+1-i} \cdot b_{s+1-j}$$

while the arguments of the Jacobian are evaluated at $\widehat{c}_i$. It seems advantageous to have methods where Λ_i and Y_{s+1-i} are evaluated at the same time points. In this case one could replace the forward stage values Y_{s+1-i} by interpolated forward solution values. This internal consistency condition reads

$$1 - c_i = \widehat{c}_{s+1-i} = \overline{c}_{s+1-i} = \sum_{j=1}^{s} \frac{a_{j,i} \cdot b_j}{b_i}, \quad i = 1, \cdots, s. \tag{16.24}$$

We see that the internal consistency condition (16.24) is precisely the simplifying $D(1)$ condition (Hairer and Wanner 1991, Section IV.5) under the assumption that all $b_i \neq 0$. In (Hager 2000) the condition (16.24) is used in the derivation of order conditions for control problems.

The following properties of the formal discrete adjoints have been proven in (Miehe and Sandu 2008). First, the formal adjoint has the same linear stability properties as the forward method since $\overline{R}(z) = R(z)$. Next, consider the matrix $M = (b_i\, a_{i,j} + b_j\, a_{j,i} - b_i\, b_j)_{i,j=1,\cdots,s}$. If a Runge-Kutta method is algebraically stable ($b_i \geq 0$ for all i and M non-negative definite) then its formal discrete adjoint is algebraically stable (since $\overline{b}_i \geq 0$ for all i and $\overline{M}$ non-negative definite). A stiffly accurate Runge-Kutta method is characterized by $a_{s,j} = b_j$ for all $j = 1, \cdots, s$. The formal discrete adjoint is stiffly accurate if the forward method satisfies $a_{i,1} = b_1$ for all $i = 1, \cdots, s$ (e.g. see Lobatto-3B and Lobatto-3C families of methods). Interesting adjoint relationships happen between known Runge-Kutta methods. Gauss and Lobatto-3C are formally self-adjoint while Radau-2A and Radau-1A are each the formal adjoint of the other.

16.4 Linear Multistep Methods

Consider the linear multistep method (Hairer *et al.* 1993)

$$y_0 = y_{\text{ini}}\,, \tag{16.25a}$$

$$y_n = \theta_n\left(y_0, \cdots, y_{n-1}\right)\,, \quad n = 1, \cdots, k-1\,, \tag{16.25b}$$

$$\sum_{i=0}^{k} \alpha_i^{[n]}\, y_{n-i} = h_n \sum_{i=0}^{k} \beta_i^{[n]}\, f_{n-i}\,, \quad n = k, \cdots, N. \tag{16.25c}$$

The upper indices indicate the dependency of the method coefficients on the step number; this formulation accommodates variable step sizes. The numerical solution is computed at the discrete moments $t_{\text{ini}} = t_0 < t_1 < \cdots < t_N = t_{\text{end}}$. As usual y_n represents the numerical approximation at time t_n. The right hand side function evaluated at t_n using the numerical solution y_n is denoted $f_n = f(t_n, y_n)$, while its Jacobian is denoted by $J_n = J(t_n, y_n) = (\partial f/\partial y)\,(t_n, y_n)$.

Equation 16.25(a)–(16.25c) is a k-step method. The method coefficients $\alpha_i^{[n]}$, $\beta_i^{[n]}$ depend on the sequence of (possibly variable) steps, specifically, they depend on the ratios $\omega_{n-k+2}, \cdots, \omega_n$ where $\omega_\ell = h_\ell / h_{\ell-1}$.

A starting procedure θ is used to produce approximations of the solution $y_i = \theta_i\,(y_0, \cdots, y_{i-1})$ at times t_i, $i = 1, \cdots, k-1$. We consider the starting procedures to be linear numerical methods. This setting covers both the case of self-starting LMM methods (a linear i-step method gives y_i for $i = 1, \cdots, k-1$) and the case of initialization by Runge-Kutta methods

It was shown in (Sandu 2008) that the discrete adjoint method associated with (16.25a)–(16.25c) and the cost function (16.5) reads:

$$\alpha_0^{[N]} \lambda_N = h_N\, \beta_0^{[N]}\, J_N^T \cdot \lambda_N + \left(\frac{\partial g}{\partial y}\,(y_N)\right)^T \tag{16.26a}$$

$$\sum_{i=0}^{N-m} \alpha_i^{[m+i]}\, \lambda_{m+i} = J_m^T \cdot \sum_{i=0}^{N-m} h_{m+i}\, \beta_i^{[m+i]}\, \lambda_{m+i}\,, \qquad m = N-1, \cdots, N-k+1 \tag{16.26b}$$

$$\sum_{i=0}^{k} \alpha_i^{[m+i]}\, \lambda_{m+i} = h_{m+1}\, J_m^T \cdot \sum_{i=0}^{k} \widehat{\beta}_i^{[m+i]}\, \lambda_{m+i}\,, \qquad m = N-k, \cdots, k \tag{16.26c}$$

$$\lambda_{k-1} + \sum_{i=1}^{k} \alpha_i^{[k-1+i]}\, \lambda_{k-1+i} = J_{k-1}^T \cdot \sum_{i=1}^{k} \left(h_{k-1+i}\, \beta_i^{[k-1+i]}\, \lambda_{k-1+i}\right) \tag{16.26d}$$

$$\lambda_m + \sum_{i=k-m}^{k} \alpha_i^{[m+i]}\, \lambda_{m+i} = \sum_{i=m+1}^{k-1} \left(\frac{\partial \theta_i}{\partial y_m}\right)^T \lambda_i + J_m^T \cdot \sum_{i=k-m}^{k} h_{m+i}\, \beta_i^{[m+i]}\, \lambda_{m+i}, \qquad m = k-2, \cdots, 0 \tag{16.26e}$$

where

$$\widehat{\beta}_0^{[m]} = \omega_{m+1}^{-1}\, \beta_0^{[m]}\,, \quad \widehat{\beta}_1^{[m+1]} = \beta_1^{[m+1]}\,, \quad \widehat{\beta}_i^{[m+i]} = \left(\prod_{\ell=2}^{i} \omega_{m+\ell}\right) \beta_i^{[m+i]}.$$

The gradient of the cost function with respect to the initial conditions is $\nabla_{y_{\rm ini}} \Psi = \lambda_0$.

16.4.1 Consistency of Discrete Linear Multistep Adjoints at Intermediate Time Points

Theorem 16.1 (Consistency at Interior Trajectory Points) *In general the numerical process (16.26a)–(16.26e) (the discrete linear multistep* *adjoint with variable step sizes) is not a consistent discretization of the adjoint ODE (16.3).*

Proof: See (Sandu 2008) for a complete proof. Here we only consider an example. Consider the variable step BDF2 method (Hairer and Wanner 1991, Section III.5, page 401). The zeroth order consistency condition for the discrete adjoint formula (16.26c) is

$$\sum_{i=0}^{k} \alpha_i^{[m+i]} = 1 - \frac{(1+\omega_{m+1})^2}{1+2\omega_{m+1}} + \frac{\omega_{m+2}^2}{1+2\omega_{m+2}} = 0$$

and has a single positive solution $\omega_{m+2} = \omega_{m+1} = \omega$ (the step sizes change at a constant ratio). For a general sequence of steps the discrete adjoint BDF2 is not (zeroth order) consistent with the adjoint ODE. ■

16.4.2 Consistency of Discrete Linear Multistep Adjoints at the Initial Time

Proposition 16.3 (Consistency at the Initial Time) *Consider a linear multistep method (16.25a)–(16.25c) convergent of order p, and initialized with linear numerical methods. (This covers the typical situation where the initialization procedures $\theta_1, \cdots, \theta_{k-1}$ are Runge-Kutta or linear multistep numerical methods.) The numerical solutions at the final time are such that*

$$\left\| y^h_{N(h)} - y\left(t_{\mathrm{end}}\right) \right\|_\infty = \mathcal{O}\left(|h|^p\right) , \quad \forall h \, : \, |h| \leq H \, ,$$

for a small enough threshold H. Let λ^h_n be the solution of the discrete adjoint linear multistep process (16.26a)–(16.26e).

Then the discrete adjoint solution λ^h_0 is an order p approximation of the continuous adjoint $\lambda(t_0)$ at the initial time, i.e.

$$\left\| \lambda_0^h - \overline{\lambda}\left(t_0\right) \right\|_\infty = \mathcal{O}\left(|h|^p\right) , \quad \forall h \, : \, |h| \leq H \, , \qquad (16.27)$$

for a small enough threshold H.

Proof: The proof is based on the linearity of the method and of its starting procedures, which leads to forward mode discrete sensitivities that are order p approximations of the continuous sensitivities. The initial time reverse model sensitivities and the final time forward sensitivities are the same. See (Sandu 2008) for details. ■

16.5 Numerical Results

We consider the following test problem

$$y' = 1 - y^2 \, , \quad 0 \leq t \leq 1 \, , \quad \overline{\Psi} = y(1). \qquad (16.28)$$

Table 16.1 Experimental orders of convergence for the adjoint solution of (16.28). Shown are discrete adjoints (DADJ) of a third order Runge-Kutta scheme (SDIRK3) and of the second order backward differentiation formula (BDF2), and the continuous adjoint solution (CADJ) obtained with BDF2.

	BDF2 CADJ	BDF2 DADJ	SDIRK3 DADJ
Trajectory	1.96	−0.01	2.98
Final time	1.98	1.99	2.95

The problem admits the analytical solution $y(t) = tanh(t)$. The numerical integration uses the second order backward differentiation formula (BDF2) and a two-stage, third order, singly diagonally implicit Runge-Kutta method (SDIRK3) with coefficients $a_{1,1} = a_{2,2} = 0.788675134594813$, $a_{1,2} = 0$, $a_{2,1} = 1 - 2a_{1,1}$, and $b_1 = b_2 = 0.5$. A nonconstant step sequence is used with $h_{2i} = 2\,|h|/3$, $h_{2i+1} = |h|$.

A reference continuous adjoint solution $\overline{\lambda}^{\mathrm{ref}}$ is computed in Matlab using the rkf45 routine with tight absolute and relative tolerances ($Atol = Rtol = 10^{-10}$). We report the accuracy along the trajectory and at the initial time

$$\text{Trajectory Error} = \sqrt{\frac{1}{N}\sum_{i=0}^{N}\left(\frac{\lambda_i - \overline{\lambda}_i^{\mathrm{ref}}}{\overline{\lambda}_i^{\mathrm{ref}}}\right)^2}\,, \quad \text{Initial Error} = \frac{|\lambda_0 - \overline{\lambda}_0^{\mathrm{ref}}|}{|\overline{\lambda}_0^{\mathrm{ref}}|}$$

where λ_i is the numerical adjoint solution at t_i, $i = 1, \cdots, N$. The experimental orders of convergence presented in Table 16.1 are obtained by integrating the system (16.28) and its adjoint with various step sizes in the range $0.0067 \le h \le 0.02$ and by measuring the rate of error decrease. These results reveal that the discrete Runge-Kutta adjoint solution converges with order three both along the trajectory and at the initial time. In contrast, the discrete BDF2 adjoint solution converges with order 2 at the initial time, but it does not converge along the trajectory. For comparison we include the continuous adjoint BDF2 solution which converges with order two along the trajectory as well. These numerical results are a direct confirmation of the theory presented here.

16.6 Application to Data Assimilation

The methodology of discrete ODE adjoints discussed in this chapter is an essential computational ingredient for solving variational data assimilation problems with large-scale, time-dependent systems. Data assimilation is the process by which model predictions utilize real measurements to obtain an optimal representation of the state of the physical system. The system considered here is the chemically perturbed atmosphere. The state we seek to estimate is the spatial and temporal distribution of air pollutants associated with the gas and aerosol phases. As more chemical observations in the troposphere are becoming available chemical data assimilation is expected to play an essential role in air quality forecasting, similar to the role it has in numerical weather prediction (Carmichael *et al.* 2008).

Data assimilation combines information from three different sources:

1. *The model which encapsulates our knowledge of the physical laws that govern the evolution of the system.* For example, chemical transport models (CTMs) incorporate descriptions of processes like emissions, chemical transformations, long range transport, radiation, deposition, etc. Consider the CTM

$$\mathbf{c}_i = \mathcal{M}_{t_0 \to t_i}(\mathbf{c}_0) \quad \text{for } i = 1, \cdots, N \tag{16.29}$$

   which advances the state from the initial time t_0 to future times t_i ($t_0 < t_1 < \cdots < t_N$). The state vector $\mathbf{c}_i$ represents the chemical concentrations of all tracer species at all gridpoints of the model.

2. *Prior information about the state of the system.* Here we focus on the estimation of the initial state $\mathbf{c}_0$ of the pollutant distribution. This state is not known exactly, and can be correctly represented only in a probabilistic framework that accounts for the uncertainty. The initial state is typically assumed to be normally distributed, $\mathbf{c}_0 \in \mathcal{N}(\mathbf{c}_0^{\mathrm{B}}, \mathbb{B})$, where $\mathbf{c}_0^{\mathrm{B}}$ is the 'background' state (the most likely state according to prior knowledge) and $\mathbb{B}$ is the background error covariance matrix. (The normality assumption does not account for the constraint that concentrations are non-negative. This assumption is reasonable only when errors are small (e.g. the mean is greater than three standard deviations). The issue of the choice of prior distributions and non-negativity of concentrations is discussed in (Constantinescu *et al.* 2007).)

3. *Limited measurements of the true state.* Observations $\mathbf{y}_i$ of the 'true' concentrations $\mathbf{c}^t$ of different species at selected locations are taken at times t_0, t_1, $\cdots$, t_N. The observations are corrupted by measurement and representativeness errors ε_i (assumed Gaussian with mean zero and covariance $\mathbb{R}_i$)

$$\mathbf{y}_i = \mathcal{H}_i\left(\mathbf{c}_i^t\right) + \varepsilon_i, \quad \varepsilon_i \in \mathcal{N}(0, \mathbb{R}_i), \quad i = 0, \cdots, N.$$

   Here $\mathcal{H}_i$ is an operator that maps the model state to observations (e.g. selects the concentration of a given species at the location where an instrument is placed).

The *data assimilation problem* is to find an optimal estimate of the true initial state $\mathbf{c}_0^t$ using information from the model ($\mathbf{c}_i$, $i \geq 0$), from the prior ($\mathbf{c}_0^B$, $\mathbb{B}$), and from the observations ($\mathbf{y}_i$, $i \geq 0$). In the 4D-Var approach

(Rabier *et al.* 2000) the maximum likelihood estimate $\mathbf{c}_0^A$ of the initial state (conditioned by all observations in the time window $[t_0, t_N]$) is obtained as the minimizer of the following cost function

$$\Psi(\mathbf{c}_0) = \frac{1}{2}\left(\mathbf{c}_0 - \mathbf{c}_0^{\mathrm{B}}\right)^T \mathbb{B}^{-1}\left(\mathbf{c}_0 - \mathbf{c}_0^{\mathrm{B}}\right) + \frac{1}{2}\sum_{i=0}^{N}\left(\mathbf{y}_i - \mathcal{H}_i(\mathbf{c}_i)\right)^T \mathbb{R}_i^{-1}\left(\mathbf{y}_i - \mathcal{H}_i(\mathbf{c}_i)\right)$$

$$\mathbf{c}_0^A = \arg\min_{\mathbf{c}_0} \Psi(\mathbf{c}_0) \qquad \text{subject to (16.29).} \tag{16.30}$$

The cost function Ψ measures the misfit between model predictions and observations, and also penalizes the departure of the solution from the background state. Note that the model equations appear as constraints (linking $\mathbf{c}_i$ in Ψ to the argument $\mathbf{c}_0$) and therefore the optimization problem (16.30) is 'PDE-constrained'. The covariance of the error/uncertainty in the maximum likelihood estimate $\mathbf{c}_0^A$ is approximated by the inverse of the Hessian of the cost function evaluated at the optimum $\nabla^2_{\mathbf{c}_0,\mathbf{c}_0}\Psi(\mathbf{c}_0^A)$. The use of second order information for estimating posterior errors is discussed in (Sandu and Zhang 2008).

The number of control variables in the optimization problem (16.30) is equal to the number of states in a CTM, which is typically $n \sim 10^7$. A gradient-based minimization method is usually employed (e.g., quasi-Newton). The gradient of the cost function with respect to the initial state is obtained as

$$\nabla_{\mathbf{c}_0}\Psi = \mathbb{B}^{-1}\left(\mathbf{c}_0 - \mathbf{c}^{\mathrm{B}}\right) + \sum_{i=0}^{N} \mathcal{M}'^{*}_{t_i \to t_0} \mathcal{H}'^{*}_i \, \mathbb{R}_i^{-1}\left(\mathbf{y}_i - \mathcal{H}_i(\mathbf{c}_i)\right), \tag{16.31}$$

where $\mathcal{M}'$ is the tangent linear model associated with $\mathcal{M}$, $\mathcal{M}'^{*}$ is the adjoint of $\mathcal{M}'$, $\mathcal{H}'$ is the linearized observation operator, and $\mathcal{H}'^{*}$ its transpose. Formula (16.31) reveals the importance of adjoint models. The methodology of discrete adjoints discussed in this chapter is an essential computational ingredient for solving the large-scale variational data assimilation problem (16.30).

The numerical tests use the state-of-the-art regional atmospheric chemical transport model STEM (Carmichael *et al.* 2003). The three-dimensional chemical reaction and transport equations are solved using an operator splitting approach; therefore the discrete adjoint model solves in succession the adjoints of individual science processes. STEM uses a second order spatial discretization of advection and diffusion operators (Sandu *et al.* 2005). The time integration of advection-diffusion is based on the Crank-Nicholson scheme in the forward model, and on the discrete adjoint Crank-Nicholson in the adjoint model. Atmospheric chemical kinetics result in stiff ODE equations that use a stable numerical integration that preserve linear invariants.

The SAPRC-99 (Carmichael *et al.* 2003) gas phase mechanism accounts for 93 chemical species, and involves in 235 chemical reactions. The chemistry time integration is done by the Lobatto-3C Runge Kutta method, and is implemented using the Kinetic PreProcessor KPP (Daescu *et al.* 2003; Damian *et al.* 2002; Miehe and Sandu 2008; Sandu *et al.* 2003). The discrete adjoint of the Lobatto-3C numerical integration method is used for the adjoint chemistry.

We simulate the air quality over Texas in July 2004. The computational domain (shown in Figure 16.1, left panel) covers $1500 \times 1500 \times 20$ Km with a horizontal resolution of 60×60 Km and a variable vertical resolution. The 24 hours simulation starts at 5 a.m. Central Standard Time on 16 July, 2004. The initial concentrations, meteorological fields, and boundary values correspond to ICARTT (International Consortium for Atmospheric Research on Transport and Transformation 2006) conditions. The EPA standard emission inventory NEI-2001v3 is used in this simulation. A detailed description of the setting of this data assimilation problem is given in Sandu (2007). The results reported in this chapter, however, are obtained with a different chemical solver (Lobatto-3C).

Measurements ('observations') of ground level ozone concentrations are available from the Environmental Protection Agency (EPA) network of monitoring stations AirNow (EPA 2004). In this work we use the hourly averages of ozone concentrations available (in electronic format) from AirNow for July 2004. The location of the AirNow stations in the computational domain is illustrated in Figure 16.1 (left panel).

The background error covariance matrix in (16.30) is constructed using autoregressive models to capture flow and chemistry dependent

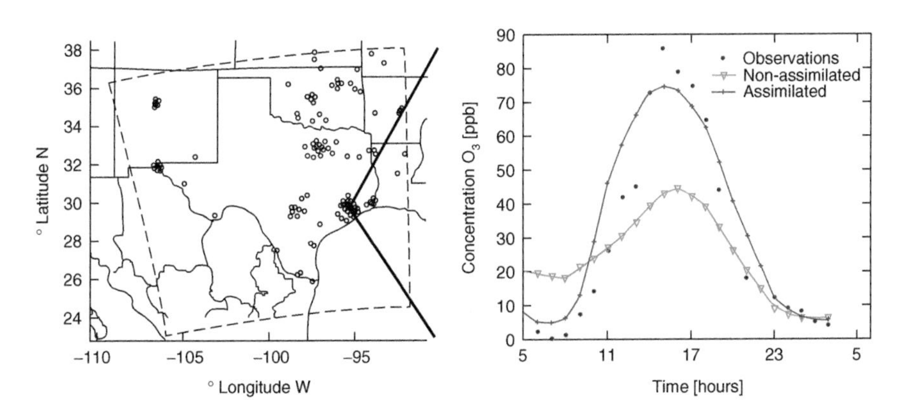

Figure 16.1 Left: The computational domain covering Texas and the location of the AirNow stations measuring ground level ozone. Right: Time series of ozone levels at a selected AirNow station in the Gulf area. The model predictions after assimilation better match the observed concentrations.

correlations of the errors (Constantinescu *et al.* 2007). The observational errors are considered independent and with a standard deviation of 0.5 ppb (parts-per-billion). The optimization algorithm used to minimize the cost function (16.30) is L-BFGS-B (Zhu *et al.* 1997). The optimization process is stopped after 10 iterations (during this process the forward and the adjoint models are each resolved 25 times to compute the cost function and its gradient; the additional calls are needed during line searches).

The overall performance of data assimilation is quantified by the R^2 ('R-square') correlation factor between observations ($Y = [\mathbf{y}_0; \cdots; \mathbf{y}_N]$) and model predictions ($X = [\mathcal{H}_0\mathbf{c}_0; \cdots; \mathcal{H}_N\mathbf{c}_N]$). The R^2 correlation factor of two series X and Y of length n is defined as

$$R^2(X,Y) = \frac{\left(n\sum_{i=1}^n X_i Y_i - \left(\sum_{i=1}^n X_i\right)\left(\sum_{i=1}^n Y_i\right)\right)^2}{\left(n\sum_{i=1}^n X_i^2 - \left(\sum_{i=1}^n X_i\right)^2\right)\left(n\sum_{i=1}^n Y_i^2 - \left(\sum_{i=1}^n Y_i\right)^2\right)}.$$

For our simulation the correlation between model predictions and observations increases from $R^2 = 0.36$ for the background run to $R^2 = 0.70$ after data assimilation.

One station located in the Gulf area is selected for a detailed analysis. The location of this station is shown by arrows in Figure 16.1. The right panel of this figure presents the time series of hourly observations and the time series of model predictions (at the location of the station) before and after data assimilation. The analysis of model predictions and data reveals that originally the model tends to over-predict ozone values. The model results after data assimilation reproduce better both the high and the low ozone values.

16.7 Conclusions

In this chapter we consider the solution of inverse problems when the underlying model is described by ODEs, and the gradients needed in the solution procedure are obtained by a 'discretize-then-differentiate' approach. Discrete adjoints are very popular in optimization and control since they can be constructed automatically by reverse mode automatic differentiation. However, the properties of the discrete adjoints are often poorly understood.

The consistency and stability properties of discrete adjoints are discussed for two families of numerical schemes: Runge-Kutta and linear multistep methods. The analysis reveals that adjoints of Runge-Kutta methods have the same orders of accuracy as the original schemes. Discrete linear multistep adjoint solutions have the same orders of accuracy as the forward methods at the initial time; however they are not consistent approximations of the adjoint ODE at the internal trajectory points. Numerical tests confirm the theoretical findings.

ODE adjoint models are an essential computational ingredient for the solution of large scale inverse problems. To illustrate this point, discrete Runge Kutta adjoints are used to solve a real chemical data assimilation problem.

Acknowledgments

This work was supported by the National Science Foundation through the awards NSF CCF–0515170 and NSF CCF–0635194, by NASA through the award AIST 2005, and by the Texas Environmental Research Consortium (TERC) through awards H-59 and H-98.

References

Environmental Protection Agency. Airnow web site: http://airnow.gov, 2004.

Alexe, M. and Sandu, A. On the discrete adjoints of variable step time integrators. Submitted, 2008.

Baguer, M.L. and Romisch, W. Computing gradients in parametrization-discretization schemes for constrained optimal control problems. *Approximation and Optimization in the Caribbean II*. Peter Lang, Frankfurt am Main, 1995.

Bonans, J.F. and Laurent-Varin, J. Computation of order conditions for symplectic partitioned Runge Kutta schemes with application to optimal control. *Numerische Mathematik*, Vol. 2006, No. 103, p. 1–10, 2006.

Carmichael, G.R., Tang, Y., Kurata, G. *et al.* Regional-scale chemical transport modeling in support of the analysis of observations obtained during the TRACE-P experiment. *Journal of Geophysical Research*, Vol. 108, Issue D21 8823, p. 10649–10671, 2003.

Carmichael, G.R., Sandu, A., Chai, T., Daescu, D., Constantinescu, E.M. and Tang, Y. Predicting air quality: improvements through advanced methods to integrate models and measurements. *Journal of Computational Physics (Special issue on predicting weather, climate and extreme events)*, Vol. 227, Issue 7, p. 3540–3571, 2008.

Christianson, B. Reverse accumulation and attractive fixed points. *Optimization Methods and Software*, Vol. 3, Issue 4, p. 311–326, 1994.

Constantinescu, E.M., Chai, T., Sandu, A. and Carmichael, G.R. Autoregressive models of background errors for chemical data assimilation. *Journal of Geophysical Research*, Vol. 112, D12309, DOI:10.1029/2006JD008103, 2007.

Constantinescu, E.M., Sandu, A., Chai, T. and Carmichael, G.R. Ensemble-based chemical data assimilation. I: General approach. *Quarterly Journal of the Royal Meteorological Society*, Vol. 133, Issue 626, Part A, p. 1229–1243, 2007.

Daescu, D., Sandu, A. and Carmichael, G. Direct and adjoint sensitivity analysis of chemical kinetic systems with KPP: II – Numerical validation and applications. *Atmospheric Environment*, Vol. 37, p. 5097–5114, 2003.

Damian, V., Sandu, A., Damian, M., Potra, F. and Carmichael, G. The kinetic Pre-Processor KPP - a software environment for solving chemical kinetics, Computers and Chemical Engineering, Vol. 26, p. 1567–1579, 2002.

International Consortium for Atmospheric Research on Transport and Transformation (ICARTT). ICARTT web site: http://www.al.noaa.gov/ICARTT, 2006.

Giering, R. and Kaminski, T. Recipes for adjoint code construction. *ACM Transactions on Mathematical Software*, Vol. 24, Issue 4, p. 437–474, 1998.

Giles, M.B. On the use of Runge-Kutta time-marching and multigrid for the solution of steady adjoint equations. Technical Report NA00/10, Oxford University Computing Laboratory, 2000.

Giles, M.B. On the iterative solution of adjoint equations. *Automatic Differentiation: From Simulation to Optimization*, p. 145–152, 2001.

Griewank, A. A mathematical view of automatic differentiation. *Acta Numerica*, Vol. 12, p. 321–398, 2003.

Hager, W. Runge-Kutta methods in optimal control and the transformed adjoint system. *Numerische Mathematik*, Vol. 87, Issue 2, p. 247–282, 2000.

Hairer, E., Norsett, S.P. and Wanner, G. *Solving Ordinary Differential Equations I. Nonstiff Problems*. Springer-Verlag, Berlin, 1993.

Hairer, E. and Wanner, G. *Solving Ordinary Differential Equations II. Stiff and Differential-algebraic Problems*. Springer-Verlag, Berlin, 1991.

Miehe, P. and Sandu, A. Forward, tangent linear, and adjoint Runge Kutta methods in KPP–2.2 for efficient chemical kinetic simulations. *International Journal of Computer Mathematics*, in print, 2008.

Rabier, F., Jarvinen, H., Klinker, E., Mahfouf, J.F. and Simmons, A. The ECMWF operational implementation of four-dimensional variational assimilation. I: Experimental results with simplified physics. *Quarterly Journal of the Royal Meteorological Society*, Vol. 126, p. 1148–1170, 2000.

Sandu, A. Reverse automatic differentiation of linear multistep methods. *Advances in Automatic Differentiation*, p. 1–12. Series: Lecture Notes in Computational Science and Engineering, Vol. 64. Bischof, C.H.; Buecker, H.M.; Hovland, P.; Naumann, U.; Utke, J. editors. Springer, XVIII, 370 p., 111 illus. ISBN: 978-3-540-68935-5. 2008.

Sandu, A. On the properties of Runge-Kutta discrete adjoints. *Lecture Notes in Computer Science, Part IV*, Vol. LNCS 3994, Alexandrov, V.N. *et al.* (Eds.), Springer-Verlag Berlin Heidelberg. International Conference on Computational Science, Reading, U.K., 2006.

Sandu, A., Daescu, D. and Carmichael, G.R. Direct and adjoint sensitivity analysis of chemical kinetic systems with KPP: I – Theory and software tools. *Atmospheric Environment*, Vol. 37, p. 5083–5096, 2003.

Sandu, A., Daescu, D., Carmichael, G. and Chai, T. Adjoint sensitivity analysis of regional air quality models. *Journal of Computational Physics*, Vol. 204, p. 222–252, 2005.

Sandu, A., Zhang, L. and Constantinescu, E.M. Sensitivity analysis and 4D-Var data assimilation. Part I: the July 2004 episode. Report for project H59 to Houston Advanced Research Center, 2007.

Sandu, A. and Zhang, L. Discrete second order adjoints in atmospheric chemical transport modeling. *Journal of Computational Physics*, Vol. 227, Issue 12, p. 5949–5983, 2008.

Zhu, C., Byrd, R.H., Lu, P. and Nocedal, J. Algorithm 778: L-BFGS-B: Fortran subroutines for large-scale bound-constrained optimization. *ACM Transactions on Mathematical Software*, Vol. 23, Issue 4, p. 550–560, 1997.

Walther, A. Automatic differentiation of explicit Runge-Kutta methods for optimal control. *Journal of Computational Optimization and Applications*, Vol. 36, p. 83–108, 2007.

Index

Large-Scale Inverse Problems and Quantification of Uncertainty Edited by L. Biegler, G. Biros, O. Ghattas, M. Heinkenschloss, D. Keyes, B. Mallick, Y. Marzouk, L. Tenorio, B. van Bloemen Waanders and K. Willcox

Printed and bound by CPI Group (UK) Ltd, Croydon, CR0 4YY

16/04/2025

14658498-0001